MUSHROOMS:
POISONS AND PANACEAS

A HANDBOOK FOR
NATURALISTS, MYCOLOGISTS,
AND PHYSICIANS

The origin of women and the power of the fungus
—A Haida myth

The canoe is paddled by Fungus Man, while Raven rides in the bow, holding a spear with which to cap-
ture the female genitalia along the shore. Raven tried other creatures to paddle the canoe, but only Fungus
Man has the supernatural powers to overcome the spiritual barriers protecting the regions where women's
genital parts are located. (Reprinted by permission from R. Blanchette et al., Mycologia (1992)
84:119–124. Redrawn from an argillite plate carved by Charles Edenshaw, courtesy of the Field
Museum of Natural History.

Mushrooms: Poisons and Panaceas

A HANDBOOK FOR

NATURALISTS, MYCOLOGISTS,

AND PHYSICIANS

DENIS R. BENJAMIN, M.B., B.Ch.

Director, Department of Laboratories
Children's Hospital and Medical Center
Seattle, Washington

Professor of Laboratory Medicine and Pathology
University of Washington School of Medicine

W. H. Freeman and Company
New York

Library of Congress Cataloging–in–Publication Data

Benjamin, Denis R.
  Mushrooms : poisons and panaceas : a handbook for naturalists, mycologists, and
physicians / Denis R. Benjamin
        p.   cm.
     Includes index.
     ISBN 0–7167–2600–9 (hard cover). — ISBN 0–7167–2649–1 (soft cover)
       1. Mushrooms, Poisonous—Toxicology. 2. Mushrooms— Composition. I. Title
     [DNLM: 1. Basidiomycetes. 2. Mushroom Poisoning. QW 180.5.B2 B468m
1995]
     RA1242.M9B46 1995
     615.9'529222—dc20
     DNLM/DLC
     for Library of Congress
94–44126

Cover photographs by Steven Trudell and Denis Benjamin.
Top: *Amanita muscaria*. Bottom: *Boletus edulis*.

1 2 3 4 5 6 7 8 9 0   VB   9 9 8 7 6 5

579.6
5

To Vivien, Craig, and Graham,

who gave me the time
to pursue a passion

HOW TO USE

MUSHROOMS:
POISONS AND PANACEAS

Mushrooms: Poisons and Panaceas is designed as a convenient and readable resource for people who want to learn more about mushrooms and the effects of eating mushrooms on human health. The informed lay public, naturalists, mycophagists, and both amateur and professional mycologists will enjoy the treatment in Part One of such topics as cultural attitudes toward mushroom eating, the use of mushrooms in criminal poisonings, and the nutritional benefits and positive health effects of eating mushrooms.

Part Two covers such topics as recognizing sickness that is *not* a result of mushroom poisoning and the likelihood of ingesting a truly poisonous mushroom. It discusses other possibilities for illness after eating food one has gathered, and it describes the most likely groups of people to suffer mushroom poisoning. Chapter 11, "Diagnosis and Management of Mushroom Poisoning," is printed on shaded pages so that it can be found quickly for reference. The chapter will be of interest to those who are symptomatic after a mushroom meal and to emergency personnel faced with a suspected mushroom poisoning. Subsequent chapters frequently refer to this general discussion for details about specific treatments or dosage recommendations.

Chapters 12–18 in Part Three present information about the recognition and treatment of specific mushroom poisonings. Each chapter opens with a drawing of a representative mushroom. The drawing is repeated at the top of each page containing emergency clinical information: It first signals the section "Clinical Features," which describes the clinical symptoms and signs of the syndrome, and later signals the section "Management," which lists clinical treatment procedures. The drawing appears in a decorative band at the top of the page so that clinical information can be accessed quickly.

The color insert shows photographs of the most commonly ingested poisonous mushrooms. Although readers are encouraged to consult standard field guides or mycologists for assistance in mushroom identification, the photographs in the insert may be handy in an emergency.

All people who eat wild mushrooms will find the clinical information useful if they are in doubt about symptoms they or others are experiencing after a mushroom meal. To emergency medical personnel and physicians, the general information will be of interest; even if it does not encourage them to share the enthusiasm of the mycophagist, it suggests critical questions to ask when mushroom poisoning is suspected.

CONTENTS IN BRIEF

PART ONE MUSHROOMS AND HEALTH / 1

1   Cultural Attitudes Toward Mushrooms .2
2   History of Mushroom Eating and Mushroom Poisoning 29
3   Introduction to the Biology of Mushrooms 51
4   Nutritional and Culinary Aspects of Edible Mushrooms` 75
5   Health Benefits and Medicinal Properties of Edible
    Mushrooms 75
6   Guidelines for Would-Be Mycophagists 100

PART TWO MUSHROOM POISONING / 107

7   Poisoning Not Caused by Mushroom Toxins 108
8   Spectrum of Poisonous Mushrooms 136
9   Incidence of Mushroom Poisoning 152
10 Sociology of Mushroom Poisoning 162
11 Diagnosis and Management of Mushroom Poisoning 171

PART THREE MUSHROOM POISONING SYNDROMES / 197

12 Amatoxin Syndrome: Poisoning by the Amanitins 198
13 Delayed-Onset Renal Failure Syndrome: Orellanine or
    Cortinarin Poisoning 242
14 Gyromitrin Poisoning: Effect of Hydrazines 264
15 "Antabuse" Syndrome: Coprine Poisoning 283
16 Inebriation or Pantherine Syndrome: Effects of Isoxazole
    Derivatives on the Central Nervous System 295
17 Hallucinogenic Syndrome: Poisoning by Tryptamine
    Derivatives 318
18 Muscarine Poisoning: PSL Syndrome or SLUDGE
    Syndrome 340
19 Gastrointestinal Syndrome 351
20 Miscellany of Toxins 378

APPENDIXES

1   Chemistry of Mushroom Toxins and Methods of Analysis . . .390
2   National and Regional Field Guides for Mushroom
    Identification 397
3   Mycological Associations and Consultants 401
4   Mushroom Cookbooks 407

CONTENTS

HOW TO USE
MUSHROOMS: POISONS AND PANACEAS / vi

FOREWORD
BY J. F. AMMIRATI /xix

PREFACE / xxi

ACKNOWLEDGMENTS / xxv

PART ONE MUSHROOMS AND HEALTH / 1

CHAPTER ONE
CULTURAL ATTITUDES TOWARD MUSHROOMS / 2

Anglo-Saxon Cultures 4
Africa 14
India 16
Native North American Cultures .17
Europe 19
Russia 21
South America 23
Asia and Australasia 24

CHAPTER TWO
HISTORY OF MUSHROOM EATING AND MUSHROOM
POISONING / 29

Mycophagy in Western Cultures .30
  Poisonings and False Allegations 32
  Regulations and Remedies 37
  Modern Criminal Poisonings 40
Mycophagy in Non-Western Cultures 42
Mushrooms in Medicine 42
  Use of Mushrooms as Panaceas: Agarick and Suilli 43
  Use of Mushrooms in Cautery and Hemostasis 45
  Miscellaneous Medical or Paramedical Uses 47
Historical Uses of Truffles 48

CHAPTER THREE
INTRODUCTION TO THE BIOLOGY OF MUSHROOMS / 51

What Is a Fungus? 51
What Is a Mushroom? 55

CHAPTER FOUR
NUTRITIONAL AND CULINARY ASPECTS
OF EDIBLE MUSHROOMS / 62

Nutrients in Mushrooms 62
    Moisture and Energy 64
    Protein 65
    Amino Acids 65
    Fat Content 66
    Carbohydrate and Fiber 66
    Minerals and Vitamins 67
    Bioavailability and Nutritional Value of Nutrients 67
Taste and Flavor 68
Cooking and Digestibility 71
Mushrooms as a Significant World Food Source 72

CHAPTER FIVE
HEALTH BENEFITS AND MEDICINAL PROPERTIES
OF EDIBLE MUSHROOMS / 75

Two Views of the Effects of Food 75
Two Approaches to Medicine and Research 77
Mushrooms in the Asian Pharmacopoeia 78
Separating Folklore Fact from Fantasy 80
General Medicinal Properties of Mushrooms 81
    Anticancer Activities 82
    Antimicrobial Activity 84
    Antiviral Activity 86
    Other Medicinal Properties 87
Mushroom of Immortality—Ling Chi or Reishi 88
    Medicinal Properties of Ling Chi .89
    Effects on the Immune System 90
    Effects on Blood Lipids and Blood Pressure 91
Medicinal Properties of Shiitake 91
    Effects on Serum Cholesterol Levels 92
    Antitumor and Immunological Effects 93
    Antiviral Activity 94
Final Comment on the Benefits of Eating Mushrooms 94

CHAPTER SIX
GUIDELINES FOR WOULD-BE MYCOPHAGISTS / 100

How to Handle Specimens 101
What to Eat or Avoid 102
Look-alikes 103

PART TWO MUSHROOM POISONING / 107

CHAPTER SEVEN
POISONING NOT CAUSED BY MUSHROOM TOXINS / 108

Panic Reactions 108
Bacterial Food Poisoning 111
Chemical Contaminants 113
Alcohol Intoxication 115
Idiosyncratic Reactions and Intolerances 116
Allergic Reactions 118
Intestinal Obstruction 122
Heavy-Metal Poisoning 122
Radioisotope Contamination 125
Cancer 127
    Agaricus bisporus 127
    Gyromitra esculenta and Relatives 128
    Other Mushrooms 129
    Coda on the Cancer Risk 129

CHAPTER EIGHT
SPECTRUM OF POISONOUS MUSHROOMS / 136

Edibility Characteristics in the Field Guides 137
Distribution of Mushrooms in North America 138
Variation in Toxicity 139
Quantity of Toxins in Edible Species .142
Dangers of Raw Fungi 146
Biological Role of Mushroom Toxins .147
Domestic Animals and Mushrooms 148
Mushroom Poisoning of Children 149

CHAPTER NINE
INCIDENCE OF MUSHROOM POISONING / 152

Incidence of Mushroom Poisoning in the United States 154
Group or Mass Poisonings 156
Mushroom Poisoning Among Amateur Mycologists 159

CHAPTER TEN
SOCIOLOGY OF MUSHROOM POISONING / 162

Children 163
Chemically Dependent Individuals 164
Immigrants 165

Elderly 168
Mycologically Naive 168
Psychologically Disturbed 169

CHAPTER ELEVEN
DIAGNOSIS AND MANAGEMENT OF MUSHROOM
POISONING 171

Treatment of Common Clinical Presentations 173
    Asymptomatic Child Suspected of Eating a Mushroom . . .174
    Symptomatic Patient Known to Have Ingested
    Mushrooms 176
    Unconscious Patient with No History of Mushroom
    Ingestion 176
Major Clinical Syndromes 177
    Poisonings with Delayed Onset: Amatoxin, Gyromitrin,
    and Orellanine Syndromes 179
        Amatoxin Syndrome / Gyromitrin Syndrome /
        Orellanine Syndrome / Differential Diagnosis
    Poisonings with Rapid Onset 181
        Muscarine Syndrome / Inebriation Syndrome / Hallucinogenic
        Syndrome / Coprine Syndrome / Gastrointestinal Syndrome
General Management of Any Case of Suspected Mushroom
    Poisoning 183
    Elimination of Further Toxin Absorption 183
        Activated Charcoal / Cathartics
    Enhancement of Excretion 186
    Maintenance of Vital Functions 187
    Supportive and Symptomatic Care 187
        Fluid and Electrolyte Balance / Pain / Convulsions / Anxiety or
        Hallucinations / Nausea and Vomiting / Monitoring
Handling and Identification of Mushroom Specimens or
Fragments 189
    Is the Specimen an *Amanita?* 190
    Is the Specimen a *Galerina?* 190
    Is the Specimen a *Lepiota?* 191
Testing for the Presence of Amatoxin 191
    Weiland Newspaper Test, or Meixner Test 191
    Spore Identification 193

PART THREE MUSHROOM POISONING SYNDROMES / 197

Color plates of poisonous mushrooms appear in an insert
between pages 294 and 295.

CHAPTER TWELVE
AMATOXIN SYNDROME: CYCLOPEPTIDE
POISONING BY THE AMANITINS / 198

Incidence of Amatoxin Poisoning 199
Distribution and Habitats of Amatoxin-Containing Mushrooms 202
    Amanita species 102
    Galerina species 205
    Lepiota and *Conocybe* species 206
Metabolism and Mechanisms of Action of Amatoxins 207
    Phallotoxins 207
    Amanitins 208
    Other Amatoxins 208
    Absorption, Excretion, and Transport 209
Toxicity of Amatoxins: How Much Will Kill You? 211
History of the Treatment of Amatoxin Poisoning 212
Clinical Features of Amatoxin Poisoning 216
    Onset 216
    Signs and Symptoms 217
Laboratory Findings 219
Prognosis 220
    Dosage / Age / Length of Latent Period / Laboratory Results
Management of Amatoxin Poisoning 223
    Early, Presymptomatic Treatment 224
    Removal of Amatoxins from the Duodenum 225
    Enhancement of Renal Excretion 225
    Supportive Medical Care 226
    Pharmacological Therapy 226
    Intravenous Penicillin G / Silibinin and Silymarin
    Liver Transplantation 228
Other Management Options 230
    Controversial Treatments 230
        Hemoperfusion / Plasma Exchange Transfusion /
        Hyperbaric Oxygen Therapy / Steroids
    Treatment Methods No Longer Advocated 230
Long-Term Outcome for Survivors 231
Pathology of Amatoxin Poisoning 232
Unanswered Questions about Amatoxin Poisoning 232

CHAPTER THIRTEEN
DELAYED-ONSET RENAL FAILURE SYNDROME:
ORELLANINE OR CORTINARIN POISONING / 242

Discovery and Incidence of Orellanine Poisoning 243
      Poisonings in Europe 243
      Poisonings in North America 244
Distribution and Habitat of Mushrooms Causing Orellanine
Poisoning 245
Cortinarius Toxins 246
      Cortinarins 247
      Orellanine 249
Toxicity of *Cortinarius* Poisons 251
Clinical Features of Orellanine Poisoning 252
      Onset 252
      Signs and Symptoms 252
Laboratory Findings 253
Detection of Orellanine in Mushrooms and Body Fluids 253
      Tests for Orellanine in a Mushroom Specimen /
      Tests for Orellanine in Animal or Human Specimens
Prognosis 254
Management of Orellanine Intoxication 256
      Therapies with No Proven Benefits / Contraindicated Therapy
Pathology of Orellanine Poisoning .256
      Human Biopsy Material / Animal Studies
Poisonings by *Cortinarius splendens* 258
Unanswered Questions About the Orellanine Syndrome 258

CHAPTER FOURTEEN
GYROMITRIN POISONING: EFFECT OF
HYDRAZINES / 264

Incidence 265
      Poisonings by Gyromitra Species 265
      Poisonings by Related Species 267
Distribution and Habitats of Mushrooms Causing Gyromitrin
Poisoning 268
Toxicology of Gyromitrin and Hydrazines 268
Enigmatic Toxicity of Gyromitrin 271
Clinical Features of the Gyromitrin Syndrome 273
      Onset 273
      Signs and Symptoms 273
      Gastrointestinal Phase / Hepatorenal and Neurological Phase
Management of Gyromitrin Poisoning 276

Preparation of False Morels 277
Adverse Effects of Morels 278
Unanswered Questions about Gyromitrin Poisoning 279

CHAPTER FIFTEEN
"ANTABUSE" SYNDROME: COPRINE
POISONING / 283

Distribution and Habitats of Species Responsible for Coprine
    Poisoning 285
Toxicology of Coprine 286
Clinical Features of Coprine Intoxications 288
    Onset 288
    Signs and Symptoms 288
Prognosis 289
Management of the Antabuse Syndrome 290
Unanswered Questions about Alcohol–Mushroom Interaction .290

CHAPTER SIXTEEN
INEBRIATION OR PANTHERINE SYNDROME:
EFFECTS OF ISOXAZOLE DERIVATIVES ON THE
CENTRAL NERVOUS SYSTEM / 295

Historical Reports of the Effects of Amanita muscaria 296
    Amanita muscaria and Religion 296
    Use of Amanita muscaria as an Inebrient 297
    Counterculture's Experimentation with Amanita muscaria . . .299
    Role of Amanita muscaria in Aggression and Warfare 301
    Amanita muscaria in English Literature 302
Incidence of Isoxazole Poisoning 303
Distribution and Habitats of Isoxazole-Containing
Mushrooms 305
Toxicology of Isoxazole Derivatives 305
    Myth of Muscarine 305
    Ibotenic Acid and Muscimol—The Real Toxins 306
Toxicity of the Isoxazole Derivatives 309
Clinical Features of Isoxazole Poisoning 310
    Onset 310
    Signs and Symptoms 310
Prognosis 312
Management of the Inebriation Syndrome 313
Unanswered Questions about the Pantherine Syndrome 314

CHAPTER SEVENTEEN
HALLUCINOGENIC SYNDROME: POISONING BY
TRYPTAMINE DERIVATIVES / 318

Historical Background 322
Incidence of Tryptamine Poisoning 323
Distribution and Habitats of Hallucinogenic Mushrooms 324
Species Containing Hallucinogenic Compounds 325
Toxicology and Pharmacology of the Hallucinogenic
Mushrooms 325
Toxicity (or Potency, Depending on One's Point of View) 328
Clinical Features of the Hallocinogenic Syndrome 329
    Onset 329
    Signs and Symptoms 329
Management of the Hallucinogenic Syndrome 333
Legal Status of the Psilocybin-Containing Mushrooms
in the United States 334
Unanswered Questions about the Hallucinogenic Syndrome . . .335

CHAPTER EIGHTEEN
MUSCARINE POISONING: PSL OR
SLUDGE SYNDROME / 340

Distribution and Habitats of Muscarine-Containing
Mushrooms 341
Mushrooms Implicated in Muscarine Poisonings 342
    Mushrooms Containing High Concentrations of
    Muscarine 342
    Other Species Suspected of Causing Muscarine Poisoning 343
    Species Containing Insignificant Amounts of Muscarine . .344
Toxicology and Pharmacology of Muscarine 344
Toxicity of Muscarine 346
Clinical Features of Muscarine Poisoning 346
    Onset 346
    Signs and Symptoms 346
Prognosis 348
Management of Muscarine Poisoning 348
Unanswered Questions about Muscarine Poisoning 349

CHAPTER NINETEEN
GASTROINTESTINAL SYNDROME / 351

Clinical Features and Prognosis **354**
Management of the Gastrointestinal Syndrome **354**
Chlorophyllum molybdites 355
Agaricus species 356
Armillaria mellea Complex 358
Red-Pored Boletes 359
Entoloma (=Rhodophyllus) Species 360
Hebeloma crustuliniforme 362
Lactarius Species: The Milky Caps 363
Laetiporus sulphureus 365
Omphalotus Species 366
Pholoiota squarrosa 367
Russula Species 369
Suillus Species 370
Tricholoma pardinum and Relatives 370
Rarely Eaten Toxic Mushrooms 371

CHAPTER TWENTY
MISCELLANY OF TOXINS / 378

Amanita smithiana (=Amanita solitaria): Renal Failure 378
Auricularia auricula: Easy Bruising and Excessive Bleeding 380
Hypholoma fasciculare (=Naematoloma fasciculare) 381
Panaeolus (=Panaeolina, =Psathyrella) foenisecii 382
Paxillus involutus and the Paxillus Syndrome: Immune Hemolytic
Anemia 382
Two Other Toxins 384
    Acromelic Acid 385
    Hydrocyanic Acid 385

EPILOGUE / 388

APPENDIXES

ONE: CHEMISTRY OF MUSHROOM TOXINS AND
METHODS OF ANALYSIS / 390

Cyclopeptides: Amatoxins 390
Coprine 391
Gyromitrin and Its Hydrazine Metabolites 391
Psilocybin and Psilocin 392

Isoxazole Derivatives 392
Muscarine 392
Orellanine 393

TWO: NATIONAL AND REGIONAL FIELD GUIDES FOR
MUSHROOM IDENTIFICATION / 397
General Field Guides 397
Regional Field Guides 398

THREE: MYCOLOGICAL ASSOCIATIONS AND
CONSULTANTS / 401

FOUR MUSHROOM COOKBOOKS / 407

Cookbooks by One or Two Authors 407
Mycological Society Cookbooks 410

INDEX / 411

FOREWORD

Mushrooms are everywhere! They have intrigued us for centuries, have become part of our culture, and delight us with their flavors and aromas. They are mysterious, we are often suspicious of them, and some are surely poisonous. At the same time we find them fascinating and seek their nutrition and health benefits, their panaceas. Each year thousands of mycophagists search for edible mushrooms in fields and forests, purchase them at the local market, and toil to cultivate them. Mushrooms, both edible and poisonous, are becoming part of our everyday lives. Have we truly become mycophiles?

Hundreds of books and papers on poisonous mushrooms and mushroom toxins have appeared over the years, and legions of mushroom hunters have loyally procured them for that moment when a poison victim cries for help or a reference is needed to separate fact from fiction about mushroom poisoning. Finally a marvelous book has appeared, craftily assembled to lure us farther into the exciting world of poisonous mushrooms and at the same time exploring for us mycophagy, the health benefits of mushrooms and the history of mushroom gathering.

This book is the best of its kind, combining lore and facts in a delightful and informative treatment. The main substance of the book is its excellent chapters on the groups of poisonous mushrooms, including details of toxicology, symptomatology, and treatment and management of mushroom poisoning. The text is rounded out with an account of our cultural attitudes toward mushrooms and an overview of mushroom biology. Thoroughness and accuracy are liberally laced with the author's enthusiasm for mushrooms, which knows no bounds!

Mushrooms: Poisons and Panaceas is both a gem and a cornerstone of modern mushroom literature. It is equally useful for the mushroom hunter and the mycophagist, the mycologist and the physician. It can be enjoyed by the casual reader and will delight all who venture into its pages to gain knowledge of fungi and mushrooms.

J. F. Ammirati

PREFACE

In the pantheon of foods, mushrooms have always played a slightly strange role. Mushrooms evoke more passion than most other foods. In some cultures they are the epitome of gastronomy; in others they provide life-enhancing sustenance. In some cultures they are feared, despised, and rejected, except for a single tame and domesticated variety; in a few they are part of the magic of life, used to touch the unknown universe. This then is the story of the mushroom: its history in the foodways of mankind, its role in health and spirituality, an exploration of the potent toxins possessed by a few, and finally, the way we currently deal with anyone unfortunate enough to be poisoned.

Mushrooms: Poisons and Panaceas is about the health effects of mushroom eating—the bad and the good, the toxins and the tonics. It is neither the traditional nor the standard book on mushroom poisoning. It is not a treatise on taxonomy or identification. It is not designed for the chemist or the toxicologist interested in the synthesis of a toxin or the formulas of the many poisons. Rather it is meant for people intrigued by food, for people who eat mushrooms, and for physicians who have to treat the occasional dietary indiscretion or potential tragedy. It contains the most current information available through 1994. It also reflects on how our knowledge of mushrooms accumulated and on the role of mushrooms in various cultures.

Mushroom poisoning is not a major public health problem, at least not in the United States. Nor is there much evidence that the incidence is increasing, although wild mushrooms are now more widely available in the marketplace and many more people are being exposed to the new gastronomic possibilities. Many authors have used increased public familiarity with mushrooms as justification for writing a book or an article on the subject. My justification is no more than the immense fascination provoked by a potentially lethal food. This is partly reflected in the way each occurrence of poisoning garners so much space in the media. Following a report of mushroom poisoning, the public is usually regaled with dire warnings regarding the dangers of wild mushroom eating, and the effect is to elevate the mystique still further.

Mushrooms: Poisons and Panaceas is a deliberate attempt to achieve two goals, which, if not inherently incompatible, are certainly a challenging combination. The first is to provide everyone, from the scientifically naive to the dedicated amateur mycologist, with a broad overview of mushroom poisoning that is both instructive and entertaining. The second is to ensure that all the appropriate technical details, up to our current state of ignorance, are made available to the physician who has to deal with a case of possible mushroom poisoning. Stated simply, the challenge is to bridge the jargon gaps in the languages of the mycologist, toxicologist, physician, and naturalist. Since my background is in medicine, this is the vantage point from which I approach the subject, but I have no intention of slighting the other experts who have played such a crucial role in illuminating the subject. Except in the technical portions relating to symptoms, treatment, and toxicology, which can be conveniently skipped by readers with little need for clinical rigor, I have attempted to translate the jargon into understandable English. Chemical formulas and techniques for identification are in appendices.

Numerous reports on the effects of mushroom toxins on a variety of animal species have been published, but I include only those having a direct bearing on human physiology and health. Extrapolating animal information to man is always a treacherous enterprise—profound differences exist in the absorption, metabolism, tissue sensitivity, and excretion of the toxins from one animal species to the next—but in certain instances it is all we have to go on.

Mushrooms are shrouded in mystery and mythology. Some of it is quaint, some is dangerous, and all is based on ignorance. This book attempts to dispel a number of the common myths and misconceptions: "most mushrooms are poisonous"; "deadly mushrooms are instantly lethal"; "it is dangerous to touch a poisonous mushroom"; "mushrooms have no calories"—"they are nutritionally useless"; and the most pervasive, "even the experts cannot tell the good from the bad mushrooms." All these myths and more are discussed and exposed in the book.

The literature on mushroom poisoning is diffuse and widely scattered, with a great deal of it in rather obscure or difficult-to-obtain sources. Only a fraction is published in English. Much of the seminal work and many of the observations have been performed in Europe and Asia. Not all of these studies are easy to evaluate, although I have attempted to include all the pertinent foreign references.

I have employed a device that has all but disappeared from the medical literature in the last century, namely the anecdote. Anecdotes fell into disfavor because experience was rightly supplanted by observation and experiment. Unfortunately the literature became blood-less, detached from the pain and anguish of the very people it was designed to help. The phenomenon is both unfortunate and unneces-sary: A case history can help highlight important features of a problem, as well as return humanity to the field; in no way need it detract from the science. So I use anecdotes with no other apology than that they are the flesh and blood of an intensely human story. No reason exists for a medical or a scientific book to be unreadable.

This book has a very specific focus: human health associated with the eating of mushrooms. Fungi play an enormous role in many other aspects of human activity, from being responsible for bread, wine, and cheese (all examples of controlled rot) to destroying our crops and our homes. Fungi, many of which are microscopic, produce a host of toxins that affect our daily lives. Mycotoxins, the products of a myriad of minute fungi growing on and contaminating our food supply, are not discussed here. Even the great scourge of the Middle Ages, St. Anthony's fire, caused by an ergot-producing fungus growing on grain, is barely mentioned. Hundreds of different compounds have been described in the higher fungi, many having interesting biological effects. They are not catalogued here unless they have some relation-ship to human health. This book is not encyclopedic in scope. Rare mushrooms responsible for human suffering are not included, either through oversight or beause their contribution to human illness was judged to be trivial.

I avoided taxonomy and identification for a number of reasons. Most important, it would have been an incompetent attempt, because it is not my professional field. Second, a variety of excellent texts are uni-versally available on the subject, many of them specific for a particular locality. This is important, since toxicity can vary significantly from one region of the world to another. In addition, both amateur and pro-fessional mycologists are as widely scattered as many mushrooms and can provide a helpful identification service. Moreover, successful man-agement of a poisoned patient seldom requires expert identification. This heresy is further discussed in Chapter 11, perhaps to the distress of some professional mycologists. However, it does appear to serve the best interests of the patient. It is a very pragmatic approach to poi-soning. Naturally, attempts at identification should be made in all cases so that we may add to our knowledge and understanding. Third, the science of taxonomy is constantly refining the classifications. It is a dynamic science, best advanced by those who have made it a career. So we will let the professional mycologists debate the taxa, rearrange the genera, and analyze the toxins. We will continue to eat wild mush-rooms and treat those who failed to follow the dicta for survival.

Librarians will have a difficult time deciding where to file this book—medicine? mycology? ethnobotany? toxicology? history? The simple solution is to purchase five copies so that one is available wherever it is sought.

Denis R. Benjamin
March 1995

ACKNOWLEDGMENTS

Dr. Joe Ammirati first stimulated my interest in updating our knowledge of mushroom poisoning because so much has changed in the decade and a half since the subject was last explored in book form. He made the mycological book collection of the Stuntz Library freely available to me, together with mountains of personal notes, extensive offprints, and great knowledge. He was always available for consultation and consolation.

Medicine and botany were traditionally closely linked. All the early herbals were designed to teach as much about the medicinal uses of plants as about the plants themselves. Not that many years ago, when I began medical school in Johannesburg, South Africa, botany was a required course in the first-year medical school curriculum. Today the fields are far apart. Modern physicians know little of the subject, and mycologists are rightly intimidated by the sophistication of current medicine. So it was with a sense of closing an ancient circle that I went to Edinburgh, Scotland, to write this book, to once again link knowledge of medicine with knowledge of mycology. What better place than Edinburgh, a city whose medical school has been associated with many who have influenced the world of botany. What better place than where a herbal store sits across the road from the medical school and a health food store adjacent to the Royal College of Surgeons. For the legacy and the tradition I am grateful. To all those at the Royal Botanic Gardens, especially Dr. Roy Watling, who provided help and inspiration I am indebted.

I am also indebted to all those who have explored the subject before me. Their influence has been pervasive, shaping much of my understanding. I have borrowed generously from their ideas and have disagreed when my experience or the current literature is to the contrary. Of the recent pioneers, Mr. Gary Lincoff and Dr. D. H. Mitchel, who produced the first major monograph on the subject in 1977, are to be especially commended. A liberal sprinkling of "Lincoffisms" are found throughout *Mushrooms: Poisons and Panaceas*, since his wisdom and wit are eternal. Dr. Emanuel Salzman and Dr. Barry Rumack did much to put the subject on a solid scientific foundation with the conferences they organized in the Rocky Mountains in the 1970s. Although recently revised, their original book on the subject, remains a classic in the field.

Dr. Rhona Jack and Ms. Cheryl Herdon, both scientific illustrators, provided the artwork, and the majority of the excellent photographs are the work of Mr. Steven Trudell, a dedicated amateur mycologist and photographer. Additional photographs were kindly provided by Dr. Michael Beug and Paul Stamets. Steve Bobbink of the Regional Poison Center in Seattle, Professor W. O. Robertson of the University of Washington and Director of the Regional Poison Center, Dr. Sandra Schneider of Strong Memorial Hospital at the University of Rochester, Dr. John Trestrail III, Chairman of the North American Mycological Association Toxicology Committee and Director of the Regional Poison Center in Grand Rapids, Michigan, and Dr. Gregory Mueller of the Field Museum in Chicago kindly reviewed selected chapters. Ms. Jan Lindgren, Dr. Nancy Turner, and Dr. Nancy Smith Weber all provided unique expertise. Ms. Lorelei Norvell generously permitted the use of an illustration. And to all those involved in the production of the book, especially my editor, Deborah Allen, who believed in the project, special thanks. It was a privilege to have the professional advice of Jodi Simpson and Penelope Hull; I learned a great deal from them.

Uncountable mycologists, naturalists, and physicians have, over the years, made the careful observations or performed the experiments on which this work is entirely dependent. Their contributions are more extensive than their listings in the references suggest.

We cannot forget all those who unwittingly sacrificed their lives. I hope we have learned much from them.

D.R.B.

Mushrooms and Health

CHAPTER ONE

CULTURAL ATTITUDES
TOWARD MUSHROOMS

Beware of muserons, moch purslane,
gourdes and all other things
whiche wyll sone putrifie.

Thomas Elyot, The Castel of Helth (1541)

My voice
Becomes the wind
Mushroom hunting.

Shiku (nineteenth-century Japanese poet)

I have the same opinion of dances that physicians have of mushrooms: the best of them are good for nothing.

Saint Francis de Sales, Introduction to the Devout Life (1609)

IN MANY RESPECTS the history of mushroom poisoning and its treatment reflects the attitudes of various cultures toward the consumption of mushrooms. Cultures with a tradition of picking and eating wild mushrooms have a much higher incidence of poisoning and a far greater interest in its prevention and management. Not surprisingly, the overwhelming number of scientific studies have been done in Europe, with France and Germany leading the way, followed by many of the eastern European countries, Russia, Japan, and China. In societies with well-developed written records, the evidence for fungus eating is well preserved. However, in the preliterate cultures—those many small groups scattered in the remote corners of the globe, who are the favorite subjects for anthropological research—the evidence is very sketchy. Many ethnobotanists seem to have little interest in fungi, and their surveys rarely mention mushrooms. Frequently their trips to these remote areas do not coincide with the main fruiting or picking season; and because there is often a clear division of labor between the sexes, ethnobotanists may inadvertently choose to interview the one that is the least knowledgeable about mushrooms.

It has been suggested that the world can be conveniently divided into two great camps—the mycophiles and the mycophobes. R. Gordon Wasson, an American banker-ethnomycologist, is sometimes credited with being the first to make this distinction. In fact, references to it can be found in the old herbals, and the term *fungophobia* was coined by a British mycologist in 1887.[1] The English did not need an American banker to inform them of one of their little foibles. This division of the world into camps based on a food preference is reminiscent of an old saw: When it comes to eating cilantro, there are three kinds of people—those who love it, those who hate it, and those who have never tried it. Similarly, the division of the world's peoples into those who love and those who disdain fungi may be convenient, but it is overly simplistic; some cultures have a fairly neutral attitude toward fungi. They may not use mushrooms extensively in their diets or for medicine, but neither do they actively despise them. These people are the "mycoindifferent."

Most of the world eats and collects mushrooms with a passion varying from a national pastime to an occasional indulgence. Local prejudices and tastes have selected out the favorite species, which are distinctive for each area. Mushrooms, regarded as perfectly edible and even highly desirable in one region, may be shunned in another region for other species that most groups would not consider eating. The edible species frequently acquire common names in the marketplace. There is a mycodictionary with descriptions of all the edible species, their common names, and all the appropriate references as to the edibility and safety of the various species.[2] It includes the names of 628 species that are eaten in one or more regions of the world.

Of comparable interest is the use of mushrooms for psychic rather than physical sustenance. This practice is best illustrated by the ritualistic use of hallucinogenic mushrooms in Mesoamerica, brought to the attention of the Western world in the 1950s by R. Gordon Wasson. His article in *Life* magazine introduced the concept of the "magic mushroom" to the American public. This subject is dealt with in much greater detail in Chapter 17. Even before the 1950s, however, the use of certain mushrooms for chemical recreation was widely known. This topic is explored in some depth in Chapter 16.

ANGLO-SAXON CULTURES

The mycophobia of the Anglo-Saxons is widely recognized—and is duly recorded in each and every modern mushroom guidebook. The French have disparaged the British for this attitude—"*les habitants des îles britanniques sont l'archetype des peuples mycophobes*"[3]—although, as will be seen, their own, more positive attitudes toward fungi are of rather recent origin. There is no denying that the British and the populations they spawned in their many colonies during the heyday of the Empire are not great mushroom eaters. Indeed, most people in the English-speaking world are terrified by the thought of eating a wild mushroom. The mere mention of the word *toadstool* conjures up a whole panoply of fearful images, including witches, death, and decay. The term *toadstool* originated in the Middle Ages and is still used by some to indicate a poisonous mushroom. It has no utility today other than common usage. The specific origin of the term has been debated for centuries without resolution. Because I make no claim as an etymologist, I will avoid inflaming the argument further. For those interested in such things, a chapter in *The Romance of the Fungus World*[4] and Baker's recent contribution to the debate should provide a certain amusement.[5] Exploring the term's origins may be instructive, for those origins may be associated with the fear Anglo-Saxon descendants feel when the words *mushroom* and *toadstool* are mentioned.

Although it is true that the mushroom holds a special place of fear in the hearts of the British, most Americans, Canadians, Australians, and others who trace their descent from the subjects of Richard I Coeur de Lion also seldom use other foods not found in the local supermarket. Gathering and using wild plants of any variety is a lost skill in all these countries. The Anglo-Saxon cultures have become thoroughly domesticated. The use of wattle and woad is long forgotten. The average American's knowledge of wild food is usually limited to a dandelion salad once eaten in a sophisticated French restaurant; they never realize that their forebears brought the dandelion from Europe as a cultivated salad green. Indeed, in modern times our distance from the natural world of plants and animals, food gathering, and animal husbandry is considerable; our distance from mushrooms and fungi is a yawning chasm.

In the parks, meadows, and hills of every town or on the beaches of English-speaking, industrialized countries, however, one commonly sees immigrants from Asia collecting a large variety of different plants for home consumption. And descendants of the great Slavic cultures of eastern Europe and Russia cannot wait to get out of the city and into the woods each spring and fall. In addition, the offspring of certain other immigrant groups, such as Italians, have maintained a small cadre of dedicated mushroom hunters. These are exceptions in the English-speaking cultures, tolerated as being mildly eccentric.

Wasson indicted all Celtic and Germanic peoples as being myco-phobic, although he did admit that travel and education have had some impact in recent generations[6]—so much so, in fact, that certain food companies in western Germany, for example, have been exploit-ing the wild mushroom resources of places like the Pacific Northwest in the United States to satisfy the appetites of their population, their own resources having become exhausted, polluted, or otherwise inade-quate. It is also probably true that the current passion the French have for the mushroom is relatively recent. Some reports during the late nineteenth century indicate more wild mushrooms being offered for sale in English markets than in the markets of Paris.[7] This may have been due to the regulations banning mushrooms for a while from Paris—because of a serious outbreak of poisoning—rather than to any intrinsic dislike. However, the following quotation reflects the French attitude only 200 years ago:

> But whatever dressing one gives to mushrooms, to whatever sauce our Apiciuses put them, they are really good but to be sent back to the dung heap where they were born.
>
> Louis de Jacourt, Champignons (1753)

Earlier than this, the antipathy of both peoples had been well expressed:

> Fungi ben mussherons . . . There be two manner of them; one manner is deedly and sleeth them that eateth of them and be called tode stooles, and the other dooth not. They that be not deedly have a grosse gleymy [slimy] moysture that is dysobedyent to nature and dygestyon and be peryllous and dredfull to eate & therefore it is good to eschew them.
>
> The Grete Herball (1526)

Such opinions would not have endeared the authors to the mush-room growers' association and marketing board. This judgment was further reinforced by the influential herbal written by Gerard less than a century later:

> Many wantons that dwell neere the sea, and have fish at will, are very desirous for change of diet to feede upon the birds of the mountains; and such as dwell upon the hills or champion grounds, do long after sea fish. Many that do have plenty of both do hunger after the earth's excrescence's called mushrooms, whereof some are venomous, others not so noisome, and neither of them very wholesome meat. Whereof for the avoiding of the venomous qualities of the one, and the other which is less venomous

may be discovered, I have thought good to set forth their pictures with their names and places of growth.

John Gerard, Herball or Generall Historie of Plantes (1597)

They are all very cold and moist and therefore do approach unto a venomous and muthering faculty and ingender clammy and cold nutriment if they be eaten.

Gerard, Herball

Few mushrooms are good to be eaten and most do suffocate and strangle the eater. Therefore I give my advice unto those that love such strange and new fangles meates to beware licking the honey among the thorns lest the sweetness of the one do not countervaile the sharpness and pricking of the other.

Gerard, Herball

Parkinson's *Theatre of Plants* (*Theatrum botanicum*), published a few years after the second edition of Gerard's *Herball* (1633), was only marginally kinder toward mushrooms. Already he noted that mushroom eating was more popular in other countries:

> And because our country neither produceth much variety of good or bad, to like or mislike, our Nation also not being so addicted to the use of them, as the Italians and other nations are, where they grow more plentifully. I will therfore but runne them over briefly and not insist so much on them as in other things of better respect.

John Parkinson, Theatre of Plants (1640)

In fact, the early authors frequently copied drawings and phrases from previous works, and it is easy to trace the lineage of both correct ideas as well as misconceptions. In some respects this tradition still continues, even with this present study, although I would like to believe that I am both objective and critical. However, I no doubt demonstrate my prejudices in the passages I choose to reproduce.

Both Gerard and Parkinson reproduced drawings of the stinkhorn mushroom (*Phallus impudicus*), copied from Clusius, the very cosmopolitan herbalist who did part of his work in Holland. Interestingly, Gerard reproduced the drawing upside down, an error suggesting that he never personally saw the mushroom. (Or perhaps it is an early example of a printer's gremlin.) Gerard referred to this mushroom as "fungus virilis penis effigie" or the "pricke mushroom," whereas Parkinson called it "phallus hollandicus," or the "Hollanders workingtoole," and described it as being somewhat like a "dogges pricke." The reason for mentioning this particular mushroom, which, as can be gathered from the name,

Phallus impudicus

has a remarkably phallic appearance, is that one of the great stories of British mycophobia is embodied in this species. It also reflects the morals of Victorian England and involves the daughter of Charles Darwin. The following excerpt gives a flavor of the attitude toward what was clearly regarded as obscene:

> In our native woods there grows a kind of toadstool called in the vernacular The Stinkhorn (though in Latin it bears a grosser name). The name is justified for the fungus can be hunted by scent alone; and this was Aunt Etty's great invention. Armed with a basket and a pointed stick, and wearing a special hunting cloak and gloves, she would sniff her way through the wood, pausing here and there, her nostrils twitching when she caught a whiff of her prey. Then with a deadly pounce she would fall upon her victim and poke his putrid carcass into her basket. At the end of the day's sport the catch was brought back and burnt in the deepest secrecy on the drawing room fire with the door locked—because of the morals of the maids!!

Gwen Raverat, *Period Piece* (1952)

Where and how did these attitudes arise? The usual approach to this question is to quote from the old herbals and from many of the great works of English literature, where numerous examples of mycophobia are readily found, and to assume that the attitudes of the British were influenced and shaped by the pen. However, these writings may be the result of the cultural bias rather than its cause. It is unlikely that the mass of humanity was greatly affected by much of this early writing, because most had no access to it, nor could they read. It may have touched the few with a classical education. Public education, which is a rather recent phenomenon, is also unlikely to have played a major role. This argument can be applied equally well to the entire European continent, whose population had even closer links with many of the early works but did not develop the same fears. Moreover, a few hints in the medieval literature suggest that mushrooms were at least a part of the diet and culture in old England, even appearing on a few coats of arms. Rather than the cause, it seems as if a deeply rooted fear merely became validated by the British literature and reflected in their art. Even the originators of the mycophobic epithet, such as Wasson, do not really give a good indication of the place, time, or cause of the origins of this dislike. To continue to claim that this attitude is a part of the culture's mythology is to beg the question entirely.

Some have speculated that the sexual connotations associated with some mushrooms led to the mushrooms' rejection by a prudish Anglo-Saxon culture. Undoubtedly, the phallic aspects of many mushrooms did not go unnoticed. Certain species, in addition to truffles, were regarded as aphrodisiacs. Although this argument may seem reasonable in the context of Victorian England, there is little supporting evidence. It is also evident that the negative attitude predated the prudish years. Furthermore, Puritanism is alive and well in America, but few people seriously believe that this ethic is the basis of American mycophobia.

What about some of the associations mushrooms have in Western mythology? Mushrooms were closely allied with toads, snails, snakes, spiders, and witches, all things regarded as inherently dangerous and evil. Once again, the problem of cause and effect arises. But the relationship of the English and the Germanic peoples in general to their witches does deserve some consideration. Although witches have never been beloved in any of the countries or cultures in which they have plied their skills, it was largely the Germanic peoples who developed a deep-seated fear of them. This aversion may have been related partly to the association of witches with the Devil and the outlawing of witchcraft by the church in the Middle Ages. It was primarily in the Germanic, Celtic, and Anglo-Saxon countries such as Germany,

Switzerland, Scotland, England, and subsequently in America, that witches were actively and systematically persecuted. Through the rest of eastern Europe, they continued to play an important quasireligious-medical-magical role. Is it possible that the association of the evil witch, a product refined and polished by the English, together with all her accoutrements, resulted in condemnation of the mushroom? Guilt by association—a highly speculative scenario.

Whatever its origins, the results of this phobia have been profound. Some of the most pejorative prose and poetry in English literature is directed against the mushroom. Shelley, for example, described the final dissolution of a beautiful garden that had become neglected after the death of its guardian. It was not enough for him to allow it to return to the wildness of nature. Instead, he used mushroom images to underscore its total disintegration and decay.

> Plants to whose names the verses feel loath
> Filled the place with a monstrous undergrowth,
> Prickly and pulpous and blistering and blue,
> And agarics and fungi and mildew and mould
> Started like mist from the wet ground cold,
> Pale fleshly as if the decaying dead
> With a spirit of growth had been animated!
>
> Their moss rotted off them, flake by flake,
> Till the thick stalk stuck like a murderer's stake,
> Where rags of loose flesh yet tremble on high,
> Infecting the winds that wander by.

P. B. Shelley, *The Sensitive Plant* (1820)

Indeed, the relationship of mushrooms to death and decay is a constantly recurring theme. This association comprises many of the negative attitudes toward mushrooms found in the literature of all English-speaking areas. Longfellow used the images that fungi conjure up to highlight the despondency of a fallen hero:

> Wounded, weary and desponding,
> With his mighty war-club broken,
> With his mittens torn and tattered,
> And three useless arrows only,
> Paused to rest beneath a pine tree,
> From whose branches trailed the mosses,
> And whose trunk was coated over
> With the Dead-man's Moccasin-leather,
> With the fungus white and yellow.

H. W. Longfellow, *The Song of Hiawatha* (1855)

The images of death and decay evoked by mushrooms were exploited by other great writers, for example, Arthur Conan Doyle and D. H. Lawrence.

> The rain had ceased at last, and a sickly autumn sun shone upon a land that was soaked and sodden with water. Wet and rotten leaves reeked and festered under the foul haze which rose from the woods. The fields were spotted with monstrous fungi of a size and colour never matched before—scarlet and mauve and liver and black. It was as though the sick earth had burst forth into foul pustules; mildew and lichen mottled the walls, and with that filthy crop, Death sprang also from the water soaked earth.
>
> Arthur Conan Doyle, *Sir Nigel* (1906)

> Nicely groomed, like a mushroom
> standing there so sleek and erect and eyeable
> and like a fungus, living on the remains of bygone life,
> sucking his like out of the dead leaves of greater life than his own.
>
> And even so, he's stale, he's been here too long,
> touch him, and you'll find he's allgone inside
> Just like an old mushroom, all wormy inside, and hollow
> under a smoothskin and an upright appearance.
>
> Full of seething, wormy, hollow feelings
> rather nasty—
> How beastly the bourgeois is!
> Standing in their thousands, these appearances, in damp England
> what a pity they can't all be kicked over
> like sickening toadstools, and left to melt back, swiftly
> into the soil of England.
>
> D. H. Lawrence, *How Beastly the Bourgeois Is* (1923)

The popular British viewpoint toward mushrooms was well expressed by W. D. Hay before the turn of the century and has changed little since:

> Among this vast family of plants, belonging to one class, yet diverse from one another, comprising more than a thousand distinct species indigenous to the islands, there is but one kind that the Englishman condescend to regard with favour. The rest are lumped together in one sweeping condemnation. They are looked upon as vegetable vermin, only made to be destroyed. No eye can see their beauties; their office is unknown; their varieties are not regarded; they are hardly allowed a place among Nature's lawful children, but are considered something abnormal, worthless and inexplicable. By precept and example children are taught from earliest infancy to despise, loathe, and avoid all kinds of "toadstools." The individ-

ual who desires to engage in a study of them must boldly face a good deal of scorn. He is laughed at for his strange taste among the better classes, and is actually regarded as a sort of idiot among the lower orders. No fad or hobby is esteemed so contemptible as that of the "fungus-hunter" or "toadstool-eater."

This popular sentiment, which we may coin the word "fungophobia" to express, is very curious. If it were human—that is, universal—one would be inclined to set it down as an instinct and to reverence it accordingly. But it is not human—it is merely British. It is so deep and intense a prejudice that it amounts to a national superstition. Fungophobia is merely a form of ignorance, of course; but its power over the British is immense, that the mycologist, anxious to impart the knowledge he has gleaned to others, often meets with scarcely credence or respect.

W. D. Hay, *British Fungi* (1887)

People have even accused that great writer of children's stories, Enid Blyton, of contributing to the fear of mushrooms. Her characters Noddy and Big Ears have recently come under scrutiny, not only regarding mushrooms, but also regarding other politically incorrect attitudes toward homosexuality, feminism, and racism. Conversely, the French story of Babar the Elephant, by Jean de Brunhoff, which begins with the death of the king by mushroom poisoning, appears to have done little to dampen the enthusiasm of the French children for fungi. It is clearly much too easy to indict the books of childhood as the cause of the fear. The attitudes of the parents, however, may be the key element determining how the children view the natural world as well as how they interpret and integrate the stories they hear. There is no doubt that the English are unanimous on this subject: "Don't touch that mushroom!" "Toadstools will kill you."

During the Middle Ages, when superstition superseded the existing knowledge of the natural world, mushrooms were firmly embedded in a mythology of the supernatural. The myths may have arisen because of mushrooms' strange habits—their sudden appearance, as if from nowhere, and their extraordinarily rapid growth, the rings they form, the bioluminescence of some, their bizarre shapes, and their ephemeral nature. They became part of the lives of fairies, elves, and witches. This mythology lingers on in some of the common names for mushrooms, like the "fairy ring" for *Marasmius oreades* and "witches butter" or "fairy butter" for *Tremella mesenterica*. In *The Romance of the Fungus World*, the Rolfes concluded their chapter on fungi in mythology with quaint nostalgia:[4]

> In these decadent latter days, when fairies no longer haunt the groves as of yore, and when no more do witches career merrily on broomsticks over-head, it has been pleasant to consider them for a moment, and with them the queer beliefs associating them with the toadstools, for, since a knowledge of the early history of a subject is necessary before one can do full justice to its present state, one should be neither scornful nor unmindful of such fantasies as we have endeavoured to chronicle, and which may other-wise become lost in the vast quantity of material dealing with these plants which has now accumulated.

Some brave efforts on the part of a few mycoevangelists have attempted to reverse the trend of mycophobia, but the current against which they have had to swim has been much too swift. Charles Badham made one of the best attempts in 1863.[8] He began his treatise with a familiar plaint:

> No country is perhaps richer in esculent Funguses than our own; we have upwards of thirty species abounding in our woods. No markets might therefore be better supplied than the English, and yet England is the only country in Europe where this important and savory food is, from igno-rance or prejudice, left to perish ungathered.

Badham was obviously aroused by the time he spent in Italy, where he appears to have developed his passion for mushrooms. The power of travel in educating a palate is reflected by the many people influenced by the flavors of Italy and France:

> The extremely limited time during which the Funguses are to be found, their fragility, their infinite diversity, their ephemeral existence, these, too, add to the interest of an autumnal walk in the quest of them. . . .
>
> It is a pleasant remembrance to have plucked the crimson Amanite, that ministered to a Caesar's decease, in the very neighborhood of the Palatine Hill; to have collected mushrooms amidst the meadows of Horace's farm, where he tells they grew the best, and to have watched along the moist pastures of the Cremera a stand of the stately *Ag. procera* nodding upon their stalks.

He concluded the book with an ardent plea to the populace, a plea that fell on prejudiced ears:

> He may at first alarm his friends' cooks, but their fear will, I promise him, soon be appeased, after one or two trials of this new class of viands, and he will not long pass for a conjurer or worse, in giving directions to stew toadstools. As soon as he is initiated into this class of dainties, he will, I am persuaded, lose no time in making the discovery known to the neighborhood; while in so doing he will render an important service to the country at large, by instructing the indigent and the ignorant in the choice of an ample, wholesome, and excellent article, which they may convert into money, or consume at their own tables, when properly prepared throughout the winter.
>
> C. D. Badham, *A Treatise on the Esculent Funguses of England* (1863)

A century later, the English food writer Jane Grigson wrote one of the great mushroom cookbooks, *The Mushroom Feast*.[9] Admittedly, all her anecdotes and experiences were based in France, and her approach to the subject was nowhere tainted by the traditional British mind-set.

Stephen Jay Gould recently best expressed our continuing dislike of the fungal world.[10] His essay followed the widely acclaimed discovery of an enormous underground fungal organism, which challenged some fundamental tenets of biology as well as the idea that a fungus could compete with more socially acceptable creatures such as whales and redwood trees.

> When an animal achieves disembodied immortality by becoming a verb, human speakers usually honor its behavior: we hawk our wares, gull or buffalo our naive competitors, hound our adversaries, and clam up in the face of adversity; we have also been known to man the barricades and kid around with our companions. But plants and other rooted creatures do not feature so great a range of overt action, and our botanically based verbs therefore tout growth and appearance as sources of metaphor.

Consider the two most prominent examples, citing comparable phenomena but with such different meanings—for one usually expresses our joy and the other our fear. Art and prosperity "flower"; taxes and urban violence "mushroom." The burden of difference reflects an obvious source in our culture and legends. We love the bright colors of the "higher" plants, either radiant in the sunlight or jewellike in the quiet darkness of the forest. We loathe the spongy, fruiting bodies of "lowly" fungi, growing in dank dampness, sprouting in cancerous formlessness from rotting logs. (Even a colorful mushroom usually strikes us as sinister rather than lovely.) I well remember a common schoolyard taunt, often cruelly directed at unloved classmates: "There's a fungus among us"—a cry that always inspired the ritual retort: "Kill it before it multiplies."

Stephen Jay Gould, *A Humongous Fungus Among Us* (1992)

Perhaps the greatest illustration of our Western mycophobia is to be found in a book by S. A. Friedman about the pleasures of mushroom hunting and gourmandism. Published in 1986, *Celebrating the Wild Mushroom: A Passionate Quest* is an ode to overcoming fear.[11] Its message brings into stark relief the centuries of distrust and dread.

AFRICA

"Mushrooms!" exclaimed Kamba the Tortoise, joyfully. "Do I see mushrooms? REAL mushrooms?"

Yes, they were real mushrooms, little, white, satiny buttony mushrooms, with lovely pink underneaths; little white mushrooms that had pushed all night at the dark brown earth above them, and had struggled through its hard crust just in time to see the sun rise, just in time to make a fine breakfast for a hungry Tortoise.

From a Malawi folk tale recounted in
G. Elliot, *The Long Grass Whispers* (1957)

A considerable amount of information is available on the use of edible fungi on the African continent. In some regions, such as Nigeria, it is evident that mushrooms are part of daily life as food, charms, and remedies in traditional medicine.[12,13] The most favored species are those associated with termite nests; these species fruit after the heavy rains and all belong to the genus *Termitomyces*. Numerous myths and legends surround these fungi, and a variety of unreliable folklore exists to differentiate the poisonous from the edible. The origins of such tales are unknown. The most common involve giving the mushroom to a fowl, usually a chicken. If the fowl eats the mushroom, it is regarded as edible; if the fowl rejects the mushroom, it is regarded as poisonous. If the fowl eats the specimen and then later vomits, it is considered to be "edible with caution"—which makes as much sense to the Yoruba as it does to the American reading some of the guidebooks in the United States. In addition to the fungi above the ground, the sclerotium of *Pleurotus tuber-regium* is of considerable economic importance as both a food and a medicine.[14]

Sir H. M. Stanley's journal, *In Darkest Africa* (1890), contains an interesting historical reference to the culinary use of mushrooms in Africa, a note written while he was on his way to interview Livingston: "Our Nyanza people were provident and eked out our stores with mushrooms and wild fruit." Unfortunately, not all the mushroom gathering proved to be entirely safe, because when he returned to camp one day, he found that "six people had succumbed, a Madi from a poisonous fungus."

Mushrooms also play a major role in the diet of rural Zambia and Zaire. They are most widely eaten during the "hunger" months, from late November through early April. A nutritional survey in the 1950s demonstrated that they were second only to insects (mainly caterpillars) as a food source during the rainy period.[15] Despite a wide range of mushrooms fruiting in the region, the selection is relatively restricted, and a number of commonly edible species, including boletes, are ignored. Once again, species of the genus *Termitomyces* are the most favored, although members of *Lactarius, Russula, Cantharellus,* and *Amanita* are also eaten. During the rainy season, impromptu markets spring up along the roads, each offering a wide variety of edible mushrooms. The traditional way of eating them in Zambia is to boil them in salted water, after which onions, tomatoes, or groundnuts may be added. A little cooking oil may also be included to turn the mixture into a type of mushroom stew, eaten with the staple, thick, maizemeal (cornmeal) paste called *nshima*.

References to mushrooms also abound in the folklore and language of Zambia and Zaire. Perhaps most illustrative is the Tonga proverb, *Sibbuzya takolwi bowa,* which literally translated means, "The one who asks is the one who does not get poisoned by mushrooms"—implying that one should always consult the elders or the more experienced members of the community.[16]

Neighboring Malawi is also home to a mushroom-appreciating people, whose women have a wide knowledge of both the edible and the poisonous varieties. One recent study recognized over 60 edible species, belonging predominantly to the genera *Amanita, Cantharellus,* and *Termitomyces.*[17] Unlike the peoples of Madagascar, who comsume boletes with great relish, the peoples of central and eastern Africa largely ignore them, favoring the chanterelles instead.[18] Both the knowledge and the responsibility for foraging rests with the women. During the period of fruiting, a considerable portion of the day may be devoted to collecting. Mushrooms are also sold at the roadside as well as in the town and village markets. Despite the vast quantities of mushrooms eaten each year, few poisonings are recorded. In 1984 Morris could report only a single known episode involving a group of forestry workers—all men, who probably did not have the experience—who were hospitalized after eating some fungi from a pine plantation.[17] The poisoning species was never identified, but it could have been *Amanita muscaria*.

The paucity of poisoning reports is intriguing. It could reflect the rarity of seriously toxic species in the region, such as *Amanita phalloides*, which is more common further south. Or it could indicate an extraordinarily high level of knowledge and folklore about mushrooms in the local, stable population. It will be interesting to observe any changes that occur in the next few decades as these regions industrialize and urbanize. It is also possible, of course, that many cases of poisoning are not recognized for what they are, because of a lack of medical resources, or have not been recorded in the literature.

Not all of Africa is enthusiastic about the local fungi. The peoples of northern Africa are far more reticent about using fungi, although certain groups still prize the desert truffle. South of the fungus-loving peoples of central Africa, in the region bordering Angola and Namibia—the land of the Ovambos—a few mushrooms are still used as food, but not in great variety or number.[19]

Although Africa leads the world in its consumption of the genus *Termitomyces*, these mushrooms are eaten wherever termites farm their fungal crop. The fungi, carefully tended by the termites in subterranean gardens, grow on and break down the cellulose and lignin of the plant material that the termites have gathered from the vegetation surrounding the nest. This is a fascinating group of fungi, especially in regard to the relationship that has evolved between two such disparate organisms—an example of mutual support rather than selfish exploitation. It is an evolutionary strategy that has proved beneficial for both organisms. For anyone intrigued by this topic—and able to read French—the monograph by the great mycologist Roger Heim is recommended.[20]

INDIA

Although mushrooms are eaten on the Indian subcontinent, they do not achieve the dietary prominence observed in the rest of Southeast Asia. The primary regions of consumption appear to be in the more mountainous areas north toward Kashmir and the Himalayas. There, morels are a treasured resource. Mushroom eating is also common in western Bengal. A survey of 11 markets in 1980 recorded 123 tons of wild mushrooms, including *Termitomyces* sp., *Tricholoma striatus*, *Tricholoma gigantium*, *Calvatia* (the giant puffball), and *Volveriella* sp. Species of the genus *Lepiota* are eaten in the south, where they fruit in abundance. Another interesting edible mushroom is *Calocybe indica*; attempts have been made to cultivate it commercially. A fascinating report in the *Edinburgh Review* (April 1869) stated, "We have been informed by a gentleman who has lived for many years in India that the natives seem to eat fungi promiscuously, chopping up the different species together, without any ill effects." Despite this historical reference, few of the current ethnobotanical surveys comment much on mushrooms. Whether only the surveyors or most of population are disinterested in mushrooms can be determined only by more focused ethnomycological study. Certainly the impression is of a culture lacking a major culinary interest in fungi. Although this attitude antedates the British colonization of the region, it is nevertheless difficult to gauge what impact the British fungophobia had on Indian attitudes. The British have had the habit of leaving behind many of their institutions and beliefs, including the bad ones.

A curious phenomenon has been reported from the Kerala district of India. Burial monuments, built by an ancient people between 2000 and 1000 B.C., are often composed of large stones resembling mushrooms. These monuments are worshipped by some of the present-day rural people, who also eat some of the hallucinogenic mushrooms growing in the area. They believe that the mushrooms confer on them the power to communicate with the Deity and the spirits of the dead. However, ceremonial, ritualistic, and religious uses of the hallucinogenic mushrooms do not seem to have evolved in this region, as they developed in Mesoamerica. The monuments still remain a mystery, and their relationship to mushrooms needs to be investigated further.[21]

NATIVE NORTH AMERICAN CULTURES

Despite the wide availability of fungi across the North American continent, native Americans have never used mushrooms as an important food source. Most ethnobotanical surveys failed to mention mushrooms as a significant or even occasional component of the diet. This omission is especially surprising in the Pacific Northwest and California, which have very rich mycoflora.[22–24] The native groups in British Columbia used names like "ground ghost" or "corpse" for the puffball. A general term used by the Naxalk people meant "having hats on the ground." The Athabascans and Kootenay groups shunned mushrooms. However, some mushrooms were eaten by nations across the Great Plains and in the East, including the Omaha, the Iroquois (even though they called puffballs "devil's bread"), the Miwok, and the Zuni. The interior Salish peoples ate both the matsutake, a mushroom greatly prized in Japan, and another related species (*Tricholoma populinum*), which grows under cottonwood trees.[25] It is possible that ready access to abundant first-class protein and other foods decreased the need for a food only available on a seasonal basis. The relative antipathy toward mushrooms and the limited use of mushrooms in the diet of the indigenous peoples of Canada, the Far North, and the immediate neighboring United States has recently been summarized.[26]

One widespread fungal food, at least in the East and Southeast was the "tuckahoe." In some of the earliest descriptions of fungi, which were all called "Indian bread" or "Indian potato," the authors also included roots, bulbs, and tubers. However, the name was eventually applied primarily to the sclerotium of a particular group of fungi, which was dug up and roasted. In the Southeast, the sclerotium was the product of *Poria* (= *Wolfiporia*) *cocos*.[27]

A sclerotium (from Greek *skleros*, "hard") is a mass of fungal element—sometimes incorporating soil, organic material, and other debris, sometimes consisting of the fungal hyphae alone—that usually grows under the ground and can resist some of the extreme environmental conditions to which the mushroom is exposed. Its production is a survival mechanism, because when the conditions improve and become propitious, the sclerotium can produce fruiting bodies and the fungus can reproduce. Sclerotia sometimes grow to be the size of large Idaho potatoes and are very nutritious.

Although mushrooms were not commonly eaten by native Americans, they were used in a variety of other ways, for either medicine or religious, spiritual, and ceremonial purposes. One medicinal practice—fairly universal among the native nations of North America—was the use of puffballs as a styptic (hemostatic substance) to staunch the flow of blood and to dry out wounds. (It is surprising how common this practice was on all continents.) The Rocky Mountain groups used the "prairie mushroom" to heal the umbilical cords of newborn infants. Peoples from the region of the Missouri River also used *Geastrum* and *Lycoperdon* for obstetrical purposes. It is from this application that the Dakota are said to derive their name for the puffball: *Hosli chepka*, meaning "baby's navel."[28]

The puffballs were used in a number of different ways as a hemostatic agent. One common method was to blow the spores of a mature specimen into the wound; another was to dry and pulverize a young specimen for later application. Fears existed about this fungus, also called "devil's snuff" by some. One was the belief that spores would cause blindness, a fear common around the world. Another was that inhaled spores would cause illness. The fear of blinding is not based in fact, although the spores can be very irritating to the eyes. Nevertheless, the Navajo were so convinced that spores had this property that their name for *Geastrum* and *Tulostoma* was "no-eyes." The second belief had a legitimate basis, because the inhalation of massive quantities of spores—of any variety, but especially those of the puffballs, which are released in such astounding quantities that they look like smoke—can cause a most unpleasant reactive airways disease, initially mimicking asthma. Puffballs were also used medicinally in poultices, a powder, and a lotion to treat various skin conditions.

The second use of fungi was for spiritual, religious, and ceremonial purposes. The Blackfoot called puffballs growing in typical fairy rings on the plains *ka-ka-toos*, meaning "fallen stars." They were used as an incense to ward off unwanted spirits. At times, tepees were decorated with images of puffballs, and medicine men were known to use dried puffballs filled with stones as rattles. There is no evidence that any of the hallucinogenic fungi were employed in any ceremonies as they were in Mesoamerica. In the Pacific Northwest, mushrooms were employed by individual as totems for hunting or by shamans for healing. It is intriguing that one of the common mushrooms selected for this purpose was the polypore fungus, *Fomes officinalis*, the same mushroom that was the "agarick" of the pharmacists in Europe. Not only was this fungus ground up and administered as a cure for many illnesses, but its supernatural powers were embodied in carvings of the fungus, which were used in ceremonies to heal and to protect both individuals and the community. After the death of the shaman, images carved out of the large sporocarps were placed at the head of his grave. This symbolic use of the mushroom seems to have been as potent a power as the use of hallucinogenic mushrooms was in Central America or in Eurasia.[29] Nancy Turner, an ethnobotanist from Victoria, British Columbia, reports that the Coast Salish of Vancouver Island and the adjacent mainland imbued some of the bracket fungi on trees with echo powers. Families who owned the right to the protective power of these tree fungi could reflect any "evil or malicious thought directed towards members of the family back to the person who sent them."[23] The Squamish peoples considered the tree fungus to have protective powers and would hang them up inside their homes to protect the inhabitants from evil thoughts.

EUROPE

Apart from Holland, where the attitude toward edible fungi is similar to that of the English, but to a lesser degree, European cultures are rather fond of wild mushrooms. The species favored by each group vary; for example, the Italians prefer the porcini (*Boletus edulis*) and the white truffle; the Germans and Swiss, the chanterelle; the Catalonians, the delicious milky cap (*Lactarius deliciosus*). In each of these places, however, the markets contain dozens of different varieties, because tastes for wild mushrooms are fairly catholic.

Switzerland provides an interesting example of the level of sophistication that has developed in both the picking and the marketing of wild mushrooms. In many non-European countries, especially those in developing countries, foraging and gathering are a woman's tasks, whereas in Switzerland they are generally male activities.[29] In most of the other European countries, they are a family affair.

During the spring and fall seasons, Swiss open-air markets reserve a place for the sale of wild mushrooms. The markets are under the supervision of an inspector, a local mycologist responsible to the state government. Only truffles can be sold outside these wild-mushroom markets. All the other mushrooms must pass the scrutiny of the inspector. (The inspector will identify any amateur's mushrooms as well.) Minors are not permitted to sell mushrooms. Species are labeled, taxes are paid, and all the regulations are enforced. The list of species officially sanctioned for sale numbers 54. Nevertheless, despite the care and surveillance, occasional deceptive practices have been detected. One of the most pernicious is to soak the mushrooms in water to increase their weight. To eliminate this problem, certain mushrooms, such as chanterelles, are sold by volume rather than by weight. The boletes have to be cut in half so that the customer can see the quantity of animal protein (insect larvae) inside. This is a wise regulation, because some boletes are so infested with the larvae of a variety of organisms that they could literally escape from the car before reaching home.

An interesting example of how the tastes of a people can be changed is illustrated by the situation in Finland. Mushroom eating was not a traditional practice there, except in the southeastern parts of the country, closest to Russia and its Slavic influence. In this area, mushrooms had been gathered for centuries by the members of the Orthodox Church, who dry large quantities of *Boletus edulis* for their Lenten food. During World War II, mushroom usage increased as other food stuffs became scarce and supplemental nutrition proved essential. Some years later, the Finnish government, together with a number of other organizations, launched an educational training program. More than 1600 advisers and 50,000 pickers were trained from 1969 to 1983.[31] The effort was coordinated by 22 inspectors who provided the expertise in identification. Every citizen in Finland had the right to pick both mushrooms and berries on any land, provided no damage was done. By 1979, an estimated 72% of the population was picking mushrooms. The income from the sale of the fungi, both locally and abroad, is untaxed; and the income generated during a good year can be substantial. Poisoning has been very infrequent, with only six deaths recorded between 1936 and 1978, despite the presence of both *Cortinarius speciosissimus* and *Amanita virosa*. (*Amanita phalloides* is very rare.)

As one moves into eastern Europe, the passion for mushrooms grows more intense, with a sizable fraction of the population participating in collecting fungi, a valued social and recreational activity in the fall at the peak of the fruiting season. In contrast to the field guides in North America, which repeatedly warn about the dangers of mushroom eating—some to the point of being funereal in character—a Czechoslovakian guide begins its section on edibility with a rather positive statement: "In comparison with the number of edible or harmless species, the number of poisonous mushrooms is quite insignificant."[32] Regulations in this country permitted the marketing of 75 species, with a detailed set of rules governing their pricing and selling. These rules are beautifully designed to protect the public and ensure that quality mushrooms reach the market. Four price groups were established according to "quality," which really translates into "desirability." At the top of the prime list is *Boletus edulis*, followed by the meadow mushroom, the morel, *Gyromitra esculenta*, the truffle, and St. George's mushroom (*Tricholoma georgii*). Way down in group three are the chanterelle and the oyster mushroom, once again illustrating how the tastes of nations and peoples vary.

RUSSIA

If you think you are a mushroom, jump into the basket.

Russian proverb

The fallen leaves are already smelling like spice-cakes. The white mushrooms are uncommon, but if you find them, you pounce on them like a black kite, cut them off and remember, that you promised yourself not to cut them off right away when you saw them, but to admire them first. Again I promised myself and forgot.

Mikhail Mikhailovich Prishvin (1873–1954), *The Eyes of the Earth*

One mushroom hunter comes home with small mushrooms, another with big ones. One is attentive, and using the power of attention, sees the mushrooms. The other doesn't see the small things around him and does-n't direct his attention on the mushroom, but rather the mushroom itself attracts his attention. This kind of mushroom hunter has mainly big mushrooms.

Prishvin, *The Eyes of the Earth*

If the British Isles are at one end of the spectrum, then Russia is at the opposite end. For those of us raised in the British heritage, it is difficult to even imagine the role mushrooms play in the Slavic culture. The word *mycophilic* does not do justice to the passion and pleasure that these peoples take in their fungal flora. A hint of this may be found in Wasson's first contact with mushrooms through his Russian-born wife, Valentina:

We had been married less than a year and we were off on our first holiday, at Big Indian in the Catskills. On that first day, as the sun was declining in the west, we set out on a stroll, the forest on our left and a clearing on the right. Though we had known each other for years we had never discussed mushrooms together. All of a sudden she darted from my side. With cries of ecstasy she flew to the forest glade, where she had discovered mushrooms of various kinds carpeting the ground. Since Russia she had seen nothing like it. Left planted on a mountain trail, I called to her to take care, to come back. They were toadstools she was gathering, poisonous, putrid, disgusting. She only laughed the more: I can hear her now. She knelt in poses of adoration. She spoke to them with endearing Russian diminutives. She gathered the toadstools in a kind of pinafore that she was wearing, and brought them to our lodge. Some she strung on threads to hang up and dry for winter use. Others she served that night, either with the soup or the meat, according to their kind. I refused to touch them.

R. Gordon Wasson, *Soma: Divine Mushroom of Immortality* (1968)

In three separate visits to the Russian republic, I have yet to meet anyone who did not pick and eat wild mushrooms. Fungi appear in salted or pickled form at the beginning of meals during most of the year. Mushroom hunting is not restricted to the country folk. Every urbanite has favorite spots outside the cities in the vast birch and pine forests that still cover much of the land. In fall, trains make impromptu stops along the way to pick up or let off women or families on a day's mushroom outing. It is a form of national recreation. Small brochures and pamphlets can be purchased at stores and kiosks along the streets, with drawings or photographs and descriptions of the commonly edible species.

Naturally, the Russian literature reflects this passion. In Tolstoy's *Anna Karenina* (1877), mushroom hunting plays a part in at least three scenes. Always it is a joyful and pleasurable experience. In one scene the children are squabbling, having been rebuked by their strict governess, who is naturally English. Their mother comes in and suggests that they get into their old clothes and go out mushroom gathering. The atmosphere immediately changes as the nursery is filled with yelps of delight and anticipation. A love scene in the woods is interrupted when the couple gets sidetracked into a discussion about mushroom identification.

SOUTH AMERICA

For the most part, fungi are not a major food item for the majority of the native peoples of this continent. It is, however, interesting to recall Charles Darwin's observation that the peoples of Tierra del Fuego sustained themselves on a fungus, *Cyttaria darwinia*.

> [It] has a mucilaginous, slightly sweetish taste, with a faint smell like that of a mushroom. . . . Tierra del Fuego is the only country in the world where a cryptogamic plant affords a staple article of food.
>
> Charles Darwin, *Journal of Researches into the Geology and Natural History of the Various Countries Visited by H.M.S. Beagle* (1839)

The natives call this fungus "summer fruit." A local, when asked what they had to eat, replied, "Plenty of fish and too much summer fruit."[33] The *Cyttaria* species belongs to the class Ascomycetes. It is most frequently found growing on beech trees (*Nothofagus*) in South America and is often available in the markets of the regions where beeches are common. Because Australasia was joined with South America in the ancient days of Gondwanaland, a similar group of fungi live on the beech and myrtle-beech trees of New Zealand, Tasmania, and southern Australia, although they do not appear to have been used for food in these regions.

A recent study of the ethnobotany of a group of Indians in the Brazilian Amazonian forest, the Waimiri Atroari people, failed to identify fungi as part of the diet.[34] Similar observations have been made in other surveys.[35,36] There are a few exceptions, however. In northern Brazil the Yanomamo are said to eat a wide variety of fungi.[36] As in many cultures, both the knowledge of the fungi and the responsibility for harvesting rest with the women. It is also intriguing that this people has two words for eating, one for eating meat and the other for eating all other dietary items. They use the word for eating meat when referring to the fungi. They have an elaborate taxonomic system for the edible fungi and can distinguish varieties that look almost identical, but they do not have a similar system for the inedible ones, calling them all by a single disparaging word that means "no-good." This linguistic practice resembles that of other cultures whose mushroom knowledge is based on a folk tradition; in such cultures, no reason exists to name those things that are of no value for food, medicine, or other uses. The majority of the species eaten by the Yanomamo grow on fallen logs and standing stumps, fitting in well with their agricultural tradition. As an emergency food, they also use the sclerotium of the *Polyporus indigeus*, which was called Indian bread by the white settlers and is very similar to the tuckahoes in North America. Even so, one gets the impression that this is not a favored food source, despite its wide availability at most seasons of the year.

ASIA AND AUSTRALASIA

It is perhaps unreasonable to lump the major cultures of Asia together, except for the fact that they are all generally mycophilic. However, the expressions of this attitude are strikingly different in the various countries. China has undoubtedly the longest tradition—not only do mushrooms play an important culinary role, but fungi also make up a substantial proportion of the traditional pharmacopoeia. The practice of this traditional medicine is discussed in Chapter 5. It involves a large number of different fungi, some of them quite exotic. A good example is *Cordyceps sinensis*. This fungus parasitizes and kills caterpillars. The fungus, still attached to the dead caterpillars, is made up into small bundles and used in a variety of ways, the most interesting of which is roasting in the stomach of a duck.

The Chinese mycophilia is different from that in Russia, where the population at large picks wild mushrooms. In China, most of the mushrooms now used in the diet are cultivated, domesticated varieties. So much of the population is concentrated in the intensively farmed areas that access to natural habitat for the average Chinese is very limited. The mushrooms for medicinal purposes have to be sought in regions a long way from the densely populated agricultural areas and urban centers. And some of the most prized species, like the ling chi (*Ganoderma lucidum*), are now cultivated, much to the consternation of many of the herbalists. Because the Chinese use a number of different mushrooms in their diet, they have developed considerable skill in cultivation, especially for those species seldom used in the West. Reports of poisoning are very infrequent in the Chinese medical literature, or at least in the part translated or abstracted into English.[37,38]

Japan, too, has a sophisticated mushroom ethic. Once again cultivation plays an important role in supplying mushrooms to a huge population living in limited space. But the Japanese approach the Russians in enthusiasm for the hunt itself. Perhaps the most prized edible mushroom is the matsutake or pine mushroom (*Tricholoma matsutake*). This mushroom has a unique scent, which once smelled is never forgotten. The aroma is difficult to describe and not liked by everyone exposed to it. In fact, I have the distinct advantage of having a mushroom-hunting companion who loves to find the matsutake but does not like the bouquet or the taste. All the spoils of the day's hunt are handed to me in exchange for the chanterelles and boletes. However, my wife has informed me that the only mushroom I need bring home is the matsutake. In the United States, the matsutake is probably a different species and goes under the appellation of *Tricholoma magnivalere*. In *Mushrooms Demystified*, David Arora, a master at describing textures, scents, and flavors, is almost at a loss for words for this particular mushroom—his portrayal of it as a "combination of red hots and old socks"[39] does not do justice to the spicy aroma but is better than most.

Unfortunately, the habitat for the matsutake in Japan has come under a great deal of pressure in the past few decades, and the mushroom has suffered a serious decline. In part, this is due to an infestation of a nematode in the great pine forests in which this mushroom forms a mycorrhizal relationship with the pine trees. Along with the trees, the mushroom is also starting to disappear. Each fall, millions of Japanese make pilgrimages to the forest to find this mushroom. The event is both social and aesthetic, one that is eagerly anticipated.

Unless a reversal in the fortunes of this mushroom occurs sometime soon, its future looks bleak indeed. A time may come when the only remaining matsutake are in preserves where people can look but cannot pick. However, entrepreneurs in Korea and the Pacific Northwest, from Oregon up through British Columbia, have developed a lucrative business supplying the Japanese market. Prices depend on the scent of the mushroom and its size. Most highly prized and commanding the top price are the unopened buttons, with long tapering stems, which have a decidedly phallic appearance. These mushrooms are now being exported in vast quantities, and harvesting techniques have become quite sophisticated. In British Columbia, pickers are flown by float planes to small remote lakes in the interior, where they may spend up to a week collecting before being airlifted out. Even satellite images have been used to pinpoint likely habitats. Many other mushrooms are also eagerly sought by the Japanese, some of which are not commonly eaten in the West. Incidents of poisoning are recorded from Japan, although not with great frequency.

The medicinal role of *Ganoderma lucidum* is also highly valued in Japan, where it goes under the name of reishi. Maitake (*Grifola frondosa*) is the medicinal mushroom of the 1990s, while shiitake (*Lentinus edodes*) and others still play important culinary and therapeutic roles.

Unlike the Anglo-Saxons and later Puritans, Asian cultures actively admire the phallic nature of some fungi. One of the common Chinese names for penis is the "swelling mushroom." In Japan, the more phallic the appearance, the greater the price commanded by the mushroom. Many mushrooms are imbued with the power of enhancing potency and performance.

Australasia merits but a few brief remarks, despite its size, because the original Anglo-Saxon settlers brought their mycophobia in their baggage. Few mushrooms are picked and used by the white population here, and the incidence of poisoning is consequently very low.[40] The aboriginal use of fungi is said to be limited to the sclerotium of *Polyporus myllitae*, also known as "black fellow's bread." In New Zealand, the Maoris traditionally ate a number of different fungi; for a time, large quantities of *Auricularia* species were grown and sold to China. Mushrooms were never significant food items for any of these native peoples.

REFERENCES

1. W. D. Hay, *British Fungi* (London: Swan Sonnenschein, Lowry, 1887), p. 6.

2. A. Chandra, *Dictionary of Edible Mushrooms: Botanical and Common Names in Various Languages of the World* (New York: Elsevier Science, 1989).

3. D. Thoen, G. Parent, and L. Tahiteya, "L'usage des champignons dans le Haut-Shaba," *Bulletin du CEPSE* (1973) 100–101:69–85.

4. R. T. Rolfe and F. W. Rolfe, *The Romance of the Fungus World* (London: Chapman Hall, 1925).

5. T. Baker, "The word 'toadstool' in Britain," *The Mycologist* (1990) 4(1):25–29.

6. R. G. Wasson, *Soma: The Divine Mushroom of Immortality* (New York: Harcourt Brace Janovich, 1968).

7. J. Ramsbottom, *Mushrooms and Toadstools* (London: Collins, 1953).

8. C. D. Badham, *A Treatise on the Esculent Fungeses of England* (London: Lovell Reeve, 1863).

9. J. Grigson, *The Mushroom Feast* (New York: Alfred A. Knopf, 1979).

10. S. J. Gould, "A Humongous Fungus Among Us," *Natural History* (1992) July:10–17.

11. S. A. Friedman, *Celebrating the Wild Mushroom: A Passionate Quest* (New York: Dodd, Mead, 1986).

12. S. K. Ogundana, "Nigeria and the Mushroom," *Mushroom Science X (Part II): Proceedings of the 10th International Congress on the Science and Cultivation of Edible Fungi* (France, 1978), pp. 537–545.

13. B. A. Oso, "Mushrooms of the Yoruba People of Nigeria," *Mycologia* (1975) 67:311–319.

14. B. A. Oso, "The Fungi as Understood by Local People in Nigeria," *2nd International Mycological Congress* (Tampa, FL, 1977).

15. B. P. Thomson, "Two Studies in African Nutrition: An Urban and a Rural Community in Northern Rhodesia," *Rhodes- Livingstone Papers*, No.24 (1954).

16. G. D. Piearce, "Zambian Mushrooms—Customs and Folklore," *Bulletin of the British Mycological Society* (1981) 15(2):139–142.

17. B. Morris, "Macrofungi of Malawi: Some Ethnobotanical Notes," *Bulletin of the British Mycological Society* (1984) 18(1):48–56.

18. G. Parent and D. Thoen, "Food Value of the Edible Mushrooms from the Upper Shaba Region (Zaire)," *Economic Botany* (1977) 31:436–445.

19. R. J. Rodin, *The Ethnobotany of the Kwanyama Ovambos*, Missouri Botanical Garden Monographs in Systemic Botany, Vol. 9 (Kansas City, MI: Allen Press, 1985).

20. R. Heim, *Termites et champignons* (Paris: Boubée, 1977).

21. K. S. Manilal, "An Ethnobotanic Connection Between Mushrooms and Dolmens." In *Glimpses of Indian Ethnobotany*, ed. S. K. Jain (New Delhi: Oxford and IBH, 1981), pp. 321–325.

22. N. J. Turner and M. A. M. Bell, "The Ethnobotany of the Coast Salish Indians of Vancouver Island," *Economic Botany* (1971) 25:65–105.

23. N. J. Turner, J. Thomas, B. F. Carlson, and R. T. Ogilvie. *Ethnobotany of the Nitinaht Indians of Vancouver Island*, British Columbia Provincial Museum Occasional Papers No. 24 (Victoria, BC, 1983).

24. E. Gunther, *Ethnobotany of Western Washington*, University of Washington Publications in Anthropology, Vol. 10 (Seattle, WA, 1945), pp. 1–62.

25. N. J. Turner, H. V. Kuhnlein, and K. N. Egger, "The Cottonwood Mushroom (*Tricholoma populinum* Lange): A Food Resource of the Interior Salish Indian Peoples of British Columbia," *Canadian Journal of Botany* (1987) 65:921–927.

26. H. V. Kuhnlein and N. J. Turner, *Traditional Plant Foods of Canadian Indigenous Peoples: Nutrition, Botany and Use*, Food and Nutrition in History and Anthropology, Vol. 8, ed. S. Katz (Philadelphia: Gordon and Breach, 1991).

27. A. Weber, "The Occurrence of Tuckahoes and *Poria cocos* in Florida," *Mycologia* (1929) 21:113–130.

28. W. R. Burk and T. K. Fitzgerald, "Puffball Usages Among the North American Indians," *McIlvainea* (1981) 5(1):14–17.

29. R. A. Blanchette, B. D. Compton, N. J. Turner, and R. L. Gilbertson, "Nineteenth-Century Shaman Grave Guardians Are Carved *Fomitopsis officinalis* Sporophores," *Mycologia* (1992) 84(1):119–124.

30. C. Weber, "The Popularity of Edible Mushrooms in Geneva," *Economic Botany* (1964) 18:254–255.

31. J.-P. Jappinen, "Wild Mushrooms as Food in Finland," *The Mycologist* (1988) 2(3):99–101.

32. A. Pilát and O. Ušák, *Mushrooms* (London: Spring Books, 1954).

33. W. Milliken, R. P. Miller, S. R. Pollard, and E. V. Wandelli, *Ethnobotany of the Waimiri Atroari Indians of Brazil*, Royal Botanic Gardens Publications (Whitstable, UK: Whitstable Litho, 1992).

34. O. Fidalgo, "Conhecimento micologico dos Indios Brasileiros," *Rickia* (1965) 2:1–10.

35. O. Fidalgo and G. T. Prance, "The Ethnomycology of the Sanama Indians," *Mycologia* (1965) 68(1):201–210.

36. G. T. Prance, "The Use of Edible Fungi by Amazonian Indians." In *Ethnobotany in the Neotropics*, New York Botanical Garden Publications, Vol. 1, ed. G. T. Prance and J. A. Kallunki (Kansas City, MI: Allen Press, 1984), pp. 127–139.

37. W. P. Liu and H. R. Yang, [Investigation on mushroom poisoning in Ninghua County during the last 20 years] (Chinese), *Chung Hua Yu Fang I Hsueh Tsa Chih* (1982) 16(4):226–228.

38. Y. J. Si, [A report of 25 cases of deteriorated tremella poisoning] (Chinese). *Chung Hua Yu Fang I Hsueh Tsa Chih* (1988) 22(5):293–294.

39. D. Arora, *Mushrooms Demystified* (Berkeley, CA: Ten Speed Press, 1986).

40. R. V. Southcott, "Notes on Some Poisonings and Other Clinical Effects Following Ingestion of Australian Fungi," *South Australian Clinics* (1975), pp. 443–478.

CHAPTER TWO

HISTORY OF MUSHROOM EATING
AND MUSHROOM POISONING

MUSHROOMS HAVE been an article of diet and commerce in many cultures for centuries. The few highlights covered here represent but a fraction of their history. Although they have never attained more than an ancillary role in the kitchen or the clinic, they have been associated with more than their share of magic and amazement. Mushrooms have been worshipped and prized, loathed and feared. Seldom have they been ignored.

When did human beings begin to eat, and be poisoned by, mushrooms? No one knows. Fossilized fungi have been described, although their study lags considerably behind that of the plant world. In general, the fungal fossil record is quite poor, especially compared with that of plants and invertebrates. Fungal fossils may be found either on the surface cuticles or inside the remains of the higher plants, as well as associated with the petrified remains of root systems. Sclerotia have been identified in coals. Even tiny fruiting bodies of both the ascomycetes and basidiomycetes are known.[1] Fungal hyphae and spores have also been well preserved in amber.[2] The ability of the spores as well as the hyphae to withstand extreme conditions suggests that coprolites and other remnants of Neolithic human activity should contain evidence of fungi, if they were eaten then. Recently, studies have begun on the presence of spores in medieval middens and may pave the way for further investigations.

None of the current books on paleoethnobotany even consider the fungi. But the history of mushroom poisoning is inextricably linked to that of mushroom eating, or mycophagy, the origins of which seem to be lost in the proverbial mists of time. We have to resort to the extant art and literature of various times and cultures to learn how mushrooms were used in the diet and how much of a problem their toxins might have been. At times, what is not said or written may be as revealing as what is. For example, the Bible makes no mention of mushrooms,[3] although scholars have claimed that the manna in the Book of Exodus may have been a lichen, which is a symbiotic organism between a fungus and an alga. (Other explanations for manna include the exudate of an insect that feeds on the tamarisk trees and the sweet liquid present on the branches of the white hammada tree. None of these speculations satisfactorily accounts for this "miracle.")

MYCOPHAGY IN WESTERN CULTURES

In the Western world, some of the earliest recorded dietary uses of mushrooms come, not surprisingly, from the Greeks and the Romans. Much of our knowledge about this period was well summarized many years ago. Very little has been added since the work of the Reverend William Houghton, Rector of Preston, Wellington, Shropshire, who scanned the classical literature for any and all references to fungi. It will take the persistence of another classical scholar to discover if he missed or misrepresented anything. What follows is adapted freely from his work.[4] All the books published in this century trace their allusions to the classical period from this single study,[5-7] a situation reflecting the current deficiencies in our classical education.

A number of references to mushrooms crop up in the Greek classics, including one of the first well-reported poisonings. Euripides (480–406 B.C.) mentions a family in Icarus that was poisoned by mushrooms. This reference has frequently been interpreted as one to his own family, despite the lack of evidence:

> O Sun, that cleavest the undying vault of heaven, hast thou ever seen with thine eyes such a calamity as this—a mother and maiden daughter and two sons destroyed by pitiless fate in one day?
>
> Euripides

Hippocrates (ca. 460–ca. 377 B.C.) refers to medicinal uses for mushrooms. He also records a case of a young lady who developed nausea, abdominal pains, and a sense of suffocation following a mushroom meal. She was purged with hydromel (honey and water) and given a hot bath. After vomiting the mushroom, she made an uneventful recovery, except for excessive perspiration. Whether this symptom was due to the hot bath or to the presence of muscarine in the mushroom is unknown, and we are unable to make a positive identification of the species because of insufficient details.

The antiquity of the concern for distinguishing the good mushrooms from the bad is revealed by a book written by Diphilus in the third century B.C. and entitled *Diets Suitable for Persons in Good and Bad Health:*

> The wholesome kinds appear to be those that are easily peeled, are smooth and readily broken, such as grow on elms and pines; the unwholesome kinds are black, livid and hard, and such as remains hard after boiling; such when eaten produce deadly effects.

One of the earliest taxonomic divisions of fungi, by Clusius in 1601, was into these two broad groups: edible and poisonous. One hundred and fifty years later, the botanist Battarra was cautious enough to include a motto in Greek on the title page of his 1755 monograph, *Fungorum Agri,* stating, "We study fungi, we do not eat them," a sentiment shared by many professional mycologists to this day—although for different reasons.

In Rome mushroom eating blossomed, at least among the rich and famous. And in modern times, Italy continues to be a country in which mushrooms, especially porcini and truffles, are considered great delicacies. Even Plutarch (A.D. ca. 46–ca. 120) in Greece referred to the "Roman" mushrooms, a reference suggesting that even back then Italy was the center of mycophagy. We assume that the general population shared in the passion, although the most favored mushrooms were expensive. The lavish and exotic meals indulged in by the patricians eventually led to the Sumptuary laws, which were intended to limit the use of certain rare—and rapidly becoming rarer—birds and animals. However, these laws did not apply to vegetable matter, in which category mushrooms were classified. (These ancient regulations bear some resemblance to our current environmental laws, which provide a little protection for a few animals and the occasional plant but largely ignore the fungi.) Some mushrooms were so favored that amber knives were reserved for their preparation and special silver vessels, *boletaria,* were set aside to hold and cook them. Martial (A.D. ca. 40–ca. 104) comments on the unfortunate, degrading fate of one of these vessels:

> Although Boleti have given me so noble a name,
> I am now used, I am ashamed to say, for Brussel Sprouts.
>
> Martial, Epigrams

The great value placed on these mushrooms was well illustrated by the same author:

> Gold and silver and dresses may be trusted to a messenger,
> but not a boletus, because it will be eaten on the way.
>
> Martial, Epigrams

The "boletus" of the Romans is generally believed to be *Amanita caesarea*. This was not the only prized mushroom, however. Truffles, too, have been known from antiquity, although no one seemed to have a better idea about their origin than to ascribe them to lightning bolts and thunder. Pliny thought the best truffles came from northern Africa, a view the French would undoubtedly debate. Martial also had some kind words to say about Pliny's "excrescences of the earth":

> We who, with tender head, burst through the earth that nourishes us, are Truffles, a fruit second only to the Boleti.
>
> Martial, *Epigrams*

One of the oldest cookery books, credited to Apicius, although probably a compilation of recipes from multiple sources, includes a number of suggestions for cooking fungi. The following is a typical Apician recipe, this one for cooking truffles (called a "Tuberia"):

> Boil and sprinkle salt, transfix with twigs, partly roast, place in a cooking vessel with liquor, oil, greens, sweet boiled wine, a small quantity of unmixed wine, pepper and a little honey, and let it boil; while boiling beat up with fine flour; prick the tubers that they may absorb, take out the twigs and serve.
>
> Coelius Apicius, *De opsoniis et condimentis* (ca. A.D. 200)

Cicero (106–43 B.C), a known gourmand (and who was not in those days, if one could afford it?), suffered from his overindulgence at the house of Lentulus. He developed such severe diarrhea that for 10 days he was barely able to stand. He had not eaten some of the forbidden foods, such as oysters and eels, in accordance with the proscriptions, but had eaten the highly seasoned vegetables, including the mushrooms:

> I who willingly abstained from delicate fishes, found myself taken in by vulgar herbs. . . . After this I shall be more cautious.
>
> Cicero, *Letter to Gallus*

It is interesting to speculate on what provoked such a severe gastrointestinal upset, because the fungal toxins normally affecting the gastrointestinal tract seldom act for more than two or three days. It is sad to think he may have given up mushrooms because of a bout of viral gastroenteritis or a bacterial infection or some other unrelated illness or toxin.

Poisonings and False Allegations

Pliny's *The Natural History of the World* (A.D. 22) contains at least two major references to fungi. Book XIX, delightfully subtitled "a discourse on the nature of Flax and other wonderful matters," contains a brief mention of fungi. Mushrooms were discussed in much more detail in Book XXII, where Pliny commented:

> Among all those things that are eaten with danger, I take the mushromes may justly be ranged in the first and principal place; true it is that they have a most pleasant and delicate taste, but discredited much they are and brought to an ill name, by occasion of the poyson which Agrippina the empress conveighed unto her husband Tiberius Claudius the Emperor, by their means a dangerous precedent given for the like practice afterwards. And verily by that fact of hers she set on foot another poison, to the mischief of the whole world and her owne bane especially (even her own sonne Nero, the Emperor, that wicked monster).

Pliny was referring to perhaps the most famous of all mushroom poisonings, the carefully planned murder of the Emperor Claudius I by his fourth wife, Agrippina. Pliny lived through this difficult period of Roman history and had the wisdom to retire from public life, only to perish in the eruption of Vesuvius in A.D. 79. Claudius, who had succeeded Caligula in A.D. 41, married Agrippina after dispatching his third wife, Messalina, for adultery and sundry other crimes against his person. Agrippina came to this marriage with a son, Nero, sired by a previous husband. Although Brittanicus (Messalina's son) was first in line for the laurel wreath and clearly favored by Claudius to succeed him, Agrippina persuaded Claudius to adopt her son and then hatched a plot to murder Claudius, which she felt was the only way to ensure her son's succession. She is said to have employed a number of people in this plot, including a woman, Locusta, known to be skilled in the use of poisons, the emperor's favorite and trusted eunuch, and his personal physician.

One of Claudius's most beloved foods was *Amanita caesarea*, a longtime choice of the Roman nobility. A dish of Caesar's mushrooms was prepared, liberally laced with the juice of *Amanita phalloides* (the death cap). Presumably Claudius would have detected the difference had whole mushrooms been used, because Caesar's mushrooms have an orange-red color, whereas the death cap is white, with hints of green and brown. In some accounts of the story, he became very drunk at the banquet, vomiting up a considerable fraction of the feast. That must have concerned Agrippina and Locusta, but they need not have worried. When he became symptomatic on the following day—the first symptoms of *Amanita phalloides* poisoning take 8 to 24 hours to develop—Xenophon, his physician, whom Agrippina had also seconded to be a participant in her scheme, was called in to minister to him. Xenophon used a dose of colocynth, an extract of a bitter gourd that contains a toxic alkaloid usually used in small doses as a purgative. He adroitly administered it as an enema, thus avoiding the possibility that Claudius might detect its very bitter taste. This treatment was the probable coup de grâce, dispatching Claudius and ending his reign.

For her troubles, the poisoner Locusta was given some large estates, and she continued to use her skills to dispose of other enemies of Agrippina and Nero. She herself was killed in A.D. 68 by Galba, the next emperor.

How close the details of this story are to the truth will never be known. At least three different versions are known from writers of the time, including Tacitus, Suetonius, and Seneca the Elder (who was Nero's mentor). The precise method of how the poison was administered remains in some doubt:

> And verily it is agreed upon generally by all, that killed he was by poison, but whereit should be, and who gave it, there is some difference. Some write that as he sat at a feast at the Capitol with the priests it was presented to him by Halotus, the eunuch, his taster; others report that it was at a meal in his house by Agrippina herself, who had offered unto him a mushroom empoisoned, knowing that he was most greedy of such meats.
>
> Suetonius (ca. A.D. 80)

That Claudius was poisoned by a mushroom seems to be beyond doubt. Wasson carefully reevaluated Claudius's death and came to the inevitable conclusion that the combination of *Amanita phalloides* poison and colocynth were the agents that paved his way to becoming a god.[8] Nero is known to have remarked to a guest who described the mushrooms being served at a banquet as "the food of the gods", "Oh, yes indeed, it was fungi that made my father a god!" He was certainly an accessory to his mother's plot. Despite the early accounts, however, a more recent commentator has suggested that Claudius's death was an act of nature and not the result of a nefarious plot.[9]

A number of authors on the subject of the poisoning of the rich and famous have alluded to other notables who supposedly lost their lives through the inadvertent, or more frequently intentional, eating of poisonous mushrooms. Unfortunately, they never cited their sources, and it is evident that many have perpetuated myths by copying from previous manuscripts or, worse, by misinterpreting the original reference. The supposed victims include luminaries such as Pope Clement VII (also known as the antipope), Emperor Jovian, Tsar Alexis (Aleksei) or his widow, and Charles VI of the Holy Roman Empire.

About Jovian, whose rule lasted but a year (A.D. 363–364), vague allusions to mushrooms are found in the writings of the time. He was discovered unexpectedly dead in his bed one morning, and it was surmised by some who witnessed the events that his death may have been due to an excess of mushrooms or the fumes of a charcoal fire. The suspicions of foul play were not supported by any evidence, however, so carbon monoxide intoxication seems a much more likely proposition.

Pope Clement VII, who died in 1534, had a fluctuating illness for many months before his final exit. No clinical evidence of mushroom poisoning was ever presented. His biographer noted that it was a custom in those days to attribute many famous deaths to poisoning, but there was no reason to suspect it in his case.

The Holy Roman Emperor Charles VI, the last male Hapsburg, whose death precipitated a war, may well have died of inadvertent mushroom poisoning. William Coxe tells us in *History of the House of Austria* (1906) that in October 1740 "at night his complaint was increased by indigestion, occasioned by a dish of mushrooms stewed in oil, of which he ate voraciously." He died suddenly 10 days later. This is a reasonable time course for *Amanita phalloides* poisoning, but the details of his last days are insufficient to determine whether he really suffered from acute hepatic failure. No suspicion of murder surrounded his demise, nor were allegations ever made. However, Voltaire commented that "a pot of mushrooms changed the history of Europe."

The story of Tsar Aleksei of Russia and his wife has been carefully dissected and reconstructed by R. Gordon Wasson.[8] It is well known that the tsar himself did not die of mushroom poisoning. All the misinterpretation about the role of mushrooms in their demise appears to have followed the autopsy on his wife, who actually died some years later than her husband. An English translation of an eighteenth-century text describes the circumstances of her death:

> The czarina-Dowager, Relict of the late Czar Alexius, dying in the year 1715, during Lent, her body was opened, and it was found, that her Indisposition was chiefly occasioned by too much of those pickled Mushrooms. Out of devotion of strictly observing her Fast.
>
> *The Present State of Russia* (1723)

Intestinal obstruction due to mushrooms has occurred in this century and has caused intestinal perforation. It is evident, therefore, that she did not die of mushroom poisoning. Despite her piety in observing Lent, she committed the deadly sin of gluttony and died of overindulgence.

Pliny described one of the prime characteristics of mushroom poisoning—its social nature. Eating is a social activity—except perhaps in the United States, where family meals have become limited to Thanksgiving and other holidays. It is not surprising, therefore, that poisonings commonly involve a number, sometimes a large number, of people.

It is not long since that in one place there died thereof, all that were of one household; and in another as many as met at a feast and did eat thereof at the same bourd. Thus Anneus Serenus, captaine of the Emperor Nero his guard, came by his death, with divers coronels and centurions at one dinner.

Pliny, *The Natural History of the World* (A.D. 22)

As one might anticipate, both the Greeks and the Romans sought ways to determine whether mushrooms were toxic, ways to cook them that would ensure their safety, and in the event that both failed, ways to treat the unfortunate victims. In none of these enterprises were they very successful. Once again Pliny supplied advice:

And verily all such are pernicious and utterly to be rejected near to which they come new out of the ground, there lay either a grieve-stud or leg harneisnaile or some rustie iron, or so much as an old rotten clout: for looke what naughtiness foever was in any of them, the same they draw and convert into venome and poyson . . . For dangerous they be otherwise, and meet with more means to make them deadly, namely, if a serpents hole or nest be neare by, or if at their first discoverie and comming forth, a serpent chance to breath and blow upon them.

Pliny, *The Natural History of the World* (A.D. 22)

The idea of mushrooms becoming poisonous by absorbing things in the immediate environment, such as iron or rotten cloth, or because of snakes in the vicinity, became the underpinning of the beliefs of later centuries. It was, of course, all summarily discarded when the poisonous compounds were shown to be endogenously synthesized. However, the beliefs have come full circle with the discovery that some fungi do indeed concentrate heavy metals and radioactive elements and that the environment in which they grow may have a profound effect on the concentration of many of the toxins.

The ancients recognized that mushrooms growing in association with certain trees are more likely to be poisonous than others. For many years, the hazelnut, poplar, and fig trees were believed always to be associated with edible mushrooms, whereas the olive tree has, since the days of Nicander (ca. 150–200 B.C.), had its fungal partners maligned:

The rank in smell, and those of livid show,
All that at roots of oak or olive grow,
Touch not! But those upon the fig tree's rind
Securely pluck—a safe and savoury kind!

Nicander (ca. 200–150 B.C.)

Regulations and Remedies

Pliny, voicing one of the original concerns about consumer protection, noted that buyers really do not know what they are getting in the marketplace:

> For my own part, as I have said before, I hold those good that the Beech, Oke and Cypress trees doe yeeld. But what assurance can a man have hereof, from their mouths who sit in the market to sell them?
>
> Pliny, *The Natural History of the World* (A.D. 22)

This fear has, in fact, not disappeared, at least in the United States. At present, control over the labeling and distribution of wild mushrooms in many markets is totally lacking. In the relatively sophisticated markets of Seattle, I have seen mushrooms misidentified. Unfortunately, most customers are too naive to know the difference and mistakenly trust their produce manager.

In Europe and many other countries in which wild mushrooms are picked and sold in the markets, regulations have been instituted to ensure the safety of the public. In France, all the pharmacists (apothecaries) are sufficiently trained in basic mycology to identify the important edible and poisonous mushrooms. In Paris, after one major epidemic of poisonings, an ordinance was passed in 1754, prohibiting both the sale of any fungi in the markets and the collecting of mushrooms in and around the city. This law was changed in 1808 to allow the sale of seven different species. The number has been increased substantially since then, although the regulations are still in place. In Rome, regulations for the inspection of all fungi went into effect in 1837. The provisions included the following points:[10]

> All the funguses brought into Rome by the different gates should be registered . . . a certain spot should be fixed upon for the Fungus market. And that nobody under penalty of fine and imprisonment should hawk them about the streets. . . . At seven o'clock A.M., precisely, the inspector should pay his daily visit, and examine the whole, the contents of the baskets being previously emptied on the ground. . . . The stale funguses of the previous day, as well as those that were mouldy, bruised, filled with maggots, or dangerous . . . should be sent under escort and thrown into the Tiber. . . . the Inspector should be empowered to fine or imprison all those refractory to the above regulations.

Pliny, like others, recommended cooking mushrooms with pear branches or combining them with meat. If all else failed, a wild pear after the meal was said to "kill or dull the malice that they may have." This is perhaps as good an excuse as any to end a meal with *eau de poire*.

Dioscorides echoed many of Pliny's sentiments and also suggested that

> they are helped being drenched with Nitre, and oil, or lie with ye decoction of sharp brine or of Thymbra or drinckifyed with Origanum or with acetum.
>
> Dioscorides, *De materia medica* (A.D. 65)

These recommendations are of more than passing interest. One of the common ways that the more bitter or peppery varieties of mushrooms are prepared in eastern Europe is to marinate or pickle them in salt, with or without vinegar. *Lactarius torminosus*, routinely listed as a poisonous milky cap—or, at best, inedible—in the English field guides, succumbs to this treatment and becomes eminently edible. The use of oregano may simply be fortuitous, being the most readably available herb, but many would agree that this herb has a special affinity for many mushroom dishes. The continuation of the comment from Dioscorides is equally revealing: " . . . or should be followed with a draught composed of bird's dung and vinegar . . . for even the edible sorts are difficult of digestion and generally pass whole with the excrement." Observe that he noted the relative indigestibility of fungi and introduces us to the then widespread use of chicken dung as more than just a manure for the garden.

The idea of removing the poisons by extracting the mushrooms in vinegar or water reached its peak in France in the nineteenth century. Frédéric Gérard, an assistant at the Jardin des Plantes, Paris, sent a memoir to the Conseil de Salubrité of Paris in 1851, asserting that he had eaten all kinds of poisonous fungi without any health problems. This claim may have been inspired by Jean Jacques Paulet, who in *Traite des champignons* (1790–1793) claimed that poisonous fungi became innocuous to animals after being cut into pieces and steeped in water containing salt, vinegar, or alcohol. In the presence of the commission, Gérard prepared and ate 500 g of *Amanita muscaria* one day and at least 70 g of *Amanita phalloides* on another, showing no ill effects at all. All he had done was to cut up portions of the mushrooms, soak them in vinegared water for a few hours, wash them, and then boil them for half an hour. (There is no report on how they tasted after this maltreatment.) Suspicion still lurks that he eventually succumbed to mushroom poisoning as a result of his experimental ardor. The French have a penchant for such public displays of bravado: A century later, another Frenchman went before the public to demonstrate his cure for *Amanita phalloides* poisoning. But that story will have to wait (see Chapter 12).

Nowhere did Pliny mention Dioscorides, although they made very similar statements about both the plants they described and the medicines or techniques they discussed. They were roughly contemporary with one another and may well have used similar sources. It is said that Pliny consulted over 2000 manuscripts in the compilation of his opus.

But our sympathy must be reserved for those who were truly poisoned. Some would say that the therapy for poisoning has changed little over the centuries, and in some respects this is true. The initial management, in the majority of cases, is to rid the stomach of the mushrooms. If the old herbalists could do nothing else, the one thing they could do well was purge a patient. An early remedy was proposed by Nicander:

> Let not the evil ferment of the earth which often cause swellings in the belly or strictures in the throat, distress a man; for when it has grown up under a viper's deep hollowtrack it gives forth the poison and hard breathing of its mouth; an evil ferment is that; men generally call the ferment by the name of fungus, but different kinds are distinguished by different names; but do thou take the many heads of cabbage, or cut round the twisting stems of the rue, or take the efflorescence which has accumulated on old corroded copper, or steep the ashes of clematis in vinegar, then bruise the roots of the pyrethrum, adding a sprinkling of lye or soda, and the leaf of cress which grows in the gardens, with the medic plant and pungent mustard, and burn wine-lees or the dung of domestic fowl into ashes; then putting your right finger in your throat to make you sick, vomit forth the baneful pest.
>
> Nicander, *Alexipharmaca* (ca. 185 B.C.)

Today's syrup of ipecac pales by comparison with these concoctions. And Nicander's remedy was by no means the most exotic purging formula.

Galen, who was to influence many centuries of medical practice and malpractice, had very little good to say about the mushroom kingdom:

> Of fungi the Boletus, when well boiled, must be counted among the insipid things; it is generally eaten with the various kinds of spices, as is done with other insipid food. These fungi, after being eaten in large quantities, yield cold, clammy, noxious juices as their nourishing qualities; the Boleti are the most harmless, and after them the Amanita; as for the rest, it is safer to have nothing to do with them.
>
> Galen, *De alimentis facultatibus* (ca. A.D. 150)

Galen, too, seems to have adopted the advice of Nicander in the use of chicken dung, a remedy for poisoning that lingered on for many years:

> I have heard of a physician of Mysia who administered Fowl's dung to persons suffering from fungus poisoning, and I myself have experimented with the remedy. I have used finely powdered dung mixed with water or

mixed with honey and vinegar. The patients immediately on drinking the mixture vomited and recovered. One must observe that the dung of a fowl at liberty is more efficacious than one in confinement.

Galen, *De simplicium medicamentorum temperamentis et facultatibus*
(ca. A.D. 150)

As Findlay noted,[7] this is a wonderful endorsement for the free-range chicken!

Lest we feel complacent about current therapy, it is illuminating to read of the treatment in the 1930s of two patients who had eaten *Amanita pantherina* and who were only moderately intoxicated at the time of admission:

A pint of permanganate of potash solution was immediately given, followed by one-tenth grain of apomorphine, and one-fiftieth grain of atropin. Following emesis and complete cleaning out of the stomach, each patient was given one ounce of castor oil and as soon as this worked they were given one-quarter of morphine and one-fiftieth of atropin. During all this eliminative treatment each was given a teaspoonful of a solution containing ten drops tincture agaricus muscaris in a half a glass of water, every half hour. The man did not lose consciousness, but lay in a semi-conscious condition for several hours. The woman regained consciousness in about eight hours. By the next morning both appeared normal and were sent home, and the following day they were practically recovered except that they were both very weak.[11]

Fortunately for the patients, the doses of the medications were reasonable, if entirely unnecessary. The outcome, with no therapy at all apart from emesis, would have been the same. In another 50 years, our descendants will look upon our therapeutic maneuvers with equal disdain and probably a hint of embarrassment.

Modern Criminal Poisonings

Considering the potency of some of the toxins, it is rather surprising that they have not been used more widely in the arcane art of subtle murder. Of course, many cases may have gone undetected, because the average physician would not suspect mushroom poisoning in the standard, everyday case of acute liver failure. At one trial at the Central Criminal Court in London, on September 24, 1873, a gardener was accused of having poisoned a barmaid with mushrooms. The Grand Jury dismissed the charges.

One celebrated case is described by Findlay and Ramsbottom.[7,12] It involved a Frenchman by the name of Girard, who was finally convicted of murder in 1918. He lived in Neuilly with a wife and a mistress, both of whom participated in his plots and benefited from his successes. He was apparently reasonably knowledgeable about the insurance industry, and he must have had at least a nodding acquaintance with pharmacy and mycology. His modus operandi was to make friends with individuals or couples who were relatively wealthy and were about the ages of himself and his wife. He would then take out insurance policies in their names, with himself or his wife (or mistress) impersonating the future victims at the required medical examination. Naturally, he named himself as the sole beneficiary on the policy. He would entertain these acquaintances and at the appropriate time serve them a poisoned dish. Evidently, some of his earlier efforts used bacteria, such as anthrax and typhoid, but for a while *Amanita phalloides* was his preferred toxin. The mushrooms were collected by a vagrant, *le père* Théo, who was given some brief instruction about the field characteristics and a drawing of the fungus. Evidently Théo was not always successful in picking quite the right mushroom, for at least some of the guests survived. Whether this was due to misidentification of some more harmless species such as *Amanita citrina* or to insufficient doses is not known. However, the French field guides of the time were quite conservative in their listing of edible species, so it is highly likely that Théo picked edible species such as *Amanita citrina* (also known as *Amanita mappa*), believing them to be toxic. It would appear that Girard's insurance knowledge was better than his mycological expertise.

Girard's end arrived, as is so often the case with soap opera crime, when he became a little too greedy. On the life of one woman, he took out insurance policies with four different companies. Three weeks later he murdered her. He collected the insurance money from three of the companies, but the fourth launched an inquiry when the company's physician questioned why an otherwise healthy woman should have died so suddenly and unexpectedly. The physician went to view the body, only to find that it was that of someone entirely different from the woman he had so recently examined. Instead of a well-nourished woman, he discovered a rickety one. Further investigations revealed the entire, gruesome story. All three were convicted of murder. Girard was sentenced to death, and his two accomplices were sentenced to life imprisonment. However, this saga ends in a rather operatic style— Girard died of tuberculosis before the sentence was carried out. He was interviewed just prior to his death, still claiming his innocence. Many years later, the details were published by Camille Fauvel as a supplement to the issues of June and August 1936 *Revue de mycologie*.

MYCOPHAGY IN NON-WESTERN CULTURES

In other cultures, the history of both mycology and mycophagy is also well recorded. The Chinese have perhaps the longest tradition in the use of mushrooms for both food and medicine, in addition to utilizing microscopic fungi in other forms of food preparation. Historians have suggested that this practice may extend back as far as the Yangshao culture, 6000 to 7000 years ago, in the Neolithic period.[13] A complete work devoted exclusively to mushrooms was produced in 1245 by Chen Jen-yu and called *The Mycoflora*. In this document he described in detail the development, appearances, harvest, and method of preparation of 11 species of fungi from the Zhejiang Province.

References to mushrooms had, of course, appeared in earlier literature. The earliest printed records of fungi can be found in the *Book of Songs*, which was produced some time during the Han Dynasty (202 B.C.–A.D. 220). Even some of the cultivation methods had become well established by A.D. 1300. Appreciation for the mushroom was also expressed in literature and poetry. China, unlike England, where references to the fungi were invariably pejorative, produced poetry that usually praised the beauty or flavor or value of mushrooms. In *Lu's Spring and Autumn Annals*, it is stated that "the best in taste is the mushroom from the southwest" (probably *Termitomyces*). Zhu Bian wrote the poem "Many thanks to colleague Tseh for his entertainment of the Mushroom of Flower of Heaven [*Pleurotus*]."

MUSHROOMS AS MEDICINE

The use of mushrooms for medicinal purposes was quite common in antiquity and across many cultures. In the West, the custom largely disappeared as more potent agents that had more limited and predictable ranges of effects became available. However, in Asia, as well as in scattered communities on the other continents, mushrooms still play a role in medicine. The nutritional and health benefits of mushrooms are explored in greater depth in Chapter 5. It is an area about which we still have a great deal to learn.

Allusions to the medicinal value of mushrooms, especially from Greek and Roman times, date back many years. Three major therapeutic uses are recorded:

◆ Panacea: the agarick (*Polyporus = Fomitopsis = Fomes officinalis*)
◆ Cauterizing agent, hemostatic agent, styptic: *Polyporus* (=*Fomes*) *fomentarius*; puffball spores

◆ Treatment for various and sundry minor complaints: suilli (*Boletus edulis*)

In addition to these uses, a few other applications are worth mentioning. The Jew's ear fungus (*Auricularia auricula*) was used in a decoction for treating sore throats; the truffles in general and *Elaphomyces granulatus* in particular have long been regarded as aphrodisiacs. This notion may have stemmed from the belief that the mushrooms were the cast-off testicles of stags, analogous to the antlers lost at the end of the rutting season. Because no one knew how these fruiting bodies developed, the wrinkled appearance and marbled, spore-filled interior could indeed be interpreted as an animal testis.

Use of Mushrooms as Panaceas: Agarick and Suilli

Dioscorides, who did not particularly favor the food use of fungi, did attribute some remarkable characteristics to a polypore fungus known to the ancients as "agaricum" (*Polyporus* = *Fomitopsis* = *Fomes officinalis*) or the agarick. He says:

> Its properties are styptic and heat producing, efficacious against the colic and sores, fractured limbs, and bruises from falls: the dose is two obols weight with wine and honey to those who have no fever; in fever cases with honeyed water; it is given in liver complaints, asthma, jaundice, dysentery, kidney diseases when there is difficulty passing water, in cases of hysteria, and to those with a sallow complexion, in doses of one drachma; in cases of phthisis it is administered in raisin-wine, in affections of the spleen with honey and vinegar.
>
>   Dioscorides, *De materia medica* (A.D. 65)

This passage continues with almost the entire litany of human complaints duly listed. The sense is that the fungus was regarded as a panacea. Ah, for the days before "double-blind" clinical trials and the Food and Drug Administration! Because of the helpful nature of this fungus, it was also known as the "female" agarick. It is almost exclusively found on larch trees.

The uses of these panaceas became even more elaborate during the Middle Ages as the physicians followed the dictates of Galen, Gerber, Avicenna, Andromachus, and others with unquestioning belief. Gerard's *Herball*, written in 1597, still had an amazing list of illnesses that were expected to respond to the use of the agarick. Only toward the end of the sixteenth century did physicians like Paracelsus develop alternative forms of therapy that had a more rational basis, being founded on observation and even experiment.

The use of the agarick persisted throughout the centuries, and, as we shall see later, variations on the polypore preparations play an important role in the materia medica of the Asian cultures today. In the West, the "agaric" was mentioned in a standard pharmacology textbook as late as the turn of the twentieth century.[14] Some of the early studies of its effects showed that extracts acted as a powerful purgative and had some actions similar to those of atropine. The observation that it diminished the night sweats associated with tuberculosis may have been behind its use in treating phthisis.

The second fungus to find widespread application, although it never achieved the popularity of the agarick, was most probably *Boletus edulis*, which in those times was called suilli (the pig or hog fungus). Today in Italy this fungus is called porcini, or "little pig." The conditions supposedly susceptible to the benefits of this mushroom were not as serious as those treated with the agarick; these conditions included hemorrhoids, perianal warts, and sore eyes.

> Suilli are dried and hung up, being transfixed with a rush, as in those that come from Bithynia. These are good as a remedy in fluxes from the bowels which are called *rheumatismsi*, and for fleshy excrescences of the anus, which they diminish and in time remove: they remove freckles (*letiginem*) and blemishes on women's faces; a healing lotion is also made of them, as of lead for sore eyes; soaked in water they are applied as a salve to foul ulcers and eruptions of the head and to bites inflicted by dogs.
>
> Pliny, *The Natural History of the World* (A.D. 22)

As the days of the panaceas faded and the dictums of Galen were being questioned, mushrooms were progressively eliminated from the pharmacopoeias, with the exception of the agarick, whose uses became more and more limited. In 1653 the new edition of Nicholas Culpeper's *The Complete Herbal* failed to mention mushrooms at all. This work was a rather remarkable venture, because Culpeper largely discarded the tradition of Dioscorides, Gerard, and Parkinson. Unfortunately, he had a profound belief in the astrological nature of disease and proposed a system even more bizarre and fanciful. He was careful to explain himself, however, and wrote in the preface—"epistle to the reader"—that

> I always found the disease vary according to the motion of the stars; and this is enough, one would think, to teach a man by the effect where the cause lies. Then to find out the reason of the operation of Herbs, Plants &c., by the stars went I; and herein I could find but few authors, but those as full of nonsense and contradiction as an egg is full of meat. This, not been pleasing, and less profitable to me, I consulted with my two brothers,

Dr. Reason and Dr. Experience, and took a voyage to visit my mother Nature, by whose advice, together with the help of Dr. Diligence, I at last obtained my desire; and being warned by Mr. Honesty, a stranger in our days, to publish it to the world, I have done it.

Modern pharmacopoeias, such as *Potter's New Cyclopedia of Botanical Drugs and Preparations*, published in 1988, still mention the three classic fungi: *Fomes fomentarius* (with its common names of amadou, surgeon's agarick, German tinder, and oak agaric), *Fomes officinalis* (agaricus), and the puffball.[15]

Use of Mushrooms in Cautery and Hemostasis

Today most Westerners would regard the use of counterirritants as a bizarre medical practice; yet it survives in a number of cultures. "Cupping" and moxabustion are still widely utilized in Asia. In bygone days this was a very common procedure and accompanied bleeding and purging as some of the mainstays of therapy. Hippocrates recommended that "one should cauterize the osseous and nervous parts with fungi; . . . quickly cauterize in eight places so as to intercept the extremities of the spleen."

The precise interpretation of what he intended is not fully clear, but most authorities believe that pieces of *Fomes fomentarius*, a polypore fungus that earns it livelihood by rotting wood and has been used on a large scale to make tinder (touchwood or punkwood), would be ignited and as it was smoldering applied to the affected part or to the appropriate site. Because this fungus was used to produce heat and fire, it was sometimes referred to as the "male" agarick. Employment of this procedure continued at least into the nineteenth century:

> The Laplanders have a way of using funguses, or common toadstools as we call them, as the Chinese and Japanese do the moxa, to cure pains. They collect the large funguses which they find on the bark of the beech and other large trees, and dry them for use. Whenever they have pains in their limbs, they bruise some of this dried matter, and pulling it to pieces with their fingers, they lay a small heap of it on the part nearest to where the pain is situated and set it on fire. In burning away, it blisters up the part, and the water discharged by this means generally carries off the pain. It is a coarse and rough method, but generally a very successful one, especially when the patient has the prudence enough to apply in time, and resolution enough to bear the burning to the necessary degree.
>
> Rees, *Cyclopaedia* (1819)

A similar technique has been recorded in a number of other regions of the world. In Sikkim, puffballs were used, because they have a tendency to smolder rather than to burst into flame:

> My servant having sprained his wrist by a fall, the Lepchas wanted to apply a moxa, which they do by lighting a piece of puffball, or Nepal paper that burns like tinder, laying it on the skin, and blowing until a large open sore is produced; they shook their heads at my treatment, which consisted in transferring some of the leeches from our persons to the inflamed part.
>
> J. D. Hooker, *Himalayan Journals* (1891)

Stopping the flow of blood has always been of prime concern to patients and their physicians, especially surgeons—even though it is well known that the flow always ceases eventually. The use of cautery has a long tradition in this regard, as has the use of sutures. In a number of situations, however, cautery and suturing are not possible nor are hot irons or suturing materials always available. From at least the eighteenth century, the spores of puffballs found favor in these cases. This use was developed independently by various groups in different parts of the world, including native North Americans.

No modern studies on the blood-clotting effects of the puffball spores have been reported. They may act by increasing the surface area for initiating the coagulation cascade and by absorbing large quantities of blood. The other mushroom used in a similar capacity was *Fomes fomentarius*, which has already been mentioned for its role in cautery and as a panacea. This mushroom and the wood it decomposes together form "dry rot," a material that can be very finely powdered and has great absorptive qualities. Consequently, *F. fomentarius* and the puffball have similar hemostatic actions. Another way of preparing *F. fomentarius* was to cut thin strips from young specimens, after the hard pellicle and the short tubes on the undersurface had been removed. The strips were then beaten repeatedly and occasionally wetted down until they were thin, soft, and pliable. They would then be laid on the wound as a dressing. The possible mechanism of action of this application, sometimes called the surgeon's agaric, was alluded to in the following statement:

> Agaric and sponge entangled the blood and retained the coagulum on the spot.
>
> Todd, *Cyclopedia of Anatomy and Physiology* (1836)

Fomes fomentarius was also widely used for the production of tinder. It was softened, immersed in a solution of nitre (saltpeter) to increase its flammability, and then dried and shredded. Common names for this product were amadou and German tinder.

The use of puffballs and polypores as hemostatic agents has all but disappeared, although the practice was still around at the turn of this century. A 1910 textbook recommended the giant puffball as

> a soft and comfortable surgical dressing. The dusty powder is a powerful hemostat.
>
> Whitla, *Pharmacy, Materia Medica and Therapeutics*, 9th ed. (1910)

Puffballs had another use, in the loosest culinary sense, a use even mentioned in Gerard's *Herball*: "The country people do use it to kill or smother Bees with Fusse-balls, being set on fire, for the purpose it fitly serveth." In other words, puffballs were used in honey production. Mrs. Hussey, always the proper Victorian naturalist, was not impressed by this use of fungi (or by beekeeping):

> The "Humane Bee-keepers" are probably not aware that this use of this Fungus to assist them in "depriving" bees is three hundred years old; a friend having lately asked for some to apply for that purpose, renders it impossible to refrain from saying a few words on this subject, questions of humanity being always worthy of investigation, and apt to be perverted by mistaken sentimentality. The country people of the present time "stifle" the bees irrevocably when they take away their honey; the "humane" and enlightened bee-keeper, instead of depriving the bees of life for ever, "temporarily stifles" them so that they may recover, and then glories over the rustic brute who massacres bees—"We have each got our honey, but my bees are alive!" Alive for what? To discover that the treasure that they worked for is gone, the food they hoarded replaced by an inferior substitute, in order that they toil through other summers to meet an equally bitter disappointment, living in the accumulation of wealth which they are never to enjoy—but "annually deprived," die annually.
>
> Mrs. J. T. Hussey, *Illustrations of British Fungi* (1847)

Miscellaneous Medical or Paramedical Uses

The Jew's ear fungus, *Auricularia auricula*, has a long history of use, dating back to Gerard, to soothe a sore throat. No obvious medical benefit accrues from its application, but it is rather soft and gelatinous and so feels comfortable. In an effort to cure the great scourge of tuberculosis, a variety of fungi were employed, including *Trametes suaveolens*, *Lactarius piperatus*, and *Lactarius torminosus*. Perhaps this was done with the belief that anything tasting so foul and peppery had to do some good. A number of other species of the genus *Lactarius* were also used as diuretics. Interestingly, *Amanita muscaria*, with its rather potent central nervous system toxin, was sometimes recommended for the treatment of epilepsy. And, of course, the use of mushrooms and many other plants as aphrodisiacs is a long and interesting story.

HISTORICAL USES OF TRUFFLES

> Ta femme, tes truffes et ton jardin, garde-les bien de ton voisin. [Your wife, your truffles and your garden; guard them well from your neighbor.]
>
> Old French proverb

The provocative scent of the underground (hypogeous) fungi have long attracted the attention of humans and other animals. The hypogeous fungi evolved this strategy for spore dispersal. The unique and permeative odors elaborated by these fungi as their spores mature prove irresistible to a variety of animals, which dig them up, enjoy a gourmet treat of truffles, and then spread the spores across the forest floor. A cause of wonder, truffles appear out of the earth without roots or other forms of attachment. This phenomenon caused great puzzlement among the early observers of nature, who proposed an array of imaginative possibilities for their origin.

Many ancient references allude to the use of truffles in gastronomy. They have been highly prized around the entire Mediterranean basin. Of their culinary properties, Pliny commented:

> There are two kinds of them, the one is full of sand, and consequently injurious to the teeth and the other free from sand and all impurities. They are distinguished by their colour, which is red, or black, or white within: those of Africa are the most esteemed.
>
> Pliny, *The Natural History of the World* (A.D. 22)

The white truffle obtained from Africa belongs to the genus *Terfezia*, a genus widely collected across North Africa and the middle East. In the early 1600s, the marketplace in Damascus was a popular outlet:

> [There] is sold a great quantity of truffles: sometimes twenty-five to thirty camels arrive laden with them and in three or four days they are sold. They come from the mountains of Armenia and Turkey.
>
> Ludivicio de Varthema, *Travels* (1503–1508)

Truffles formed an important part of the diet of the Bedouins and are still sold all over North Africa.

The use of animals for hunting truffles is an ancient practice, dating back to at least 1481, when Platina, a papal historian, recorded that nothing equaled the sows of Notza for finding truffles. He mentioned muzzling the pigs to avoid losing the prize. Truffle hunting even developed in England, a land not noted for its love of fungi. This activity appeared to be quite widespread from the 1750s to the 1850s, although the volume of truffles picked was probably never large. Their presence in England is illustrated by an 1860 report.[12] The trufflers of the region petitioned Parliament for exemption of the dog tax of 12 shillings a year,

> being poor labouring men . . . living in a woody district of the county where there is a great many English Truffles grow, which we cannot find without the dogs, we do therefore keep and use a small pudle sort of dog wholy and solely for that and no other . . . it has been carried on by our ancestors for generations without paying tax for the dogs.

We do not know if their plea was granted. If the taxing authorities then were anything like those of today, the trufflers probably paid the tax.

Truffling in England has almost disappeared. Its decline followed the breakup of many of the great old estates and extensive logging. With the decline in habitat, trufflers have disappeared and this rural occupation has gradually died.

Truffles have always commanded premium prices and been a food of the wealthy; in the heyday of classic French cuisine, truffles were used in almost indecent quantities. It is not surprising that chefs attempted shortcuts. A variety of adulterants were added to dishes in an attempt to give them a truffle flavor, aroma, and appearance. A common substitute was slices or finely ground pieces of the thick-skinned or black puffball of the genus *Scleroderma*. Poisonings with this fungus are recorded in the *Gardener's Chronicle* as long ago as 1868. Since then, a number of reports have adequately documented the toxicity of members of this group. No one knows what effect the adulteration of the dishes with *Scleroderma* had on the unsuspecting guests. Perhaps it was used in quantities small enough to avoid serious problems.

Almost all the truffles of commerce today belong to the genus *Tuber*. They are native to regions of Italy and France, and their collection and trade is still surrounded by intrigue and deception, a situation hardly surprising in view of the demand and the potential profits.

Truffles have had a long history as aphrodisiacs.[16] At one time, the hypogeous fungus, *Elaphomyces granulatus*, was extensively used in the manufacture of love potions. Indeed, consumption of the truffles has, from time immemorial, been associated with heightened passions. Although this usage has been disparaged, it is worthwhile to note that some truffles contain α-androstenol, a compound also found in the saliva and breath of rutting pigs. It acts as a pheromone to decrease the sexual inhibitions in the young sow. It has also been described in human perspiration, contributing to the musky odor. Similar compounds are added in minute amounts to cosmetics and perfumes to enhance their attractiveness. The old beliefs may well have some foundation in fact.

REFERENCES

1. S. V. Meyen, *Fundamentals of Paleobotany* (New York: Chapman and Hall, 1987).

2. G. O. Poinar, B. M. Waggoner, and U.-C. Bauer, "Terrestrial Soft-bodied Protists and Other Microorganisms in Triassic Amber," *Science* (1993) 259:222–224.

3. M. Zohary, *Plants of the Bible* (Jerusalem: Sadan, 1982).

4. W. Houghton, "Notices of Fungi in Greek and Latin Authors," *Annals and Magazine of Natural History* (1885) 15(5):22–49.

5. A. H. R. Buller, "The Fungus Lore of the Greeks and Romans," *Transactions of the British Mycological Society* (1915) 5:21–66.

6. R. T. Rolfe and F. W. Rolfe, *The Romance of the Fungus World* (London: Chapman and Hall, 1925).

7. W. P. K. Findlay, *Fungi: Folklore, Fiction and Fact* (Richmond, Surrey: Richmond Publishing Co., 1982).

8. R. G. Wasson, "The Death of Claudius, or Mushrooms for Murderers," *Botanical Museum Leaflets, Harvard University* (1972) 23(3):101–128.

9. V. Grimm-Samuel, "On the Mushrooms That Deified Emperor Claudius," *Classical Quarterly* (1991) 41:178–182.

10. C. D. Badham, *A Treatise on the Esculent Fungeses of England* (London: Lovell Reeve, 1863).

11. J. W. Hotson, "Mushroom Poisoning at Seattle," *Mycologia* (1934) 26:194.

12. J. Ramsbottom, *Mushrooms and Toadstools* (London: Collins, 1953).

13. Y.-C. Wang, "Mycology in Ancient China," *The Mycologist* (1987) 1(2):59–62.

14. T. Sollman, *A Textbook of Pharmacology* (London, 1906).

15. R. C. Wren, *Potter's New Cyclopedia of Botanical Drugs and Preparations* (Saffon Walden, Essex: C. W. David, 1988).

16. R. Stark, *The Book of Aphrodisiacs* (New York: Stein and Day, 1981), p. 195.

CHAPTER THREE

INTRODUCTION TO THE BIOLOGY OF MUSHROOMS

IN THE WORLD of both unicellular and multicellular nucleated organisms, fungi constitute a kingdom separate from those of the animals and the plants. In the modern vernacular, the animals are the consumers, the plants are the producers, and the fungi are the recyclers. But these designations are overly simplistic. Certainly many fungi are important decomposers of organic matter, but others, such as the mycorrhizal fungi (discussed later), are consummate consumers.

WHAT IS A FUNGUS?

Fungi inhabit every possible environment, including many unlikely ones, utilizing the organic material from plants and animals and even other fungi for their nutrition and energy source. Unlike the chlorophyll-containing plants, which convert solar energy into chemical energy, fungi, like animals, are totally dependent on the available organic material for all their nourishment. Unlike the animals, most fungi are stationary and cannot pursue their food—they must absorb it from their surroundings and grow into new food sources. A few groups, such as the cellular slime molds, are able to move with slow, amoeboid motions, a far cry from a gazelle or a cheetah. A number of aquatic fungi are able to swim with the aid of one or more whiplike organelles that protrude from the cell (flagella).

The cell walls of fungi are often made of a complex polysaccharide, chitin. This large carbohydrate also occurs in the exoskeletons of insects. Its structure is different from that of cellulose, the structural polysaccharide in plants. The interaction of fungi with the surrounding environment is regulated by the permeability of their cell walls. Like animals, they store energy in glycogen rather than in starch, which is the usual storage compound in plants. And unlike most animals and a majority of the higher plants, they do not make embryos when they reproduce. Their spores can form new organisms directly.

The vast majority of fungi are minute, microscopic organisms. We encounter them in our daily lives without noticing them. They are the yeasts in our bread and beer, the fermenters of our wine, the mold in the blue cheese, and the source of many of our antibiotics. They rot houses, cause athlete's foot or worse, and damage crops. Depending on the species, they may follow one of three common lifestyles. The saprobes, or decomposers, live on dead or decaying organic material. Some of these have the capacity to utilize material that few other organisms can handle, like the lignin in wood. The saprobes are critical for recycling the vast quantities of dead organic material that accumulates each growing season. The second group are the parasites, fungi inhabiting the living tissue of many plants, animals, and other fungi. These may live in relative harmony with the host, but on occasion they may produce damage and disease. The third group comprises the symbionts, fungi that live in intimate association with the roots on many—if not most—plants, especially trees.

The association of fungi with the fine roots of plants can take a wide variety of forms, and the structures made by the combination of fungus and plant root are called mycorrhizae. In some plants, the fungal cells not only penetrate into the cortex of the root between the cells but also surround the rootlet in a sheathlike fashion. This structure is called an ectomycorrhiza. In other plants, the cells of the fungi do not form a sheath around the rootlets, but penetrate directly into the root cortex cells of the host plant, thereby forming an endomycorrhiza. These mycorrhizal fungi are crucial for the health of the tree, supplying water and nutrients such as minerals. In return the tree supplies carbohydrates and substances that the fungus itself cannot synthesize. It is a mutually beneficial relationship. The health of the world's forests depends on mycorrhizae. But this is not the only association between a fungus and another organism. One of the great cooperative relationships in the natural world is the symbiotic alliance between fungi and algae that produces the lichens.

This description of the lifestyles of the fungi is obviously superficial, because there are some fungi that can live quite well for a time in a symbiotic relationship, then, when conditions change, can alter their behavior and become either saprobic or parasitic. One of the striking aspects of all life is the complex interactions and interdependencies exhibited between organisms. One of the many biological misconceptions most of us are raised with is the idea that organisms are in constant combat with everything else around them. It stems from the

Darwinian cliché of "survival of the fittest" and is encouraged by many of the television science programs that focus largely on the violence in nature. (Science programs have to compete with network television in which violence is a sure draw. The decapitation of the male praying mantis after copulation and the killing technique of the great white shark attract viewers more readily than a picture of a fungus and an alga that have set up housekeeping together.) The reality is quite different. There are as many examples of cooperation and collaboration between organisms as there are of adversarial relationships. Most cooperation probably came about by coevolution, because the organisms that could exploit successful arrangements with other organisms in the environment would have been favored. There are as many models of mutual support as there are of selfish exploitation. Fungi have done this to a remarkable degree. They can be found everywhere, from the interior of a growing bud to the surface of a leaf to the interior of the cells of the root of wheat.

Some fungi have developed remarkable dependencies, like the fungus *Termitomyces*, a favorite food item in Africa. This fungus has an intimate association with the termites and their nests. The termites supply the wood and leaves, which they cannot digest, to fungal "gardens." From this crop of fungi and wood, the termites obtain all their food. The fungi convert the lignin and are consumed by the termites. When this genus fruits, the top of the termite mound is covered with mushrooms that have long tapering "stems" extending into the heart of the nest.

The association of fungus to insects is not limited to *Termitomyces* species. Bark beetles and ambrosia fungi, leaf-cutter ants with their fungal gardens, and wood wasps and some wood-rotting fungi are other examples of this close dependency.

Exploiting unusual food sources, ones not utilized by other organisms, has permitted fungi to expand into some unlikely places. The keratin in the skin of animals is the daily diet of the fungi that produce ringworm, barber's rash, and athlete's foot. An insect, such as a caterpillar, is the nutritional basis for the *Cordyceps sinensis*, a fungus highly esteemed in the Chinese pharmacopoeia. Some fungi are able to ensnare or capture living organisms. These include members of the genera *Arthrobotrys* and *Dactyella*, soil-inhabiting fungi that feed on minute roundworms or nematodes. Fungi have been known to eke out an existence on substrates that appear to offer little in the way of nutrition, including the surfaces of metal and glass. Mushroom-producing fungi can be found in the desert, at the seashore, and coming up through melting snowbanks.

It is impossible to overestimate the importance of fungi, not only for the role they play in every ecosystem, but also for the huge impact they have had on the history of humankind. The presence and survival of the vast temperate or boreal forests largely depend on the mycorrhizal relationship between many fungi and the conifers. The soils are often rather poor and the rainfall only marginal. The mycorrhizal fungi vastly increase the surface area of the tree rootlets, maximizing the absorption of water, as well as providing crucial nutrients. Foresters, now well aware of this relationship, infect the roots of seedlings prior to planting to ensure that this beneficial association is maintained. But this alliance extends far beyond the forests. In fact, it is rare to find a plant that does not have a fungal partner.

The immensely valuable relationship between mycorrhizal fungi and plants, little known by the general public, has often been overshadowed by the darker side of fungi, the destroyers of our crops or our health. Included in the group of fungi that have plagued mankind are the smuts and the rusts. These microscopic fungi have produced famine and have changed the historical course of nations. Best known is the Irish potato blight (*Phytophthora infestans*). The reliance on a single crop and years of weather that favored the spread of this fungus culminated in widespread crop failure. The resulting starvation was an important stimulus for the emigration of countless Irish, especially to the United States. All our major food crops have at one time or place fallen victim to a fungal infection. Chestnut blight, Dutch elm disease, and the most current scourge, dogwood anthracnose, have all helped to alter the landscape. The wine industry has repeatedly had its vineyards decimated by fungi such as *Plasmopara viticola*.

Fungi have evolved to occupy almost every ecological niche imaginable. This has given rise to an immense diversity of lifestyles and strategies for survival. Some fungi are extraordinarily long-lived, like the recently described *Armillaria* in the Midwest and West. Not only have these organisms spread over an immense area to become the "humongous fungus" of the popular media, but they appear to have survived for as long as 1500 years. These fungi are now known to be some of the largest organisms on Earth, as well as some of the oldest. At the other extreme are fungi that exploit a very temporary food source. A good example are the fungi that profit from cow dung. A remarkable succession of fungi grow and reproduce in a cow pat in a matter of hours to days. There is the spectacular *Pilobolus*, which actively shoots its spores long distances to ensure that cows will eat clean grass to which the spores adhere. With this strategy, they can once again get into the digestive tract of an animal. Other fungi inhabiting the decomposing cow dung are members of the genera *Coprinus*, *Ascobolus*, and *Psilocybe*.

For those interested in learning more about this incredibly diverse kingdom, two general references are recommended.[1,2]

WHAT IS A MUSHROOM?

The mushroom-producing fungi with which we are concerned are composed of long, cellular, threadlike structures called hyphae, which form strands or a web of tissue in the substrate upon which the fungus feeds. This tangle of hyphae is termed the mycelium. It is the biologically important, vegetative, business end of the organism. It is responsible for absorption of nutrients, metabolism, and energy production. Most often, these microscopic threads are buried in the soil, around the roots of trees, beneath leaf litter, in the tissues of a tree trunk, or in some other nourishing substrate and are almost invisible to us. Turning over a pile of leaves in the fall will often reveal a mass of the silky threads of the mycelium of a fungus. For most of the year, we are unaware of the presence of these fungi in our surroundings. Only when they reproduce, by creating a fruiting body that we recognize as a mushroom, do they reveal themselves.

The fungi responsible for producing such grand fruiting bodies belong to two of the large classes of fungi, Basidiomycetes and Ascomycetes. The former develop and bear their spores on microscopic flask-shaped cells termed basidia. The latter, to which the prized morel belongs, develop their spores, usually in a series of eight cells, in long microscopic sacs termed asci. The life cycle of the ascomycetes is considerably different and somewhat more complex than that of the basidiomycetes, especially in the way the spores develop.

Reproduction in many fungi does not resemble the simple "birds and bees" concept all of us learned about. Most sex education classes ignore the fungi, and for good reason—their method of reproduction would thoroughly confuse the beginner. In addition to a variety of asexual methods for producing new organisms, fungi have also developed methods of sexual reproduction during which there is the usual exchange and recombination of genetic material. Most fungi do not display sexual dimorphism in the manner to which we are accustomed—they do not display maleness or femaleness but instead exist as various "mating types." Some species have many more than two mating types. The spores of a fungus germinate and develop into mating hyphae. When the hyphae of two different but compatible mating types touch each other, they may join; then the nuclei from one hypha are transferred into the cells of the other. These multinucleated cells then become the vegetative hyphae that grow throughout the food substrate. The organism may remain in such a state for a long time, depending on the species, the environmental conditions, and the available food supply. Under the appropriate conditions, however, sexual reproduction can begin. Specialized hyphae form a solid structure to become the earliest primordium of a fruiting body—or, a mushroom.

From the microscopic primordium, the mushroom grows larger, and specialized cells form into spores, which mature on the surface. In many of the basidiomycetes, this spore-bearing surface is arranged as plates (gills), as pores, or even as toothlike structures. All of these specialized structures vastly increase the surface area for spore production and are designed to ensure that the mature spores, once released, will be maximally dispersed. In the ascomycetes, the spore-bearing surface may be convoluted or composed of pits, as in the genera *Gyromitra* and *Morchella*, respectively, or it may be exposed to the surroundings in a way that allows the mushroom to eject its spores into the atmosphere. Once the spores are released, they germinate immediately if the environment in which they find themselves is propitious, and the life cycle begins anew. The life cycle of a generic basidiomycete is illustrated in the accompanying figure.

A mushroom, then, is the fruiting body (sporocarp) of a fungus belonging to either of two large classes, Basidiomycetes and Ascomycetes. The fruiting body is only one of the reproductive strategies that fungi have evolved. Fruiting bodies are exquisitely designed to spread the fungus's spores. In one sense, a mushroom is vaguely analogous to an apple on an apple tree. When the environmental conditions are favorable for spore dispersal and germination, the fungus makes the ephemeral structure we recognize as a mushroom. Some mushrooms are extraordinarily short lived, exploding out of their substrate in the morning, maturing during the day, releasing their spores, and disappearing by the next day. Others may exist for many seasons, such as the polypores or the conks, which appear as shelves or brackets on tree trunks. Most mushrooms are around for days to a few weeks. Their sudden appearance from the soil and the rapidity of their growth has always been a source of fascination and amazement.

On rare occasions, especially when the climatic conditions are unfavorable, a few fungi form a hard mass of mycelia, perhaps incorporating part of the substrate on which they are growing, into underground structures called sclerotia. These structures act as a food store for the fungus over the bad times. When conditions improve, the fungus can once again produce new fruiting bodies.

Each fruiting body is designed to release the spores in a way that maximizes the likelihood for successful spore dispersal and the establishment of new organisms. In some, the spores develop on the surface of gills or tubes, on the underside of the mushroom, ready to fall to the ground or to be wafted away on a passing breeze. The way in which the mushroom grows ensures that the spores are released with just

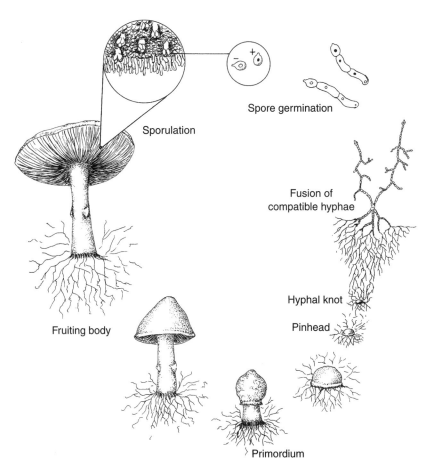

Sporulation

Spore germination

Fusion of
compatible hyphae

Hyphal knot

Fruiting body

Pinhead

Primordium

Life cycle of a typical agaric

enough force to prevent them from being trapped by the opposite gill or pore surface and that the spore-releasing surface itself is perpendicular to the force of gravity. This arrangement allows the spore to fall under its own weight between the gills until it reaches the atmosphere in which it will be passively dispersed. Other fruiting bodies, such as the morels and the cup fungi, actively shoot their spores into the air from the exposed outer surface. In some ascomycetes, the microscopic spore-containing sacs actively turn toward the light to maximize the likelihood that, when ejected, the spores will reach the surrounding air currents. The truffle relies on scent to lure and entice animals to dig it up and eat it, thereafter dispersing its spores in their droppings. Puffballs release clouds of spores when they are compressed, even by so slight a pressure as that exerted by a raindrop. The beautiful little bird's-nest fungus also relies on rain to splash its spores out of the cup. The variety in the shapes and sizes of the fruiting bodies reflects the many strategies these fungi have developed to ensure their survival.

The mechanisms precipitating and controlling the fruiting of many mushrooms in the wild are not well understood. Changes in temperature, humidity, length of day, amount of light, substrate availability all play important roles. Some fungi fruit rather regularly, season after season. Others remain in their vegetative phase for many years before suddenly producing huge numbers of mushrooms. In relatively temperate climes, spring and especially fall are the two prime seasons for mushroom fruiting. However, during most of the year, some mushrooms can be found. For some fungi, merely watering a lawn provokes a flush of new fruiting bodies. Although each species appears to have its usual season to fruit, unexpected weather patterns can significantly alter the fruiting behavior of a particular species.

The basic anatomy of the gilled mushroom is illustrated in the accompanying figure. In certain mushrooms, the prototype being those in the genus *Amanita*, the mushroom develops inside a layer of tissue, the universal veil. As the mushroom grows, it ruptures through this eggshell-like casing and expands upward. Remnants of this veil may persist in the form of one or more patches of tissue (warts) on the surface of the cap. Around the base of the stem, fragments of the veil may

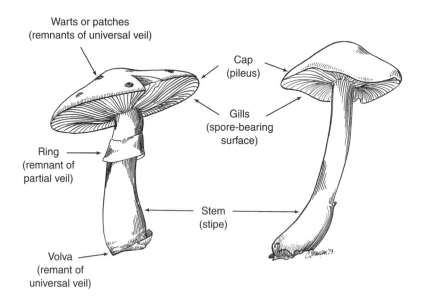

Anatomy of a typical mushroom

also persist as a cup-shaped structure (volva). In certain species the volva is not very prominent, and in others it may be entirely absent. The spore-bearing surface in both mushrooms with gills and those with pores may be covered by another thin layer of tissue (the partial veil). When this veil breaks, vestiges may remain as a ring around the stem. In certain genera, such as *Amanita*, *Lepiota*, and *Agaricus*, one or both of these structures prove most useful in classification and identification. In the majority of mushrooms, however, no part of either the universal veil or the partial veil persists in the mature mushroom. Furthermore, many mushrooms never develop either type of veil.

The characteristic shape of a mushroom's cap (pileus) depends on its species. The features of the cap that are used to identify a specimen include shape, size, margin, surface texture, and color. However, color is an unreliable feature because it also depends on environment and light conditions. Gill color, gill attachment, and spore color are also useful identifying features. Spore color, of prime importance

Mycetism and Mycotoxicoses

This book explores the history of mushroom eating and mushroom poisoning; the role mushrooms have and still play in health and nutrition; and the mechanism, symptoms, and therapy of mushroom toxicity. It looks at who gets poisoned and why, and some of the possible health benefits from mushroom eating. In short, the book examines the good and the bad sides of this ephemeral reproductive organ. But it is limited to mushrooms. The term *mycetism* has been used to describe poisonings caused by mushrooms. Mycetism is distinct from the *mycotoxicoses,* which are poisonings caused by any of the toxins produced by fungi. *Mycotoxins,* which are the fungal toxins that adulterate our crops and our food supplies, constitute an immense subject. Some were first recognized in the Middle Ages during the great epidemics of ergotism, when the fungus *Claviceps purpurea* grew on the grain. St. Anthony's fire, as it was called then, was only one of the scourges of those times. And the adulterations continue today; for example, aflatoxin contaminates many peanut products. (Of course, when fungi elaborate mycotoxins that are useful, we call them *antibiotics*.) Despite their immense economic and health implications, these mycotoxins and their associated mycotoxicoses are not covered. This story is confined to mushrooms and their effects, both good and bad.

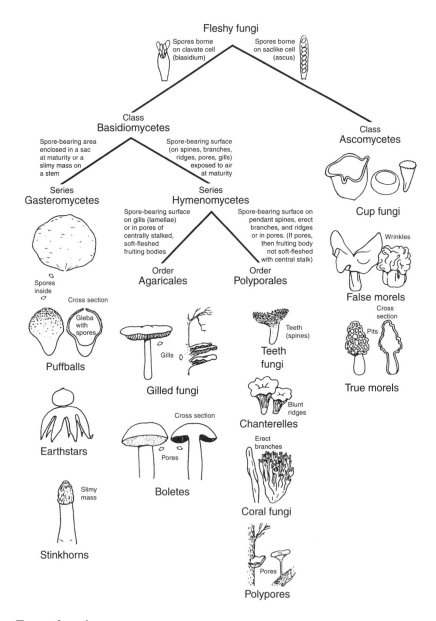

Types of mushrooms

in identifying mushrooms, can be determined by making a "spore print." This simple test is done by cutting the stem flush with the cap and placing the cap, gill or pore side down, on a piece of white paper. The specimen should be covered with a bowl or glass and left undisturbed for two or more hours. The color of the deposited spores is readily visible against the white paper.

The nature of the stem (stipe) is also of value in mushroom identification. The shape, surface texture, basal structure, and rigidity or plasticity of the stem vary from species to species.

The basic structure of the common agaric—the mushroom we buy at the supermarket—is well known to everyone, but it is only one of many. Mushrooms have evolved a remarkable diversity of forms to optimize their spore dispersal and to ensure their survival. The basic groups are represented in the accompanying figure.

REFERENCES

1. B. Kendrick, *The Fifth Kingdom* (Waterloo, Ontario: Mycologue Publications, 1985).

2. C. J. Alexopolous and C. W. Mims, *Introductory Mycology,* 3rd ed. (New York: Wiley, 1979).

CHAPTER FOUR

NUTRITIONAL AND CULINARY ASPECTS OF EDIBLE MUSHROOMS

Many phantasticall people doe greatly delight to eat of the earthly excrescences calles Mushrums . . . They are convenient for no season, age or temperament.

Venner, *Via Recta ad Vitam Longam* (1620)

Many do fear the goodly mushrooms as poysonous damp weeds; but this doth in no way abate the exceeding excellence of God's Providence, that out of grass and dew where nothing was, and onlie the little worm turned in his sport, came, as the shaking of bells, these delicate meats.

Anonymous; quoted in Findlay, *Fungi: Folklore, Fiction and Fact* (1982)

MUSHROOMS ARE COMMONLY believed to lack caloric value, to be nutritionally insignificant at best or useless and inert at worst. These views are wrong. But, then again, mushrooms are unlikely to solve the world's food shortage, as some of the mycoevangelists would have us believe. Even though no one would recommend a diet of mushrooms alone, they do contain reasonable quantities of protein and carbohydrates and a modest complement of other important nutrients. In the grand scheme of nutritional value, mushrooms occupy a place somewhere above the vegetables and legumes but below the first-class protein in meat, fish, and poultry.

NUTRIENTS IN MUSHROOMS

Studies of the nutritional value of mushrooms to humans are few in number, so considerable speculation about their "true" nutritive value remains. No formal human feeding studies have been done, so what we do know is based only on chemical analyses of the composition of many different mushroom species and a limited number of animal studies. Feeding a pure diet of wild mushrooms to weanling rats to assess their nutritional value bears little relationship to human nutrition,[1] and such experiments need to be interpreted with caution: Rats do not necessarily digest or metabolize like people.

Interpreting nutritional analyses poses difficulties similar to those faced in toxicology because of wide variations in the measurements. The factors influencing the concentration of a particular nutritional compound include

- Differences between strains of the same species
- Composition of the growth substrate
- Method of cultivation
- Stage of the life cycle analyzed
- Specific portion of the fruiting body used for the analysis
- Time interval between harvest and measurements
- Inherent inaccuracies in the method used
- Skill of the analyst

Crisan and Sands have published one of the best summaries of the nutritional aspects of mushrooms.[2] It covers more species than similar works. This reference should be consulted if more specific details for a particular fungus are required. Although less comprehensive than the Crisan and Sands analysis, other recent studies have produced similar results.[3,4] Measurements of the nutrient characteristics of mushrooms are based on the following assumptions:

- Crude protein content is calculated from the nitrogen content (usually determined by Kjeldahl analysis). Instead of using the

Nutrient content of various common edible mushrooms

Species	Water	Protein	Fat	Total CHO	N-Free CHO	Fiber	Ash	Energy (kcal)
Agaricus bisporus	89	27	2	60	51	9	10	350
Auricularia species	90	5	8	90	75	9	6	360
Boletus edulis	87	30	3	60	52	8	7.5	360
Cantharellus cibarius	91	22	5	65	54	11	8.5	350
Coprinus comatus	92	25	3	59	51	7	12.5	345
Lentinus edodes	90	17	8	75	70	7	3	390
Morchella esculenta	90	20	4	63	54	8	10	350
Tuber melanosporum	77	23	2	66	38	28	8	270

From Crisan and Sands (1978). Values are estimates based on many measurements and are expressed as g/100 g dry weight, except water content, which is expressed as g/100 g wet weight.

standard $N \times 6.25$ as the conversion factor, $N \times 4.38$ is used, because some of the nitrogen in mushrooms is present in the indigestible chitin in the cell wall. Studies on the digestibility of the protein in mushrooms have demonstrated that approximately 60–70% of the protein can be utilized by humans.[5]

♦ Fat content is determined by solvent extraction.

♦ Fiber content is defined as the residue remaining after enzymatic digestion in the gut.

♦ Carbohydrate content is calculated by difference: The total carbohydrate value includes the fiber; the nitrogen-free carbohydrate value excludes it.

♦ Ash content is defined as the material remaining after incineration.

♦ Energy value of a food can usually be estimated on the basis of the content of protein, fat, and carbohydrate by using the famous Atwater factors of 4.0, 9.1, and 4.2 kcal/g for protein, fat, and carbohydrate, respectively. However, these factors assume complete digestibility of the food. Because mushrooms are not totally digestible, the following factors have been used: protein, 70% (2.62); fat, 90% (8.37); carbohydrate, 85% (3.48) These numbers allow us to estimate the caloric content of mushrooms.

Moisture and Energy

As those of us who have cooked mushrooms are well aware, mushrooms are largely water—somewhere between 85 and 95%, depending on the environmental conditions under which they have been harvested and stored. For this reason, great shrinkage occurs during the cooking process, and water accumulates in the pan at low cooking temperatures. Because of this high moisture content, the early studies, which were done on fresh specimens, showed very few calories and low nutritional content. However, similarly low values are found for many fruits and vegetables. The high moisture content does became a limiting factor in using mushrooms as a significant contributor to the diet, because a large volume of mushrooms would have to be eaten to supply sufficient calories and protein. For example, there are approximately 350 kcal in 100 g (dry weight) of *Agaricus bisporus*. In the fresh state, this same material would weigh about a kilogram, or 2.2 lb. An average sedentary human male requires about 2000 kcal per day, so he would have to eat approximately 12.5 lb of mushrooms a day if mushrooms were his sole source of calories. This would be a formidable task. We would all have to spend our days grazing like sheep just to get enough calories. Dried mushrooms, however, contain only 5–20% moisture, so dishes based on dried mushrooms would be much more energy and protein dense.

Unfortunately, dried mushrooms have to be reconstituted to some extent to make them palatable, so the bulk that would have to be consumed to obtain sufficient calories would still be large.

The energy content of most mushrooms varies from 320 to 380 kcal/100 g dry weight. A very generous serving of mushrooms—say, half a pound—would provide only approximately 80 kcal or about the same number of calories as a slice of bread. This information tends to validate the widely held belief that mushrooms are a low-calorie (but not a no-calorie) food when consumed in the quantities to which most of us are accustomed. The butter, cream, or other fats in which we cook the mushrooms make a much more substantial contribution to our calorie intake.

Protein

Because few people have a diet deficient in either fat or carbohydrates, except in the tragic areas of famine throughout the world, the limiting factor for good nutrition, health, growth, and development tends to be the availability of protein. As has already been mentioned, only a proportion of the protein in mushrooms is digestible—somewhere in the neighborhood of 70%. The protein content may be as low as 4–9% in species such as the wood's ear or Jew's ear fungus (*Auricularia auricula*) or as high as 44% in some strains of *Agaricus bisporus*. Most species fall in the 15–40% of dry weight range. This protein content is higher than that of most vegetables and legumes except soybeans.

Protein nutrition is not a problem in the United States, where excessive protein consumption is more often the rule, but it is a crucial factor in many of the developing countries. In these areas, mushrooms can be a valuable protein supplement.

Amino Acids

In addition to total crude protein content, amino acid composition is a very important feature in assessing the nutritional value of a foodstuff. Contrary to the widespread belief that mushrooms are deficient in some of the essential amino acids—a belief recorded in most textbooks and articles on the subject—it is well documented that mushrooms contain all the essential amino acids, as well as most of the nonessential ones. It is true that the sulfur-containing amino acids are not abundant, but they are present. To some extent, they limit the nutritional value of mushrooms. The essential amino acids constitute 25–40% of the total amino acids present in mushrooms, the exact amount depending on species. In addition, about a quarter of the amino acids are present in a free form, not as a component of protein. Some of these, such as glutamate, are present in particularly high concentrations in certain mushrooms, for example, the oyster mushroom. This compound is a flavor enhancer. Glutamate also contributes to a problem that has led to a common chef's name for this species—the "baby diaper" mushroom. The abundant glutamic acid rapidly converts to ammonia, so stored oyster mushrooms develop the typical ammonia smell of the nursery.

The amino acids and amides are not the only nitrogenous compounds in mushrooms, of course. Sometimes, quite novel compounds are present, including pipecolic acid. Some of these no doubt contribute to the flavor of the mushrooms. In one study, at least 53 different nitrogenous compounds were detected.[6] A more recent analysis of many mushroom species has documented the huge variety of amino acids in mushrooms.[7] The amino acid concentrations vary throughout the life of a single mushroom and also show significant substrate-dependent variations.[5] The role of nitrogenous compounds in human toxicity, in addition to taste and flavor, has not been completely explored.

Fat Content

The average fat content of mushrooms is 5–8% of the dry weight, but this value can vary from less than 1% up to a rare 15%. All the classes of lipids are represented, including relatively large amounts of the essential fatty acids, especially linoleic acid. Some species also contain reasonable quantities of steroids, including ergosterol, which can be converted to vitamin D.

Carbohydrate and Fiber

Carbohydrate is a major component of mushrooms, averaging 40–60% of the dry weight. In fresh mushrooms, carbohydrate content varies from 3 to 28% and consists of pentose and hexose sugars; methyl pentoses, such as rhamnose and fucose; inositol and mannitol, the latter in sizable amounts of up to 13%; disaccharides, including sucrose; glucuronic and galacturonic acids; the amino sugars; and a number of currently unidentified compounds. In addition α-trehalose is present in significant concentrations in a young specimen, but as the mushroom matures, most of it is hydrolyzed to glucose.

Fiber content varies from 5 to 15% of the dry weight of a mushroom. Recall that fiber is defined as the residue remaining after enzymatic digestion in the gut.

The two important polysaccharides in mushrooms are glycogen and chitin. Unlike plants, which store energy in the form of starch, mushrooms, like most animals, tend to use glycogen as a storage compound. The major structural component of a mushroom's cell wall is chitin, which is a large polymer of N-acetylglucosamine and identical to the substance that forms the exoskeletons of insects.

Minerals and Vitamins

Whereas the fat-soluble vitamins (A, D, E, K) are present in small quantities, if at all, most of the others are well represented. The water-soluble vitamins include the B group (thiamine, riboflavin, and niacin), biotin, and ascorbic acid (vitamin C). Concentrations depend on species and growing conditions.

All the important minerals are present except iron, which occurs in negligible amounts. The kind of substrate the mushroom grows on has a significant impact on the mineral content. Sodium, potassium, and phosphorus are particularly abundant, whereas calcium is present in lower concentrations. As will be discussed in Chapter 7, mushrooms have the ability to preferentially concentrate certain heavy metals, especially when these elements contaminate the environment.

Bioavailability and Nutritional Value of Nutrients

The concentrations of various nutrients in a food source are only important if humans have the ability to digest, absorb, and utilize them. This simple truth has been ignored again and again in food science, an unfortunate circumstance leading to abundant misinformation. Probably the best example of such misguidance is the myth of useful iron in spinach; usually that iron is complexed in such a way as to render it unavailable to humans. Mushrooms, too, have suffered from these falsehoods, because no adequate human feeding studies have been performed. We can only hope that this situation will be remedied in the near future.

As has been stated repeatedly, the cell walls of fungi consist largely of chitin, a derivative of cellulose. Human beings do not have the ability to digest or absorb this compound. Hence, the availability of the crudely measured nitrogen is not total; most sources now correct for this problem. Crisan and Sands[2] developed an interesting "nutritional index" based on the amino acid composition of a food and it "biological value" to allow them to compare mushrooms with other known

Nutritional Highlights

- For overall nutrition, mushrooms fall between the best vegetables and animal protein sources.
- Mushrooms consist largely of water (85–95%).
- The protein content of mushrooms is 15–40% of dry weight.
- All essential amino acids are present in mushrooms.
- Water-soluble vitamins and all mineral are present in mushrooms.
- A generous serving (0.5 lb) of fresh mushrooms provides approximately 70 kcal.

foods. Interestingly, on the basis of this index, many of the wild mushrooms appear to be more nutritious than the cultivated varieties. The nutritional value of mushrooms in general was largely limited by the relatively low concentrations of the sulfur-containing amino acids, methionine and cysteine. When compared with other foods analyzed by the same method, however, the high-ranking mushrooms scored higher than all the vegetables and legumes except soybeans. The low-ranking mushrooms were more comparable with some common vegetables.

In summary, I quote from Crisan and Sands:[2]

> We can say that the nutritive value of mushrooms is species if not strain specific. The great compositional variation noted in the studies discussed here points out the impossibility of making general statements which are true for edible mushrooms per se. At their best, some mushrooms provide nutritive value comparable to some high protein foods, but conversely, other mushrooms are singularly undistinguished in their nutritional contribution to a diet.

TASTE AND FLAVOR

When we introduce that species [*Boletus edulis*] to our gastronomic friends it will be time to point out clearly their discrepancies; knowledge of Mycology, however, is not intended to supersede the senses of taste and smell, and those who possess such valuable gifts need not learn to consider them mere vulgar prejudices.

Mrs. J. T. Hussey, *Illustrations of British Mycology* (1847)

Not only this same fungus [the chanterelle] never did any harm, but it might even restore the dead.

L. Trattinnick, *Essbaren Schwamme* (nineteenth century)

Nutritional value is not the only reason to eat a food—although this is usually the only aspect that expert nutritionists comment on. Taste, flavor, and palatability are equally as important, a fact that is not missed by the manufacturers, purveyors, and marketers of foodstuffs. Many mushrooms shine in this arena, with a wealth of flavors and textures that not only stand by themselves but also complement many other foods. These qualities are as difficult to express in words as are the tastes of wine, but they are often both forceful and distinctive.

This is not to say that mushrooms are for everyone. One of the great joys of life is our uniqueness and individuality. Each one of us experiences smell and taste in a different way. These two very primitive senses, coded and processed in some of the most evolutionarily ancient parts of our nervous system, frequently evoke feelings and memories of which we are only dimly aware. Any discussion among mycophagy aficionados soon degenerates into passionate defense of someone's favorite mushroom or disparaging remarks about others. We know very little about these great variations in our appreciation of taste and flavor; how they develop, what influences them, or how they mature. We do know that children are not inherently attracted to mushrooms, as they are to french fries or ice cream. In fact, one definition of a mushroom is a food widely esteemed by adults and uniformly detested by children.

A substantial number of references comment on the flavor of mushrooms being more allied to that of meat than to that of vegetables.[8–12] As Pilát and Ušák said, "Mushrooms give us the only vegetable which can conjure up on the tongue the illusion of a meat dish, and besides they have an excellent taste when well prepared."[10] This should not be a surprise, because it is only an accident of Western history that mushrooms became grouped with the plants to be studied by botanists. They obviously have very little in common with plants, apart from being rather stationary. In many cultures, including the preliterate ones, the fungi have always been regarded as different from both the plants and the animals. This "meaty" flavor of mushrooms was best expressed in the late 1800s—and by an Englishman, no less:

> I have indeed grieved, when I reflected on the straitened condition of the lower orders this year, to see pounds innumerable of extempore beef-steaks growing on our oaks in the shape of *Fistulina hepatica*. ... Puffballs, which some of our friends have not inaptly compared to sweet bread for the rich delicacy of their unassisted flavour; *Hydna* as good as oysters, which they somewhat resemble in taste; *Agaricus deliciosus*, reminding us of tender lamb-kidneys; the beautiful yellow Chanterelle, that *kalon kagathon* of diet, growing by the bushel, and no basket but our own to pick up a few

specimens in our way; the sweet nutty flavoured *Boletus*, in vain calling himself *edulis* where there was none to believe him.

Reverend Charles Badham, *A Treatise on the Esculent Funguses of Britain* (1863)

It is not surprising that the flavors of individual specimens of one species of mushrooms can vary considerably, depending on where they are grown and picked. This variability is no different from the variability in toxicity noted in the poisonous species and the differences in nutritional content. Variability from year to year in the same locality, not unlike the differences in the flavor of grapes, is a common and expected phenomenon. To complicate matters, differences have also been demonstrated in the concentration of flavor compounds, depending on the stage of the life cycle at which it is examined. This variability is attributable to the constantly changing concentration of metabolites in the fruiting body as the mushroom matures and ages. In their earliest (button) stages, most mushrooms have a mild flavor. The flavor usually strengthens as the mushroom reaches maturity, and then a series of "off-flavors" develop. Some mushrooms, such as the morel, develop a very unpleasant taste. The majority of mushrooms are eaten early in their life span, because this is usually the time when the texture of the fruiting body is most pleasing. The change in aroma and associated flavor that occurs with aging happens to both fresh and dried specimens. In fact, a number of wild mushrooms are dried, not only so that they can be used throughout the year, but also because their flavors tend to become more concentrated once the moisture has been removed. With storage, many of these flavor compounds undergo a series of subtle and often delightful alterations, and the aromas change accordingly. To extend the wine analogy, this flavor enhancement is similar to that occurring in wine as it ages in the barrel or bottle.

Differences in flavor were brought home to me in the summer of 1992. I had picked and eaten *Boletus edulis* (also called the cèpe, porcini, and Steinpilz) for many years in the Pacific Northwest. I enjoyed it fresh in its button stage, and dried pieces added a depth of flavor to my soups, stews, and stocks. But I could never really understand why this was the world's most widely sought-after edible wild mushroom. It was always exciting to find one, especially if I had beaten the maggots to the prize—a beautiful, statuesque mushroom—but I always felt vaguely disappointed with the culinary results. I knew it was not my cooking. In 1992 I traveled to the Dolomites in northern Italy for a week of mountain walking. The first meal of porcini in the small town of Bassano del Grappa on the way up north was close to being a religious experience. The intensity of flavor was almost overwhelming. I realized then that the flavor of our varieties is but a shadow of the flavor of the Italian specimens. I must presume that regional flavor differences explain my experience. It is also possible that the porcini of Italy is not the same species as the American porcini but is instead another, more flavorful, bolete.

Other mushrooms also exhibit this variability. The chanterelles in the eastern part of the United States, although considerably smaller than those in the West, have much more perfume, a greater whiff of apricot, and a deeper flavor. Morels also vary considerably. Flavor differences are not always endogenous, however. Many cookbooks comment on the "smoky" flavor of morels, but this is not an inherent characteristic of morels. Many of the morels available for sale in the markets around the world, and especially in the United States, come from Asia, including India, where they are commonly dried over slow-burning smoky fires, for which cow dung often is the fuel—hence, the smoky flavor.

Not surprisingly, the majority of studies on taste and flavor have been performed on the cultivated varieties of mushrooms, although great interest in truffles and morels, both of which have extraordinarily distinctive aromas, have prompted similar investigations. Because the smell of the mushrooms contributes so heavily to the perceived flavor, much of the research has been concentrated on the volatile compounds. As many as 150 different chemicals have been identified in various species.[13,14] Some of these compounds appear to be formed from the few essential fatty acids present in many mushrooms: linolenic and linoleic acids.[15] The most important contributors to the flavor are 8- and 10-carbon compounds. In the fresh sporocarp of the chanterelle (*Cantharellus cibarius*), 1-octen-3-ol is the preponderant agent, whereas 1-octen-3-one has the highest aroma value in the cèpe (*Boletus edulis*). The common cultivated mushroom, *Agaricus bisporus*, has a benzyl alcohol as its major volatile substance, a compound that actually does not smell very "mushroomy." During heating, a number of these compounds become modified to pyrazines, pyrroles, lactones, and furans. These are responsible for the distinctive taste of many of the fungi in cooked dishes. A Japanese investigator has published a detailed study of wild mushroom tastes.[16] The connoisseur who wishes to know what chemical is tweaking her taste buds should consult this study.

COOKING AND DIGESTIBILITY

Mushrooms lend themselves to a variety of different cooking methods. But one fundamental principle applies to all methods: Mushrooms should always be well cooked. This dictum is the antithesis of the

**ALL mushrooms should be WELL COOKED.
Do not eat raw mushrooms!**

cooking fad that has swept the country in the past few years, especially in the preparation of vegetables. In a much-needed reaction to widespread overcooking, many of the more innovative and health-conscious chefs have adopted the practice of using fresh, young produce and minimal heat, cooking vegetables only until they are "tender-crisp." Although this is certainly a decent treatment for most vegetables, it is not good practice with mushrooms. The cell walls are made of chitin, a polysaccharide that humans do not have the necessary intestinal enzymes to digest. Heat is required to begin the degradation of cells walls so that we can absorb the nutrients within.

For certain mushrooms, cooking is an absolute requirement for safe eating. The most notable examples belong to the genus *Gyromitra*. If you are going to indulge in specimens of *Gyromitra*, you must blanch or parboil them, discarding the cooking water prior to final preparation. This procedure eliminates most of the gyromitrin toxin. However, I firmly recommend that this genus never be eaten (see Chapter 14).

The widely enjoyed morel should also be well cooked, because in its raw state it will produce unpleasant gastrointestinal symptoms. Unfortunately, the nature of the morel toxin is currently unknown. A story about a banquet in Vancouver, British Columbia, at which raw morels were consumed is recounted in Chapter 14.

Many of the other toxic compounds in mushrooms, including a variety of hemolysins and lectins, are heat labile and are adequately degraded during cooking. The proteins, carbohydrates, and minerals, of course, are not. Vitamin C is destroyed by high heat for a prolonged period, but no one should rely on mushrooms for their daily dose of this nutrient.

MUSHROOMS AS A SIGNIFICANT WORLD FOOD SOURCE

Although it is clear that the nutritional characteristics of mushrooms allow them to be only a supplement to humankind's food supply—albeit a very pleasant one—reasons other than nutrition and palatability argue for the production of mushrooms to complement our diets. The most important reason for cultivating mushrooms is the incredible ability of these fungi to utilize otherwise unavailable organic material and convert it into protein.[17] No other multicellular organism can match the fungus's efficiency in this regard. Vast quantities of agricultural and industrial waste are accumulating daily, along with the rest of the detritus of human living. Fungi could play the dual role of recycling some of this discarded material and simultaneously contributing to our protein requirements.[18] The waste products currently amenable to transformation by fungi are those composed largely of cellulose, hemicellulose, and lignin—the last a major component of wood and paper.

As the habitats for wild mushrooms, especially the mycorrhizal ones associated with forests, continue to be ravaged in every corner of the globe, this source of food, recreation, and gastronomic joy is likely to be only a memory within the next few generations. The cultivation of saprobic edible fungi, however, could replace declining natural habitat as a source of mushrooms. Of course, the host substrates used to grow mushrooms must be free of all heavy metals, because mushrooms preferentially concentrate them. As a corollary to this, however, fungi might be used to break down certain toxic organic compounds and might assist in the disposal of the undesirable by-products of our industrial lifestyle. Fungi such as *Phanerochaete chryosporium* have already been employed on an experimental basis for this purpose.

Worldwide, the cultivation of edible fungi has increased dramatically in the last three decades. Even countries such as the United States, which for years have cultivated only *Agaricus bisporus*, have begun a cottage industry of small "farms" for growing shiitake and oyster mushrooms. In Asia, the cultivation of a number of species, including the shiitake, the oyster mushroom, the wood's ear (*Auricularia auricula*), the paddy straw mushroom (*Volvariella volvacae*), the enoki (*Flammulina velutipes*), and the hamejitake (*Pholiota hameji*) is an ancient art now carried out on a huge scale. Because of the flavor and high nutritional value of many of the species of *Termitomyces*, some interest in the artificial cultivation of these species has recently been expressed.[19] Great interest has swirled around the propagation of morels, and patents for their cultivation and fruiting were issued in 1986 and 1989 by the U.S. Patent Office. The cultivation of the gastronomically valued mycorrhizal mushrooms, which require the symbiotic relationship with a tree for their survival and continued fruiting, will obviously be orders of magnitude more difficult to accomplish. This difficulty has not deterred some individuals from "infecting" the roots of oaks and planting entire hillsides in the hopes of producing the prized truffles. A plan to develop mycorrhizae in artificial culture, from which the edible mushrooms could be fruited, has recently been proposed.

REFERENCES

1. S. R. Adewusi, F. V. Alofe, O. Odeyemi, O. Afolabi, and O. L. Oke, "Studies on Some Edible Wild Mushrooms from Nigeria: 1. Nutritional, Teratogenic and Toxic Considerations," *Plant Foods and Human Nutrition* (1993) 43(2):115–121.

2. E. V. Crisan and A. Sands, "Nutritional Value." In *The Biology and Cultivation of Edible Mushrooms*, ed. S. T. Chang and A. Hayes (New York: Academic Press, 1978), pp. 137–187.

3. S.-T. Chang and P. G. Miles, *Edible Mushrooms and Their Cultivation* (Boca Raton, FL: CRC Press, 1989).

4. S.-T. Chang, "Cultivated Mushrooms." In *Handbook of Applied Mycology: Food and Feeds*, Vol. 3, ed. A. K. Arora, K. G. Mukerji, and E. H. Marth (New York: Marcel Dekker, 1991), pp. 221–227.

5. F. A. Gilbert and R. F. Robinson, "Food from Fungi," *Economic Botany* (1957) 11:126–145.

6. M. R. Altamura, F. M. Robbins, R. E. Andreotti, L. J. Long, and T. Hasselstrom, "Mushroom Ninhydrin-Positive Compounds: Amino Acids, Related Compounds, and Other Nitrogenous Substances Found in Cultivated Mushrooms," *Journal of Agricultural and Food Chemistry* (1967) 15:1040–1043.

7. S.-I. Hanataka, ed., *Amino Acids from Mushrooms*, Progress in the Chemistry of Organic Natural Products, Vol. 59, ed. W. Herz, G. W. Kirby, R. E. Moore, W. Steglich, and C. Tamm (New York: Springer-Verlag, 1992).

8. B. Morris, "Macrofungi of Malawi: Some Ethnobotanical Notes," *Bulletin of the British Mycological Society* (1984) 18(1):48–56.

9. G. T. Prance, "The Use of Edible Fungi by Amazonian Indians." In *Ethnobotany in the Neotropics*, New York Botanical Garden Publications, Vol. 1, ed. G. T. Prance and J. A. Kallunki (Kansas City, MI: Allen Press, 1984), pp. 127–139.

10. A. Pilát and O. Ušák, *Mushrooms* (London: Spring Books, 1954).

11. B. Morris, "The Folk Classification of Fungi," *The Mycologist* (1988) 2(1):8–10.

12. R. K. Dentan, *The Semai: A Non-Violent People of Malaya* (New York: Holt Reinhart, 1968).

13. J. A. Maga, "Mushroom Flavor," *Journal of Agricultural and Food Chemistry* (1976) 16:517.

14. H. Pyysalo, "On the Formation of the Aroma of Some Northern Mushrooms," *Mushroom Science* (1978) 10(2):669–675.

15. R. Tressl, D. Bahri, and K. H. Engel, "Formation of Eight-Carbon and Ten-Carbon Components in Mushrooms (*Agaricus campestris*)," *Journal of Agricultural and Food Chemistry* (1982) 30:89–93.

16. Y. Miyaji, "Notes on Tastes of Mushrooms," *Transactions of the Mycological Society of Japan* (1976) 17:587–592.

17. W. D. Gray, *The Use of Fungi as Food and in Food Processing* (Cleveland, OH: CRC Press, 1973).

18. S. Rajarathnam and Z. Bano, "Biological Utilization of Edible Fruiting Fungi." In *Handbook of Applied Mycology: Food and Feeds*, Vol. 3, ed. A. K. Arora, K. G. Mukerji, and E. H. Marth (New York: Marcel Dekker, 1991), pp. 241–281.

19. W. J. Botha and A. Eicker, "Nutritional Value of *Termitomyces* Mycelial Protein and Growth of Mycelium on Natural Substrates," *Mycological Research* (1992) 5:350–354.

CHAPTER FIVE

HEALTH BENEFITS AND MEDICINAL PROPERTIES OF EDIBLE MUSHROOMS

What medicine is there for hunger? . . . obviously that which appeases hunger. But food does this, hence it must contain some medicine.

Hippocrates (460–377 B.C), *About Winds*

TWO VIEWS OF THE EFFECTS OF FOOD

The belief that what we eat affects our state of well-being is ancient. Beyond either nutrition and gastronomy, certain mushrooms have been promoted as agents that have a direct, positive impact on our health, prescriptions that come predominantly from Asia, where for centuries both the Chinese and the Japanese have used a variety of mushrooms in their pharmacopoeias—solely to promote good health. The belief in the health benefits of mushrooms is certainly not confined to these cultures, but it finds its fullest expression there.

The concept of the influence of diet on health has been embraced by various distinct communities in the United States, often those with religious overtones. For example, the Seventh-Day Adventists, the Mormons, and the followers of Reverend Sylvester Graham have all elevated diet to a significant place in their respective lifestyles. Within the general population, however, there has been only a peripheral awareness of the importance of diet in maintaining health. In the early 1930s, the characterization of vitamins and the publication of the minimum daily requirements put a major stamp of legitimacy on the idea that certain nutrients prevent disease. Popular publications of the 1970s, such as Adele Davis's *Let's Eat Right for Health and Fitness,* induced contemporary Americans to view their food in a somewhat new light.

In more recent times, the focus has shifted to the role of dietary factors—sometimes unwanted contaminants, but often regular components of the Western diet—in causing disease. The current concern centers around the dangers rather than the benefits of food, perhaps because we have moved from a period of subsistence to one of overwhelming surfeit. (Unfortunately, this surplus has not been shared worldwide, for the diet of a great portion of humanity is still deficient.) In the past few decades, a week has not gone by without a report of some foodstuff being the cause of cancer or some other equally deadly scourge. But the thrust of most of the reports has been toward prevention. In other words, the most frequent reason given for eating or not eating a particular nutriment has been to prevent the occurrence of a particular disease. A large subculture and economy has grown up to exploit the fears of the public. This market is now well served by "health food" stores, which are liberally sprinkled in the many shopping malls around the country.

An underlying assumption of both traditional allopathic medicine and the health food promoters is that humans are on the brink of dissolution—that only by ridding ourselves of enemies such as viruses and bacteria and by the careful control of diet can we avoid falling victim to the myriad ills surrounding us, each waiting to exploit the slightest weakness. Food marketing has now taken on an entirely new dimension. Promoting the avoidance of many foods has become a more profitable enterprise than promoting their consumption—labels boast of "reduced calories" and "cholesterol-free," "low-fat," "sodium-free," and "decaffeinated" contents. A few items, such as calcium, fish oil, fiber, and oats, have been pushed with the message that they provide positive health benefits, but even these products target either a single disease or a laboratory measurement, such as the level of serum cholesterol. Generally, each fad passes rather quickly. In the health food ethic, prevention plays a similar role, but it is coupled to an almost mystical belief in the "natural" or the "organic." Many of these movements have been short-lived, to be replaced the next week with another "heal-all."

No one has described our current obsession with lifestyle as well as the medical essayist Lewis Thomas:

> Many of us take the view that [death] can be put off, stalled off anyway, by changing the way we live. "Life-style" moved into the language by way of its connection with getting sick and dying. We jog, skip, attend aerobic classes, eat certain food "groups" as though food itself was a new kind of medicine, even try to change our thoughts to make our cells sit up and behave like healthy cells; meditation is taken as medication by some of the most ardent mediators. We do these things not so much to keep fit,

which is healthy exercise and good for the mind, but to fend off dying, which is an effort not so good for the mind, maybe in the long run bad for the mind.

Lewis Thomas, *The Fragile Species* (1992)

The Western approach of using foods to avoid disease contrasts with the approach of many Asian cultures in which foods are often employed to actively enhance health, to improve performance, and to increase one's sense of well-being. Not that the concepts of longevity and prevention are eschewed in Asia. Those, too, are potent driving forces behind the consumption of certain foods. But the underlying philosophy is clearly different. Almost every food is imbued with some beneficial life-giving or health-promoting property. In other words, foods are consumed because of their positive effects rather than avoided because of their negative effects. Moreover, as noted earlier, the interrelation between diet and health is an ancient belief. Recently, Chinese female athletes turned in astounding, world record-breaking performances in a number of long-distance races. On being accused of using steroids, the Chinese countered with the "revelation" that the athletes had been on a diet that included *Cordyceps sinensis*, a fungus growing on caterpillars. Great skepticism still exists in the West, where some cynics believe that the fungus is a euphemism for anabolic steroids.

TWO APPROACHES TO MEDICINE AND RESEARCH

The difference in belief systems about the effects of food has some interesting consequences. The Western reductionistic approach and the simplistic "one cause, one effect" paradigm used in evaluating a particular foodstuff or drug almost preclude any serious study of the claims made by scientists in Asia. Our attitude is further tainted by the general belief that nutrients are largely for prevention. For example, we can readily demonstrate the development of particular symptoms in the absence of a vitamin in the diet or the undesirable consequences of a steady diet of a certain nutrient. However, we are almost incapable of designing studies that demonstrate a gradual improvement or enhancement of a system or a function. Indeed, we look for discrete effects in a single function rather than general changes in major organ systems or processes. The Chinese and Japanese, however, are comfortable with global changes such as "increased energy," "enhanced potency," or "a general improvement in the immune system." We disparage remedies that produce such effects and call them panaceas, because we have no easy way of measuring the changes. The results are too imprecise, too "unscientific," too general for us to deal with. One of the dictums of Western medical science is that if a phenomenon cannot be measured, then it cannot be known. (However, there is no reason to accept the corollary: Therefore, it cannot be true.) The word *panacea* has a pejorative ring and is associated with quackery and charlatanism. We tend to assign all therapies producing whole-body effects to practitioners outside the medical mainstream—herbalist, naturopaths, homeopaths, and chiropractors.

Western therapeutics have largely been reduced to two major modalities in the last few decades—remove the problem (the surgical solution) or change it chemically (the pharmacological solution). In pharmacology, the goal has been to find the most specific agent, one that has a targeted and direct effect on a particular organ or physiological function. This is the "magic bullet" approach. Conversely, Asian practitioners have a long-standing custom of classifying drugs into two major groups: the "higher" drugs, which regulate the homeostasis of the body and restore the diseased organism to a normal state of well-being, and the "lower" drugs, which act directly on the disease focus. By this classification, almost all the remedies we use in the West would be characterized as "lower" drugs. Moreover, the Chinese use small amounts of "drugs" over a long period. They expect gradual beneficial effects. We expect an instant cure. Once again, the disparity between belief systems makes the evaluation of health claims very difficult.

MUSHROOMS IN THE ASIAN PHARMACOPOEIA

The overwhelming majority of studies on the health benefits of mushroom eating have been performed in China and Japan. The Russians, too, have made significant contributions.[1] Unfortunately, most of the published reports have not been translated into English. The methodologies of the studies frequently do not conform to what Western bureaucracies and accepted scientific practice now demand of a pharmacological evaluation—namely, the double-blind study. In this kind of experiment, neither the test subjects nor the investigators know whether the subjects are getting the drug or a harmless placebo. The outcomes of therapy are measured in rather precise ways, with laboratory tests, quantitative functional measurements, or clearly defined end points. Unfortunately, these techniques, which are very valuable in assessing the effectiveness of a treatment, have generally not been followed in Asia. In fact, the whole style of Asian therapeutics precludes this approach, because their pharmacological agents have to be taken for a long time and are often combined with a specific diet or other substances in a carefully balanced and often individually customized regimen. Double-blind studies in such a setting would be both complex and expensive.

The recent publication of the translation of a major work on Chinese medicinal fungi proved to be much less informative than anticipated. The *Icones of Medicinal Fungi from China*[2] is a beautifully illustrated compendium of Asian mushrooms, with a fascinating account of their uses. However, documentation of their effects in any sort of controlled clinical trail is virtually absent. Most of the references are to the antitumor properties of the mushrooms. These studies were all performed in questionable animal models and are of limited value, because the results cannot automatically be extrapolated to humans. Enhancement of the immune function is another major topic.

For a physician trained in the Western medical tradition, some of the recommendations in *Icones* are difficult to comprehend. The three passages reproduced here illustrate the problem.

Tyromyces sulphureus (Laetiporus sulphureus)—Medicinal value

Mild in mature and sweet in taste, capable of regulating the human body, improving health and defending the body against illnesses. Taken regularly, it plays an important part in the regulation of the human body. The water extract of the sporophore inhibits Erhlich carcinoma in white mice. Burning the sporophore drives away mosquitoes and midges. The eburicoic acid produced by the sporophore of this fungus may be used to synthesize steroid drugs which play important roles in regulation of human body.

Xanthochrous hispidus—Medicinal value

It dispels endogenous wind, stops hemorrhage, counteracts noxious heat, relieves pain and cures mixed hemorrhoids, prolapse of the rectum and bleeding piles. Use appropriate amount for external application. Its inhibition rate against sarcoma 180 and Erhlich carcinoma is 80% and 70% respectively.

Ganoderma sinense—Medicinal value

Being mild in property and insipid, it is a nourishing and roborant tonic, it strengthens the brain and nourishes the stomach, is anti-inflammatory and diuretic. According to the "Shen Nong's Herbal" it cures deafness, is good for the joints, conserves the vital energy, nourishes the essence of life, strengthens tendons and bones. It remedies neurasthenia, dizziness and insomnia, chronic hepatitis, pyelonephritis and bronchial asthma.

Despite the cultural differences, it would be a great mistake to dismiss the Asian health claims out of hand. We have embraced the products of the lower fungi, the antibiotics. So why have we largely ignored the potent compounds in the higher fungi? It is noteworthy that all the reviews on this topic have appeared in the mycological literature rather than in medical reports. Western medicine is reluctant to accept the possibility that mushrooms have important benefits. This attitude is reflected in the paucity of funded studies, by either the pharmaceutical industry or governmental agencies, and mycologists and medical scientist have great difficulty in getting funding agencies to take their research proposals seriously. It may well be a subtle manifestation of mycophobia at work. Conversely, it could be a realistic assessment of mushrooms' potential.

SEPARATING FOLKLORE FACT FROM FANTASY

There is therefore no sharp distinction between food and medicine from a utilitarian standpoint. There never has been. In all the ancient materia medica only a few toxic substances were included, and most of the medicinal plants were used as food. Even food additives were selected for their medicinal value. Food, therefore, contains many necessary and useful chemicals which help in the maintenance of optimal health.[3]

In their search for food, ancient peoples tasted all that grows on the earth, that crawls and walks over it, that flies above it and lives in the waters—the plants, animals, birds, fishes and invertebrates. People noticed in the process that along with satiation, other sensations arise: different moods—and sometimes the cure of ailments.[4]

Considerable folklore about mushrooms and health has developed in many different cultures. When this lore is tested in the modern arena, many of the traditional products do show some biological activity. It is not always the result expected or anticipated, but in many instances the original use has been verified. Lucas has quoted an earlier investigation that reported a predictive value of 65% for identifying plants with antibiotic activity on the basis of the descriptions in the old herbals;[5] for a random selection of plants, it is 29%.[6]

The older therapies were frequently directed at the symptoms and not at the cause of the disease. Because the "modern" approach is to treat the cause, except for the common cold, some of these ancient remedies have limited value. A good example is the use of the agarick (*Fomes fomentarius*) for the treatment of tuberculosis. We know today that all it achieved was a decrease in the night sweats associated with the infection.

One retrospective study of plants found to have some anticancer activity in a screening program of the National Cancer Institute estimated that the yield of active species could have been increased by 50–100% if the ethnobotanical evidence had been considered. "Folklore use is an excellent prescreen for activity since if the treatment succeeds and the patient lives, the knowledge tends to get passed on to future generations."[7]

The likelihood of finding something truly useful is not the same for each therapeutic area. In those conditions in which the response is immediate and dramatic, the chances are very high that the indigenous product will prove to be effective. In chronic situations or conditions where it is much more difficult to relate the change in symptoms to the ingestion of the herb or mushroom, the odds favoring efficacy decline but are still substantial. It has been estimated that the odds of finding some anticancer activity is two times higher for indigenous medicinal plants than for plants chosen at random; for antiparasitic activity, three times higher; for a fish poison, four times higher; and for arrowhead poisons, five times higher.[8] Obviously, if a poisoned arrow does not kill its intended victim, it rapidly falls out of favor—and out of the local pharmacopoeia.

GENERAL MEDICINAL PROPERTIES OF MUSHROOMS

The subject of the relationship of mushrooms to health can be approached in two ways. The most comfortable approach for a Westerner is to list the various pharmacological effects currently regarded as important and provide a catalogue and analysis of the fungi associated with that effect. The second approach is to list each fungus, with an investigation of its proposed benefits. Neither approach is entirely satisfactory, so I have adopted a hybrid. In this section, the general benefits of mushrooms are examined. Then follows a more detailed analysis of the two fungi most frequently associated with medicinal applications, the ling chi and the shiitake.

Constituents in plants and fungi can be thought of in two broad categories. The first group comprises the key structural and metabolic compounds, with which we are all familiar, namely, proteins, carbohydrates, fats, and the critical micronutrients, minerals and vitamins. These are the compounds—the nutrients—that the nutritionists have tried to educate us about. In the second group, usually unmeasured and ignored, are the nonnutritional biochemicals, with which the nutritionists are unfamiliar and the general public entirely ignorant. These compounds include a bewildering array of alkaloids, flavonoids, terpenes, coumarins, lectins, and other plant and fungal products. They are the so-called secondary compounds, of which the various mushroom poisons are only one very small example. It is these compounds, eaten in every mouthful of food, that affect health. Many of them have very potent actions in mediating, inhibiting, or changing a wide spectrum of physiological processes. The major medicinal properties attributed to mushrooms include anticancer activity, antibiotic activity (directed against bacteria, fungi, and protozoa), antiviral activity, immune response-stimulating effects, and blood lipid-lowering effects. Much of the earlier work on these effects was summarized by Cochran.[9]

Anticancer Activities

Of all the biological activities lying untapped within the fleshy macrofungi, the possible role of mushrooms in treating patients with cancer has created the most excitement. The antineoplastic properties of fungi were initially reported in the 1950s when the giant puffball, *Calvatia gigantea*, was shown to contain a compound—labeled calvacin—that was thought to be active against tumors.[10] This report followed earlier work with other mushrooms and was published during the period when many products were being screened for antimicrobial activities.[11] Because various forms of cancer are common, many folklore remedies have been developed to treat them. It was from old stories about the use of mushrooms for this purpose that the idea to test the puffball first originated. A more complete list of the folklore cancer remedies, prepared from both plants and fungi, was published in the early 1970s. Today the search continues, and excitement is still generated by discoveries of compounds like taxol, which comes from the bark of the Pacific yew tree and is proving to be quite useful against ovarian cancer. (Of course, it has not done much for the survival of the yew trees, whose numbers are rapidly declining as many trees are harvested to provide this new drug.)

A major problem in the evaluation of claims for antineoplastic activity is the imperfect nature of the screening processes. Investigating the cancer-curing effects of mushrooms is a difficult and nontrivial business. Animal models of cancer are poor representatives of the human conditions. Both the genetics and the environmental factors leading to the development of tumors are very different among the various animal species. Furthermore, despite the superficial similarities between the different kinds of human tumors in the various organs and at different ages, we know that each has unique mechanisms responsible for its initiation and progression. It is therefore not surprising that one form of therapy may be useful in treating one form of cancer and almost useless in another. This heterogeneity makes interpretation of the animal experiments even more difficult; yet it is on these data that many of the current claims rest. Unfortunately, very few products that show promise in vitro or in nonhuman tumors ever fulfill the promises of their advance publicity. Certainly animal studies should not be abandoned, but valid evaluations can be made only with human subjects, a difficult prospect at best.

The following is an incomplete list of mushroom species for which anticancer properties have been claimed:[9,11–18]

Ganoderma applantum	Lentinus (=Lentinula) edodes
Coriolus (=Trametes) versicolor	Flammulina velutipes
Piptoporus betulinus	Pleurotus ostreatus
Lentinus (=Neolentinus) lepideus	Pholiota nameko
Morchella hortensis	Auricularia auricula
Boletus edulis	Tricholoma matsutake
Agaricus species	Pleurotus spodoleucus

A number of interesting and novel chemical compounds have been extracted from many of these mushrooms, but it must be reiterated that the activities demonstrated are purely experimental. Very little evidence is available to suggest that any of these compounds will ever play a role in the prevention or management of human cancer. At this time, reports of their actions can be regarded only as interesting, preliminary observations.

It has become clear that many of the anticancer agents do not act directly on the tumor itself but instead act through the immune system of the host. This indirect mechanism of action became evident when mushroom extracts known to inhibit transplanted tumors showed no activity against spontaneous tumors of the same type. In this case, the mushroom compounds had simply stimulated the immune system of the experimental animals to reject the transplanted tumor. This effect is very similar to that obtained with Calmette-Guerin bacillus (BCG) or Corynebacterium parvum. These bacteria and their extracts act as immunological adjuvants or stimulators. Enhancement of immune system functions is potentially a very beneficial effect in the management of a patient with cancer. However, it is not known how successful mushroom extracts will be in treating spontaneous tumors in humans. Much of the work with higher fungi and tumors has been well summarized by Chihara.[19]

More recent work has centered on the common edible species, such as *Pleurotus ostreatus,* the oyster mushroom,[20] and *Flammulina velutipes,* the enoki.[18] A number of compounds from enoki have been carefully characterized, such as a β-(1→3)-glucan, a polysaccharide called EA3, and a second, higher molecular weight polysaccharide, EA501. Again, these agents do not act directly on the tumor but somehow affect the host.[21]

Other species, such as the coral mushrooms, also have been investigated.[22] Some of the compounds in mushrooms have the ability to directly stimulate certain cells of the immune system to divide,[23] whereas others can actually decrease the responsiveness of these same cells, acting as immunosuppressants.[24]

The most contemporary entry into the health sweepstakes is the maitake (*Grifola frondosa*), a mushroom commercially produced in Japan. Both the fruiting body (the part normally eaten) and the mycelium contain antitumor glucans similar to those isolated from other edible species.[25,26] These are said to have immunity-enhancing effects on the host. A Chinese relative of this mushroom, the zhu ling (*Polyporus umbellatus* = *Grifola umbellata*), is said to have similar properties.

In addition to this indirect effect on tumor growth, recent evidence points to a more direct and possibly beneficial role in the prevention of tumors for compounds in certain mushrooms.[27] Fungi such as *Craterellus cornucopoides* ("horn of plenty" or, in France, *trompet de mort,* "trumpet of death"), which despite its gothic name is an excellent edible species, contain substances that can inhibit the development of tumors normally provoked by mutagenic agents such as aflatoxin B and benzo[*a*]pyrene. In a series of experiments, seven different mushroom species demonstrated this antimutagenic property.[28] Once again it must be stressed that these results were produced in a very artificial experimental situation, one based on bacterial cell systems and not on human cell systems. Translating such results into everyday human biology and health is not yet possible.

Antimicrobial Activity

Much of the early work on mushrooms involved the search for compounds with antibacterial activity. This search was a natural extension of prior observations of extraordinary substances, such as penicillin, produced by microscopic, often soil-dwelling, fungi. The ecological role of such compounds may be to reduce the competition between the fungus and the soil microbes in its immediate environment. From the late 1940s through the early 1960s, numerous plants and fungi were systematically screened for activity against bacteria, other fungi, protozoa, and viruses. This screening was usually done with rather crude extracts of the fruiting bodies, using simple in vitro testing conditions. As was to be expected, a large number of fungi were found to produce compounds that showed activity in selected test systems under specific conditions. However, no compound that could be used to treat all microbial infections was ever found. This work was carried out in many laboratories around the world. The details have been summarized in a number of reviews.[1,29–31] (For readers interested in pursuing the subject further, these references are the place to start, because they contain extensive literature citations.) More recently, the work has switched to characterizing the active compounds chemically and exploring the mechanism of their action.[31,32,33,34]

Although the initial surveys suggested that antimicrobial activity was more frequently present in the genera associated with the wood-rotting fungi and therefore present in the mushrooms that grow on trees, such as *Fomes*, *Trametes*, and *Polyporus*, it soon became clear that this activity was not restricted to these genera. Many mycorrhizal fungi also showed similar activity. In fact, many species of edible mushrooms were also discovered to possess some antimicrobial activity. Not surprisingly, the antimicrobial effects were not always consistent; not every collection of a particular species produced the active compound. This variation is similar to that seen in mushroom morphology and in the occurrence of other metabolites.

The types of chemical compounds identified as the active agents for antimicrobial activity are very diverse. They include polyacetylenes (phenylpropanoid derivatives), terpenoids, quinones, purines, pyrimidines, and phenolic and quinoid compounds. None has proved particularly attractive from a therapeutic standpoint, and none has undergone further investigation or development. Undoubtedly, dozens of compounds are still waiting to be discovered and perhaps exploited. With the increase in the number of pathogenic organisms now resistant to our current crop of antibiotics, the search for new drugs will once again intensify. Whether the higher fungi will contribute to this is uncertain, but at least some promise for the discovery of novel compounds remains.[34,35]

As noted earlier, many of the fungal products with antimicrobial activity were found in edible mushrooms. Their presence raises interesting questions about the role of such mushrooms in a person's regular diet. For example, could individuals be sensitized to an antibiotic as a result of small doses in their diet? Is it possible that indigenous bacteria could become resistant to an antibiotic as a result of repeated exposure to these small doses? Does the consumption of mushrooms with these properties change an individual's normal bacterial flora in the mouth, GI tract, and vagina? Does eating these mushrooms increase or

decrease an individual's incidence of infections? Is there any therapeutic or prophylactic role for these fungi in specific clinical circumstances? The answers to all such questions and dozens of others
provoked by these observations are unknown.

The biological role of many of these compounds is to reduce the
competition from other organisms for the available food supply, so it
is reasonable to anticipate that antifungal chemicals also are produced.
Such antifungal activity was noted as early as 1923.[36] Other examples—including the edible fungi *Lentinus (Lentinula) edodes* (shiitake) and
Coprinus comatus (shaggy mane)—were provided by those who summarized the available data in the 1960s.[31,37] Still other mushrooms
inhibit the protozoa and the small roundworms (nematodes) in the
soil. (Carnivorous soil fungi actually trap and digest the tiny nematodes, ensnaring them in loops of hyphae that suddenly expand by
osmosis when a nematode is "sensed" within the loop.) None of these
fungi plays any significant role in the production of antifungal or
antiparasitic compounds suitable for human consumption.

Antiviral Activity

At this time, very little is known about the antiviral properties of
fungi, because only a few fungi and a few viruses have been studied.
However, hints of such antiviral activity are intriguing. For instance,
extracts from a variety of the higher, fleshy fungi were shown to
protect mice against the effects of poliomyelitis.[38] Some serotypes of
another of the small RNA viruses (the ECHO group) also were
inhibited by fungi. Cochran and coworkers demonstrated that extracts
of a number of the edible fungi, including *Boletus edulis, Calvatia gigantea,
Suillus luteus,* and *Lentinus (=Lentinula) edodes,* could inhibit the influenza
virus, both in vivo and in vitro.[39] Of course, variation in effect from
one collection of mushrooms to the next has been observed.

For the giant puffball (*Calvatia gigantea*), the compound that inhibits
tumor growth (calvacin) is not the same as the compound that inhibits
viral replication. Indeed, inhibition of viral replication may be largely
nonspecific, that is, it may be caused by viruses that are already present within the fungus—so-called mycoviruses. They may stimulate
the immune system of the animal that ingests them or provoke the production of interferon, thereby indirectly inhibiting other viruses
attacking the animal host.

It became evident in the 1960s, largely through electron microscopic
studies, that many fungi contain large quantities of viruses composed
exclusively of double-stranded RNA.[40] Most of these viruses appear to
be latent, doing no obvious damage to the fungus itself. A few, how-

ever, cause cytolysis of the hyphae and lead to disease in the developing mushroom. Because of their economic implications, such viral diseases of fungi have been best studied in *Agaricus bisporus*.

Fungi may also act as vectors, transmitting viruses to their plant hosts.[41] No current research has implicated the mycoviruses in human disease, however.

The current AIDS epidemic has spawned another hurricane of activity, because every conceivable product is being screened for possible activity against HIV (human immunodeficiency virus). In late 1992, extracts of maitake (*Grifola frondosa*) were shown to inhibit the virus in an in vitro tissue culture system. The clinical significance and value of this observation are awaiting further investigation.

Other Medicinal Properties

A number of other effects have been ascribed to certain mushrooms, such as the hypotensive action of one species of *Pleurotus*.[42] In both Japan and China, members of the genus *Pleurotus* have been used to control high blood pressure. However, all the information is anecdotal and related by herbalists. No well-controlled clinical studies have been done. In a few animal experiments, mushroom extracts were given intravenously. Although hypotension was produced in a dose-dependent manner, experiments of this kind do not even distantly resemble the clinical situation of essential hypertension. Fragmentary evidence of vasoactive ingredients in extracts of other mushrooms has been reported; these mushrooms include maitake (*Grifola frondosa*), shiitake (*Lentinus* (=*Lentinula*) *edodes*),[43] and ling chi (*Ganoderma lucidum*).

The effects of various mushrooms on blood lipids have been well studied in experimental animal systems. *Pleurotus* decreases serum cholesterol levels in rats and Syrian hamsters,[44,45] as do extracts from certain polypores.[46] A 40% reduction in cholesterol—both the very low-density lipoprotein (VLDL) and the low-density lipoprotein (LDL) forms—was noted in female rats at the end of a seven-week trial. The reduction was postulated to be the result of decreased absorption of cholesterol from the gastrointestinal tract, because of possible cholesterol binding to the fiber (undigested residues) from the mushroom. Other studies in rats have demonstrated an increased turnover of lipoproteins.[47] In each animal system, the mechanism of the hypercholesterolemia may be different from that in humans, so the beneficial effects noted in animals may not be readily transferable to the human situation. In the human gut, the chitin in the hyphal cell wall is changed into chitosan, which is suspected of binding bile salts and may influence the absorption of lipids.

MUSHROOM OF IMMORTALITY— LING CHI OR REISHI

The preeminent mushroom from a medicinal standpoint is *Ganoderma lucidum*, called ling chi in China and reishi or mannentake ("10,000-year mushroom") in Japan.[48–50] It has been associated with health and longevity, wisdom and potency, royalty and happiness. The presence of ling chi in traditional Chinese medicine, folklore, and art goes back to the earliest days of recorded history. Gordon Wasson postulated that its adoption as an "herb of immortality" was a product of the knowledge of the Soma, which the Chinese emperors learned about from the Rig Veda.[51] The first mention of the fungus concerns its growth in the royal palace in 109 B.C., during the Han Dynasty, while Wu-ti was on the throne. Since that time, the fungus has been depicted in art, architecture, textiles, sculpture, poetry, song, and literature. It was sought by the powerful for its supposed life-giving and life-enhancing properties. Even today, it is still an expensive and highly valued product.

The ling chi is the paradigm of the Asian approach to health. This mushroom is regarded as a tonic. It is used for almost everything one could imagine—and then some. The compendium of disorders it supposedly cures, or at least helps, reads very much like the lists in the old

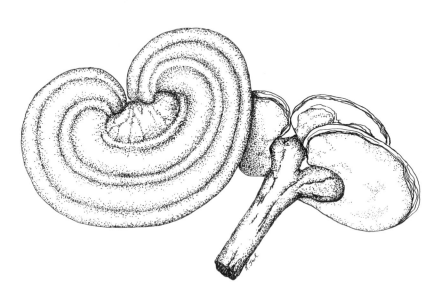

Ganoderma lucidum

herbals describing the benefits of the agarick. It is frequently employed, together with a number of other herbal products, in complex mixtures carefully prescribed and formulated. It rivals ginseng in many of its actions.

But does it work? What is it really good for? The simple answer is that we simply do not know. Although the folklore associated with it is abundant, no studies have adequately answered these questions. Hints, suggestions, and legions of anecdotes abound, but until the appropriate trials are undertaken, its actions will remain unknown.

Polypharmacy—the use of multiple drugs to treat a single problem—is generally decried in the West. But polypharmacy is not just widely practiced in Asia, it is part of the traditional medical practice. Consequently, ling chi is seldom used alone. This practice makes it very difficult to dissect out the various effects of the fungus. Because Westerners look for very specific pharmacological effects, and the Chinese hope for general effects on the whole organism, the two systems are at odds in trying to understand and maximize the benefits of this fungus.

Ling chi, a polypore, has been successfully cultivated.[52] The natural supply has always been limited, so initially only the wealthy could afford the tonic produced from it. Now that ling chi is produced in quantity, many different preparations are available to everyone; it is even available in Western health food or herbal stores. (Some herbalists believe that the wild variety is superior. Once again, we encounter a belief in the superiority of the "natural" product. Perhaps the substrate on which the wild mushroom grows does alter its chemical constituents.)

Medicinal Properties of Ling Chi

According to the *Icones of Medicinal Fungi from China*,[2] *Ganoderma lucidum* has the following medicinal properties:

In clinical practice of late, the following main products have been trially produced, such as *Ganoderma lucidum* syrup, injection, tablet, containing powdered medicine and honey, solution and mixture etc. . . . It has to various extents curative effects on the following diseases: neurasthenia, dizziness, insomnia, chronic hepatitis, pyelonephritis, high serum cholesterol, hypertension, coronary heart disease, leucocytopenia, rhinitis, chronic bronchitis, bronchial asthma, gastropathy, duodenal ulcer etc. *Ganoderma lucidum* is also used as an antidote for poisonous mushroom; in that case take a decoction of 200 gm dried *Ganoderma lucidum* in water. The sporophore contains carbohydrates (including reducing sugars and polysaccharides), amino acids, slight amount of protein, steroid, triterpene, lipid and a small amount of inorganic ions, alkaloids, glucosid, coumaringlucoside, volatile oil, riboflavin and ascorbic acid. Furthermore, according to the research institutes concerned, *Ganoderma lucidum* juice makes white mice resistant to radioactivity. Fluid from submerged fermentation of *Ganoderma lucidum* may enhance the tolerance of white mice against hypoxia under ordinary and low atmospheric pressure. It is also effective in resisting the muscarine and nicotine in white mice.

Jong and Birmingham and Su have recently published two of the most comprehensive reviews of the chemistry and possible medicinal effects of ling chi,[50,53] including well over 100 citations for the chemical compositions of the physiologically active compounds in *Ganoderma*. Unfortunately, few of the articles cited have been translated into English. It is evident, however, that the overwhelming majority of the studies did not involve human subjects; most employed rodents as experimental models, so the results may not be applicable to humans. Many effects have been identified in these animal experiments, and it is clear that the mushrooms contain compounds with potent physiological effects. These active compounds include bitter-tasting triterpenes (at least 100 different triterpenoids have been identified in *G. lucidum*) and many polysaccharides. Properties accorded to this mushroom include an analgesic effect (adenosine), antihepatotoxic activity (*R,S*-ganodermic acid and ganasterone), antiinflammatory activity (glucan), antitumor activity (polysaccharides and glucans), cardiotonic activity (alkaloids, polysaccharides), hypocholesterolemic action (ganodermic acids), inhibition of histamine release (ganodermic acids and oleic acid), hypoglycemic action (ganoderans), hypotensive action (ganderols), immunomodulatory activity (polysaccharides and proteins), stimulation of interferon production, inhibition of platelet aggregation (adenosine),[54] radiation protection, enhancement of protein synthesis, and restoration of neuromuscular activity. Despite this impressive list, the role of this mushroom in medicine and health will remain shrouded in mysticism and mystery until clinical studies are undertaken in human subjects.

Effects on the Immune System

Considerable anecdotal evidence suggests that extracts of this mushroom influence the immune system by affecting a variety of functions. A protein claimed to be an immune potentiator has been described.[55,56] This protein, termed LZ-8, consists of 110 amino acids and has a molecular weight of 12,400. It bears some similarity to the variable region of the heavy chain of the immunoglobulin molecule and may be related, in some distant way, to the ancestral proteins of the immunoglobulin superfamily. It can stimulate cell division in vitro (mitogenic), hemagglutinate sheep (but not human) red cells, and abolish the anaphylactic response in experimental situations. The mechanism by which it modulates the immune system is not currently known, but it may affect those T cells responsible for influencing IgE production. In another experimental disorder, it is able to prevent the development of the autoimmune diabetes type I, which normally develops in NOD mice.[56] It is not a lectin per se because it has no specificity for sugars.

Effects on Blood Lipids and Blood Pressure

Blood lipids are another aspect of human nutrition claimed to be beneficially affected by ling chi.[57] Because elevated serum lipids— especially cholesterol and one of its relatives, low-density lipoprotein (LDL)—are widely believed to be important in the development of atherosclerosis and the subsequent complications of ischemic vascular disease, any compound, especially one occurring naturally in the diet, that decreases the blood concentrations of lipids is regarded with much favor. The mechanism of action is unknown. The physiological effects are said to vary, depending on the strain of mushrooms tested and the method by which it was cultivated. A very mild reduction in systolic blood pressure in genetically hypertensive rats has also been demonstrated.[57,58]

MEDICINAL PROPERTIES OF SHIITAKE

The health benefits of *Lentinus* (=*Lentinula*) *edodes,* the shiitake (Shii-ta-ke) are legendary in both the Chinese and Japanese cultures. As with many other Asian mushrooms, a wide variety of benefits have been claimed for this easily cultivated and widely available species. Two of the most notable effects are the propensity of this species—or of hot water extracts of the mushroom—to lower serum cholesterol levels and to prevent and shrink tumors. The latter is usually linked to its immune-potentiating or immune-modulating activity. More research concerning health-related effects has been done on this mushroom than on any other, largely because of the interest of the growers and the distributors in Japan, who have supported and stimulated the work. In contrast, in the West almost all the work on our favorite edible mushroom, *Agaricus bisporus,* has focused on its dangers. No one has studied its possible health benefits. Once again, the fundamental difference between East and West in their approaches to medicine is underscored.

Effects on Serum Cholesterol Levels

Reproducible effects of *Lentinus* (=*Lentinula*) *edodes* in lowering the level of total cholesterol in the serum have been demonstrated. This outcome was noted in 1964 when a Japanese researcher observed that rats fed on a diet containing 5% ground, dried mushroom for 10 weeks had an average decrease in their serum cholesterol of 20–25%. With certain strains of the mushrooms, even greater reductions were seen.[59,60]

The precise mechanism by which this reduction occurs is not known, but a number of studies have indicated an increased excretion of cholesterol into the stool, an associated increase in the activity of lecithin:cholesterol acyltransferase activity, and an acceleration in the turnover of cholesterol.[61] No change in the rate of synthesis has been noted. This hypocholesterolemic effect has also been demonstrated in humans.[62] A recent study in spontaneously hypertensive rats demonstrated a reduction in free plasma cholesterol while the total cholesterol remain unchanged. This reduction was associated with a decrease in both the VLDL and HDL fractions. It was postulated that increased esterification, due to the increased activity of lecithin:cholesterol acyltransferase, was the underlying mechanism.[43] One problem with extrapolating such results to humans is that these experimental animal systems have unique genetic abnormalities, very different from the human situation.[63] For example, spontaneously hypertensive rats have a low plasma cholesterol and abnormal cholesterol synthesis.

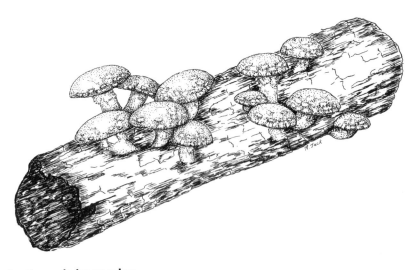

***Lentinus edodes* on a log**

One of the compounds with hypolipidemic properties is presently called eritadenine. It is also called lentysine and lentinacin.[64-66] Its effect in the rat is quite rapid, with the cholesterol level declining within 3 hours of administration. However, in human subjects, this particular compound was shown to decrease triglycerides and not cholesterol. More tests are required with human subjects to elucidate the mechanism of the lipid-lowering effects of the shiitake and its role in maintaining our long-term health.

Antitumor and Immunological Effects

The widely held belief concerning shiitake's efficacy for treating and preventing cancer appeared to be validated when a variety of studies demonstrated a striking antineoplastic effect of mushroom extracts on transplanted sarcoma-180 in mice. Several large molecular weight polysaccharides, the most potent of which was labeled lentinan, were accorded the biological property of producing tumor regression. Other murine tumors responded similarly to these compounds. What has become evident is that the outcome is not due to a direct effect of any of these compounds on the tumor cells but is entirely mediated by the host's immune system. Moreover, there is a great range of host suscep-tibilities. Certain mouse strains such as ICR, SWM/M2, and DISA/2 are responsive to this therapy, whereas C3H, Balb/c, and CBA are not.[19] This tumor model does not represent human cancers, which are autologous in nature. In all the models of spontaneously arising tumors in animals, such as the mammary carcinoma in Swiss mice, shiitake failed to demonstrate a beneficial effect.[67] No human trials have been reported in which a major beneficial role for this mushroom has been clearly demonstrated.

Despite the lack of experimental evidence in both animals and humans, it has been suggested that these compounds may still have a beneficial role in the management of patients with cancer because of their more general immune-potentiating effects. An effect on some of the T cell subsets is suggested by the lack of response in neonatally thymectomized mice. However, no measurable changes in the number or function of T cells as determined in the usual clinical laboratory assays have been demonstrated. The mechanism by which these poly-saccharides affect the T cell system is still unclear. In tumor-bearing animals whose immune systems have been depressed by chemo-therapy, these compounds can restore immune function to almost nor-mal levels.

The Japanese investigators largely responsible for this research list the following properties of the polysaccharides present in shiitake:[19]

- Restoration of various immune responses that are depressed in a tumor-bearing host (including delayed hypersensitivity and antibody production)
- Potentiation of T helper activity
- Activation of macrophages (this activity enhances cytotoxicity and phagocytosis)
- Increases in levels of biologically active serum factors (e.g., tumor necrosis factor, lysozyme, ceruloplasmin)
- Minor effects (e.g., reduces toxicity of chemotherapeutic agents, enhances effect of local irradiation)

In addition to the immunological effects of shiitake, a compound with a possible antimutagenic effect has been extracted. This compound traps substances like nitrites, thereby blocking the formation of carcinogenic N-nitroso compounds that would normally form in the body. Interestingly, this chemical is not present in the raw mushroom but is formed when the mushroom is cooked.[68]

Antiviral Activity

Considerable interest in the ability of shiitake to protect against certain viral infections has stimulated a number of investigations. Originally, this activity was attributed to the nonspecific effect of interferon, which apparently was induced in response to "extracts" of the mushroom.[69,70] It has since been shown that the effect on interferon levels is not due to any compound in the mushroom itself but to the presence of intrinsic viruses living in the fungal tissue (mycoviruses). The initial observation that the fraction stimulating interferon was double-stranded led to this idea. Single-stranded RNA had no effect; double-stranded RNA is found only in viruses.[41,69,71,72] These viruses have limited host specificity, and there is no evidence that they pose a threat to either animals or humans. Long-term feeding of this mushroom to animals had no discernible negative effect.

Interferon induction alone may not account for all the antiviral effects noted. Polysaccharide fractions in the extracts may also be important.[73,74]

FINAL COMMENT ON THE BENEFITS OF EATING MUSHROOMS

Two great cultures, those of Japan and China, have a deeply held and ancient tradition attributing many beneficial effects to the eating of mushrooms. This tradition is part of a complex cultural health system that differs from that of the West in fundamental ways. The health benefits of mushrooms have yet to be documented in convincing ways for Western scientists. The conversion of Western skeptics may be very difficult or impossible because of the differences in belief systems. Nevertheless, fragments of intriguing evidence, some based on animal experiments, suggest that many of the traditional claims may have a sound basis in fact. The challenge is for the scientists on both sides of the Pacific Ocean to find ways to bridge these differences and design studies that will either validate or negate the folklore.

REFERENCES

1. A. N. Shivrina, *Biologically Active Compounds in Higher Fungi* (Moscow: Izd. Nauka, 1965).

2. J. Ying, X. Mao, Q. Ma, Y. Zong, and H. Wen, *Icones of Medicinal Fungi from China* (Beijing: Science Press, 1987).

3. M. Iwu, "Empirical Investigations of Dietary Plants Used in Igbo Ethnomedicine." In *Plants in Indigenous Medicine and Diet: Biobehavioral Approaches*, ed. N. L. Etkin (New York: Redgrave, 1986).

4. G. Brekhman, quoted in Iwu (1986) [note 3].

5. E. H. Lucas, "The Role of Folklore in the Discovery and Rediscovery of Plant Drugs," *The Centennial Review* (1959) 3:173–188.

6. A. G. Winter and L. Willeke, "Untersuchungen über Antibiotika aus höheren Pflanzen. VIII. Die Heilpflanzen des Matthiolus (1611) gegen Infektionen," *Naturwissenschaften* (1953) 40:247–248.

7. M. Suffness and J. Douros, "Drugs of Plant Origin." In *Methods in Cancer Research*, Vol. 16, ed. V. T. De Vita and H. Busch (New York: Academic Press, 1979), pp. 73–126.

8. R. W. Spujt and R. E. J. Perdue, "Plant Folklore: A Tool for Predicting Sources of Anti-cancer Activity?" *Cancer Treatment Reports* (1976) 60(8):979–985.

9. K. W. Cochran, "Medical Effects." In *The Biology and Cultivation of Edible Mushrooms*, ed. S. T. Chang and W. A. Hayes (New York: Academic Press, 1978).

10. E. H. Lucas, R. U. Byerrum, D. A. Clarke, H. C. Reilly, J. A. Stevens, and C. C. Stock, "Production of Oncostatic Principles in vivo and in vitro by Species of the Genus *Calvatia*," *Antibiotic Annual* (1959), pp. 493–495.

11. E. H. Lucas, R. L. Ringler, R. U. Byerrum, J. A. Stevens, D. A. Clarke, and C. C. Stock, "Tumor Inhibitors in *Boletus edulis* and Other Holobasidiomycetes," *Antibiotics and Chemotherapeutics* (1957) 7:1–4.

12. M. A. Espanshade and E. W. Griffith, "Tumor Inhibiting Basidiomycetes: Isolation and Cultivation in the Laboratory," *Mycologia* (1966) 58:511–517.

13. F. J. Gregory, E. M. Healy, H. P. K. J. Agersborg, and G. H. Warren, "Studies on the Anti-tumor Substances Produced by the Basidiomycetes," *Mycologia* (1966) 58:80–90.

14. J. L. Hartwell, "Plants Used Against Cancer," *Lloydia* (1971) 34:386–437.

15. T. Ikekawa, N. Uehara, Y. Maeda, M. Nakamishi, and F. Fukuoka, "Anti-tumor Activity of Aqueous Extracts of Some Edible Mushrooms," *Cancer Research* (1969) 29:734–735.

16. J. Iwekawa, M. Nakamishi, N. Uehara, G. Chihara, and F. Fukuoka, "Anti-tumor Action of Some Basidiomycetes, Especially *Phellinus linteus*," *Gann* (1968) 59:155–157.

17. P. Sevilla-Santos and Z. P. Bernado, "Tumor Inhibitors in Philippine Basidiomycetes," *Philippine Journal of Science* (1966) 75:189–196.

18. T. Ohkuma, K. Otagiri, T. Ikekawa, and S. Tanaka, "Augmentation of Antitumor Activity by Combined Cryo-destruction of Sarcoma-180 and Protein-bound Polysaccharide, EA6, Isolated from *Flammulina velutipes* (Curt. ex Fr.) Sing. in ICR Mice," *Journal of Pharmacobiodynamics* (1982) 5(6):439–444.

19. G. Chihara, "Anti-tumor and Immunological Properties of Polysaccharides from Fungal Origin," *10th International Congress on the Science and Cultivation of Edible Fungi* (France, 1978), pp. 797–814.

20. Y. Yoshioka, R. Tabeta, H. Saito, N. Uehara, and F. Fukuoka, "Antitumor Polysaccharides from *Pleurotus ostreatus* (Fr.) Qu'el.: Isolation and Structure of a ß-Glucan," *Carbohydrate Research* (1985) 140(1):93–100.

21. T. Ikekawa, Y. Ikeda, Y. Yoshioka, K. Nakanishi, E. Yokoyama, and E. Yamazaki, "Studies on Antitumor Polysaccharides of *Flammulina velutipes* (Curt. ex Fr.) Sing. II. The Structure of EA3 and Further Purification of EA5," *Journal of Pharmacobiodynamics* (1982) 5(8):576–581.

22. I. S. Yoo, M. S. Woo, E. C. Choi, and B. K. Kim, "Antitumor Components of *Ramaria formosa*," *Han'guk Kyunkakhoe Chi* (1982) 10:165–171.

23. F. Licastro, M. C. Morini, O. Kretz, G. Dirheimer, E. E. Creppy, and F. Stirpe, "Mitogenic Activity and Immunological Properties of Bolesatine, a Lectin Isolated from the Mushroom *Boletus satanas* Lenz," *International Journal of Biochemistry* (1993) 25(5):789–792.

24. H. Fujimoto, Y. Nakayama, and M. Yamazaki, "Identification of Immunosuppressive Components of a Mushroom, *Lactarius flavidulus*," *Chemistry and Pharmaceutical Bulletin* (1993) 41(4):654–658.

25. N. Ohno, I. Kazuyoshi, T. Takeyama, I. Suzuki, K. Sato, S. Oikawa, T. Miyazaki, and T. Yadomae, "Structural Characterization and Anti-tumor Activity of the Extracts from Matted Mycelium of Cultured *Grifola frondosa*." *Chemistry and Pharmaceutical Bulletin* (1985) 33(8):3395–3401.

26. K. Adachi et al., "Potentiation of Host Mediated Antitumor Activity in Mice by ß-Glucan Obtained from *Grifola frondosa* (Maitake)." *Chemistry and Pharmaceutical Bulletin* (1989) 35:262–270.

27. A. Grüter, "Mutagens and Anti-mutagens in Mushrooms." Doctoral thesis No.8473, ETH, Zürich, 1988.

28. A. Grüter, U. Friederich, and F. E. Wurgler, "Antimutagenic Effects of Mushrooms," *Mutation Research* (1990) 231(2):243–249.

29. H. W. Florey et al., *Antibiotics* (London and New York: Oxford University Press, 1949).

30. P. W. Brian, "Antibiotics Produced by Fungi," *Botanical Review* (1951) 17:357–430.

31. D. Broadbent, "Antibiotics Produced by Fungi," *Botanical Review* (1966) 32:219–242.

32. F. S. Vogel, S. J. McGarry, L. A. K. Kemper, and D. G. Graham, "Bacteriocidal Properties of a Class of Quinoid Compounds Related to Sporulation in the Mushroom, *Agaricus bisporus*," *American Journal of Pathology* (1974) 76:165–174.

33. R. G. Benedict and L. R. Brady, "Antimicrobial Activity of Mushroom Metabolites," *Journal of Pharmacological Science* (1972) 61:1820–1822.

34. S. L. Midland, R. R. Izak, R. M. Wing, A. I. Zaki, D. E. Munnecke, and J. J. Sims, "Melleolide, a New Antibiotic from *Armillaria mellea*," *Tetrahedron Letters* (1982) 23:2515–2518.

35. K. Dornberger, W. Ihn, W. Schade, D. Tresselt, A. Zureck, and L. Radics, "Antibiotics from Basidiomycetes: Evidence for the Occurrence of the 4-Hydroxybenzenediazonium Ion in the Extracts of *Agaricus xanthodermus* Genevier (Agaricales)," *Tetrahedron Letters* (1986) 27:559–560.

36. R. Falck, "Über ein kristallisiertes Stoffwechselprodukt von *Sparassis ramosa* Shiff.," *Chemische Berichte* (1923) 56B:2555–2556.

37. G. Bohus, E. Glaz, and E. Scheiber, "The antibiotic action of higher fungi on resistant bacteria and fungi." *Acta Biologica Academiae Scientiarum Hungaricae* (1961) 12:1–12.

38. K. W. Cochran and E. H. Lucas, "Chemoprophylaxis of Poliomyelitis in Mice Through the Administration of Plant Extracts," *Antibiotic Annual* (1959), pp. 104–109.

39. K. W. Cochran, T. Nishikawa, and E. S. Beneke, "Botanical Sources of Influenza Inhibitors," *Antimicrobial Agents and Chemotherapy* (1969) 1966:515–520.

40. P. A. Lempke, ed., *Viruses and Plasmids in Fungi* (New York: Marcel Dekker, 1979), p. 653.

41. P. A. Lempke and C. H. Nash, "Fungal Viruses," *Bacteriological Reviews* (1974) 38:29–56.

42. S. C. Tam, K. P. Yip, K. P. Fung, and S. T. Chang, "Hypotensive and Renal Effects of an Extract of the Edible Mushroom *Pleurotus sajor-caju*," *Life Sciences* (1986) 38(13):1155–1161.

43. Y. Kabir, M. Yamaguchi, and S. Kimura, "Effects of Shiitake (*Lentinus edodes*) and Maitake (*Grifola frondosa*) Mushrooms on Blood Pressure and Plasma Lipid Levels of Spontaneously Hypertensive Rats," *Journal of Nutritional Science and Vitaminology* (1987) 33:341–346.

44. P. Bobek, E. Ginter, M. Jurcovicov'a, and L. Kuniak, "Cholesterol-lowering Effect of the Mushroom *Pleurotus ostreatus* in Hereditary Hypercholesterolemic Rats," *Annals of Nutrition and Metabolism* (1991) 35(4):191–195.

45. P. Bobek et al., "Effect of Mushroom *Pleurotus ostreatus* and Isolated Fungal Polysaccharide on Serum and Liver Lipids in Syrian Hamsters with Hyperlipoproteinemia," *Nutrition* (1991) 7(2):105–108.

46. H. Sugiyama et al., "Isolation of the Plasma Cholesterol-lowering Components from Ningyotake (*Polyporus confluens*) Mushroom," *Journal of Nutritional Science and Vitaminology* (1992) 38(4):335–342.

47. P. Bobek, L. Kuniak, and L. Oxd'in, "The Mushroom *Pleurotus ostreatus* Reduces Secretion and Accelerates the Fractional Turnover Rate of Very-Low-Density-Lipoproteins in the Rat," *Annals of Nutrition and Metabolism* (1993) 37(3):142–145.

48. I. K. Matsumoto, *The Mysterious Reishi Mushroom* (Santa Barbara, CA: Woodbridge Press, 1979).

49. T. Willard, *The Reishi Mushroom: Herb of Spiritual Potency and Medical Wonder* (Vancouver, BC: Sylvan Press, 1990).

50. S. C. Jong and J. M. Birmingham, "Medicinal Benefits of the Mushroom *Ganoderma*," *Advances in Applied Microbiology* (1992) 37(1):101–134.

51. R. G. Wasson, *Soma: The Divine Mushroom of Immortality* (New York: Harcourt Brace Janovich, 1968).

52. P. Stamets, "A Discussion on the Cultivation of *Ganoderma lucidum* (Curtis. Fr.) Kar. The Reishi or Ling Zhi, Mushroom of Immortality," *McIlvainea* (1990) 9(2):40–50.

53. C.-H. Su, "Taxonomy and Physiologically Active Compounds of *Ganoderma*—A Review," *Bulletin of the Taipei Medical College* (1991) 20:1–16.

54. A. Shimizu, T. Yano, Y. Saito, and Y. Inada, "Isolation of an Inhibitor of Platelet Aggregation from a Fungus, *Ganoderma lucidum*," *Chemistry and Pharmaceutical Bulletin* (1985) 33(7):3012–3015.

55. S. Tanaka et al., "Complete Amino Acid Sequence of an Immunomodulatory Protein Ling-zhi-8 (LZ-8) from a Fungus, *Ganoderma lucidum*," *Journal of Biological Chemistry* (1989) 264:16372–16377.

56. K. Kino et al., "Isolation and Characterization of a New Immunomodulatory Protein, Ling-zhi-8 (LZ8) from *Ganoderma lucidum*," *Journal of Biological Chemistry* (1989) 264:472–478.

57. Y. Kabir, "Dietary Effects of *Ganoderma lucidum* on Blood Pressure and Lipid Levels of Spontaneously Hypertensive Rats (SHR)," *Journal of Nutritional Science and Vitaminology* (1988) 34:433–438.

58. Y. Kabir and S. Kimura, "Dietary Mushrooms Reduce Blood Pressure in Spontaneously Hypertensive Rats (SHR)," *Journal of Nutritional Science and Vitaminology* (1989) 35(1):91–94.

59. T. Kaneda, K. Arai, and S. Tokuda, "The Effect of the Dried Mushroom, *Cortinellus shiitake* on Cholesterol Metabolism in Rats," *Journal of the Japanese Society of Food and Nutrition* (1964) 16:106–108.

60. T. Kaneda and S. Tokuda, "Effects of Various Mushroom Preparations on the Cholesterol Levels in Rats," *Journal of Nutrition* (1966) 90:371–376.

61. S. Tokuda and T. Kaneda, "Effect of Shiitake Mushroom on Plasma Cholesterol Levels in the Rat," *Mushroom Science X (Part II): Proceedings of the 10th International Congress on the Science and Cultivation of Edible Fungi* (France, 1978), pp. 793–796.

62. S. Suzuki and S. Oshima, "Influence of Shii-ta-ke (*Lentinus edodes*) on Human Serum Cholesterol," *Mushroom Science* (1976) 9(pt 2):463–467.

63. M. Higuchi, S. Oshima, and S. Suzuki, "The Effects of Shi-i-take on Blood Pressure and Plasma Cholesterol of Spontaneously Hypertensive Rats," *Japanese Journal of Nutrition* (1978) 36:119–125.

64. I. Chibata, K. Okumura, S. Takeyama, and K. Kotera, "Lentinacin: A New Hypocholesterolemic Substance in *Lentinus edodes*," *Experientia* (1969) 25:1237–1238.

65. T. Kamiya, Y. Saito, M. Hashimoto, and H. Seki, "Structure and Synthesis of Lentysine, a New Hypocholesterolemic Substance," *Tetrahedron Letters* (1969) 53:4729–4732.

66. T. Rokujo et al., "Lentysine: A New Hypolipidemic Agent from a Mushroom," *Life Sciences* (1970) 9(pt 2):379–385.

67. R. Tokuzen and N. Nakahara, "Die Wirkung einiger pflanzlicher Polysaccharide auf das spontane Mamma-Adenocarcinoma der Maus," *Arzniemittel-Forschung* (1972) 21:269.

68. Y. Kurashima, M. Tsuda, and T. Sugimura, "Marked Formation of Thiazolidine-4-carboxylic Acid, an Effective Nitrite Trapping Agent in vivo, on Boiling of Dried Shiitake Mushrooms (*Lentinus edodes*)," *Journal of Agricultural and Food Chemistry* (1990) 38(10):1945–1949.

69. F. Suzuki, T. Koide, A. Tsunoda, and N. Ishida, "Mushroom Extract as an Interferon Inducer. I. Biological and Physiochemical Properties of Spore Extracts of *Lentinus edodes*," *Mushroom Science* (1976) 9(pt 1):509–520.

70. A. Tsunoda and N. Ishida, "A Mushroom Extract as an Interferon Inducer," *Annals of the New York Academy of Sciences* (1970) 173:719–726.

71. R. F. Bozarth, "Mycoviruses: A New Dimension in Microbiology," *Environmental Health Perspectives, Experimental Issue* (1972) 2:23–29.

72. K. Mori and K. Mori, "Studies on the Viral-like Particles in *Lentinus edodes* (Shiitake)," *Mushroom Science* (1976) 9(pt 1):541–556.

73. Y. Yamamura and K. W. Cochran, "A Selective Inhibitor of Myxoviruses from Shii-ta-ke (*Lentinus edodes*)," *Mushroom Science* (1976) 9(pt 1):495–507.

74. Y. Yamamura, "An Antiviral Principle of Shiitake," Doctoral thesis, University of Michigan, Ann Arbor, 1973.

CHAPTER SIX

GUIDELINES FOR
WOULD-BE MYCOPHAGISTS

THE HISTORY of mushroom eating is littered with the corpses of those who followed the folklore of the day. Because of the inherent dangers accompanying this pastime, almost every culture has attempted to develop a set of rules or simple procedures to distinguish poisonous from edible mushrooms. Some are purely fanciful, others are an attempt to generalize from legitimate biological observation, and all have been fraught with hazard and death. None of them can replace knowledge gained through observation and hard work. A delightful book published at the turn of the century examined the origin of many of these myths and developed a refutation for each one of them.[1] Unfortunately, myths have long lives and are not easily dislodged, because many are still believed today.

The following list is an incomplete compendium of common myths:

- Poisonous mushrooms tarnish a silver spoon.
- Poisonous mushrooms cooked with rice will turn the rice yellow or blue (a myth from Asia).
- Garlic cooked with poisonous mushrooms turns black.
- Mushrooms can be made safe by parboiling. (This myth is based on the known volatility of certain toxins, such as the gyromitrins.)
- Poisonous mushrooms cannot be peeled.
- Cooking mushrooms with metal, such as a coin, makes them safe (an Italian myth).
- Cooking mushrooms with wild pears (Parkinson) or pear twigs (Pliny) renders them safe.
- If a fowl refuses to eat a mushroom, the mushroom is poisonous (a Nigerian myth).
- Mushrooms growing on wood are safe.

◆ If animals eat the mushroom, it must be safe. (This myth is one of the most dangerous, because, although it makes intuitive sense to the biologically naive, some of the most deadly, such as *Amanita phalloides*, can be eaten without harm by rabbits and the gray squirrel.)

Persons high and low have perpetuated these untruths. Consider the oft-quoted epigram of Horace (65–8 B.C.):

> Fungi which grow in meadows are the best, it is not well to trust the rest.
>
> Horace, *Satires*

Today, there is no reason why any mushroom should be misidentified. Many excellent field guides are readily available (see Appendix 2). Amateur mushroom clubs and societies are present in many communities and, in conjunction with universities and community colleges, offer courses in identification. Beginners should take advantage of these aids.

HOW TO HANDLE SPECIMENS

Neophytes still need a great deal of guidance in the early stages of their exploration of the world of mycophagy. At the risk of perpetuating modern myths, the following list is a set of guidelines (not rules) that have proved useful to beginners. These guidelines assume that the mushroom hunter has at least a passing knowledge of the basic terminology of mycology. The truly uninitiated—for whom folk wisdom and pure chance have been the only protection—should invest some time in learning the fundamentals of mycology. It will pay great dividends.

These guidelines are designed to limit the risk inherent in wild mushroom eating. They represent the combined wisdom of many mycology teachers and writers, and provide a very conservative approach to mycophagy. Because of that, they exclude a number of perfectly edible species. This was a conscious decision. A large number of safe, succulent, excellent esculents are readily available, so why flirt with disaster?

◆ Never eat a wild mushroom raw.
◆ Try only one species at a time. Wait at least 24 hours before eating another.
◆ Do not overeat. Have only a small meal of mushrooms when sampling a new species.
◆ Save at least two good specimens for the toxicologist and the mycologist!
◆ Eat only young, fresh specimens. Overmature, decomposing, or frankly rotten specimens are never a good test of flavor or the

fortitude of one's own digestive system. Do not challenge it unnecessarily.

- Eat mushrooms soon after picking them. If you must store them for a brief period, keep them in a paper sack or paper towels in the refrigerator. Many mushrooms deteriorate rapidly, some within hours, such as the *Coprinus* genus (inky caps). Storing mushrooms in impermeable plastic is taboo, because it accelerates the rate of decay.

- Examine each mushroom individually. It is easy to gather different species at one time, especially when mushrooms are fruiting in abundance and are close together, for example, *Marasmius oreades* and *Clitocybe* species, or *Armillaria mellea* and *Hypholoma* (=*Nematoloma*) *fasciculare*.

- When collecting, use a separate container for each species. Contaminating a day's worth of hard foraging with a poisonous specimen is wasteful and dangerous.

- Determining the color of a mushroom's spores is sometimes required for identification. If a sporeprint is made as part of the classification process, ensure that the results have been observed before sitting down to the meal. This admonition may require you to delay dinner by a couple of hours or even longer.

- Do not serve wild mushrooms to the very young, the very old, or the infirm. If a poisonous mushroom has been included in the stew inadvertently, the consequences are likely to be more severe for these vulnerable groups. Please do not subject your mycological and culinary prowess to any who are afraid of or reticent about mushrooms or who claim to be "allergic" to them.

WHAT TO EAT AND WHAT TO AVOID

There is little fear of English folk committing any rashness in respect to Funguses as an article of food, but it will assist discrimination if the decidedly dangerous are known as well as the decidedly safe; besides is it not the duty of the moralist to point out examples to deter, as well as models to imitate; and of a mycologist to warn against the involuntary emetics, as well as to recommend the dainty stew?

Mrs. T. J. Hussey, *Illustrations in British Mycology* (1847)

Be sure of the identification of the specimen. The word *absolutely* comes to mind, but this is not usually possible. However, you can

- Be absolutely sure that you are not eating an amatoxin-containing mushroom. This is certainly possible, because there are only a handful of species producing amatoxins. Learn their characteristics perfectly. It should be a mycophagist's first assignment. (See Chapter 11.)

♦ Avoid any white-capped mushroom not positively identified, especially one with white gills.

♦ Check any mushroom with an annulus (ring on stem) carefully. Be certain of its identification.

♦ Avoid all the nondescript little brown, gray, pale, or otherwise pallid mushrooms that few people on Earth can identify.

♦ If the base of the mushroom is swollen and bulbous, be sure that a volva is not present (see figure in Chapter 3) and that the specimen is not an *Amanita*.

♦ Only eat *Lactarius* species with red or orange milk (latex).

♦ Avoid boletes with red or orange pores that stain blue when bruised.

♦ Cut all puffballs in half to be sure that they are pure white inside, that they are not button stages of a gilled mushroom such as an *Amanita*, and that they are not a thick-skinned earth-ball (*Scleroderma*).

♦ Avoid mushrooms with a cerebriform (brainlike) or saddle shape. Do not eat any *Gyromitra* species. Many of these require special preparation to remove their toxins.

♦ Beware of any mushroom with warts or scales on the cap.

♦ Avoid collecting in places that could be contaminated with pesticides, herbicides, and fungicides, such as railroad and powerline right-of-ways, public parks, and golf courses.

♦ Never eat any small or medium-sized *Lepiota*.

♦ Avoid "chanterelles" with scales (*Gomphus floccosus*).

♦ Avoid all coral fungi (*Ramaria* species) that have gelatinous centers or a peppery taste.

♦ Avoid any *Agaricus* species that smells like creosote or stains yellow when handled.

In addition to these guidelines, it is well to recall two crucial "Lincoffism's":[2]

♦ NO RULE IS THE ONLY RULE.
♦ ANY MUSHROOM IS EDIBLE . . . ONCE!

LOOK-ALIKES

One of the great concerns of many mushroom hunters and would-be mycophagists is the fear instilled by teachers about the danger of "look-alikes." These specimens are the ones, you are warned, that will surely lure you to your doom. Although it is true that many mushrooms are remarkably similar and a microscope or chemical testing may be required to distinguish them, the common edible species are distinctive enough that, with a little care, they can be identified. It is also true that some individuals never develop the basic skills of pattern recognition. Mushroom hunting is not for them. As David Arora has so aptly put it, "If you can tell the difference between an artichoke and an asparagus, you can tell an edible from a poisonous mushroom."[3]

At the beginning of the newly initiated mycophagist's career, the task looks daunting indeed—hundreds of mushrooms to learn, each with look-alikes. The reality is much simpler. The beginner can start off on the right foot by studying the accompanying table. Edible

Common look-alike mushrooms

Edible mushroom	Common name	Look-alike	Potential outcome
Agaricus campestris	Meadow mushroom	Agaricus xanthodermus	Gastrointestinal problems
		Agaricus hondensis	Gastrointestinal problems
		Amanita species	Coma/hallucination or death, depending on species
Agaricus sylvicola		Amanita virosa	Death
Amanita calyptroderma	Coccora	Amanita phalloides	Death
Amanita caesarea	Caesar's mushroom	Amanita muscaria	Inebriation, hallucinations
Amanita rubescens	The blusher	Amanita muscaria	Inebriation, hallucinations
Armillaria mellea	Honey mushroom	Galerina species	Hepatic failure or death
		Gymnopilus spectabilis	Hallucinations
		Hypholoma fasciculare	Severe gastrointestinal symptoms
Cantharellus cibarius	Chanterelle	Hygrophoropsis aurantiaca	No problems
		Cortinarius species	Renal failure
		Omphalotus olearius	Gastrointestinal problems
Coprinus comatus	Shaggy mane, lawyer's wig	Coprinus atramentarius	Symptoms with alcohol
Flammulina velutipes	Enoki, winter velvet foot	Galerina species	Hepatic failure or death

Common look-alike mushrooms (continued)

Edible mushroom	Common name	Look-alike	Potential outcome
Gyromitra gigas	Snowbank mushroom	*Gyromitra esculenta* *Gyromitra* species	Severe gastrointestinal symptoms and possibly liver damage; potentially fatal
Kuehneromyces mutabilis	Two-tone pholiota	*Galerina* species	Hepatic failure, or death, if greedy
Leucoagaricus naucinus		*Amanita virosa* *Amanita bisporigera* *Amanita verna*	Death Death Death
Lepiota rachodes	Parasol mushroom	*Chlorophyllum molybdites*	Gastrointestinal problems
Lepiota procera		*Amanita gemmata* *Amanita phalloides*	Inebriation/hallucinations Death
Lycoperdon species	Puffballs (button stage)	*Amanita* species *Scleroderma* species	Death or hallucinations, depending on species Gastrointestinal problems; muscle spasms
Marasmius oreades	Fairy ring mushroom	*Inocybe* species *Clitocybe* species	Muscarinic syndrome Muscarinic syndrome
Morchella species	Morel	*Gyromitra esculenta*	Severe gastrointestinal symptoms and liver damage; potentially fatal
Pleuteus cervinus	Deer mushroom	*Entoloma* species	Gastrointestinal symptoms
Psilocybe species	Liberty caps	*Inocybe* species *Galerina* species *Conocybe* species	Muscarinic syndrome Hepatic failure or death Hepatic failure
Rozites caperata	Gypsy mushroom	*Hebeloma* species	Gastrointestinal symptoms
Tricholoma flavoviriens	Man-on-horseback	*Amanita phalloides*	Death
Tricholoma magnivalere	Matsutake	*Amanita smithiani*	Renal failure; gastrointestinal upset

mushrooms and their look-alikes are listed side by side. Although the list is not exhaustive, it does cover the most common pairs. Moreover, which look-alikes you may stumble across is very dependent on which area of the country you are rooting about in. It is crucial for you to learn about the common species on your home turf and to be especially careful when foraging on unfamiliar soil.

Over the past few years, the North American Mycological Association has attempted to keep track of the errors in identification made by amateur mycologists. The list has provided some insight into the difficulties people have in distinguishing different mushrooms. Some of these mistakes are quite surprising, because the differences between the desired specimen and the collected specimen are quite profound. These errors demonstrate the poor observational skills of some people, the lack of education of others, and probably the carelessness of many.

REFERENCES

1. W. Hamilton Gibson, *Our Edible Toadstools and Mushrooms* (New York: Harper and Brothers, 1895), p. 23.

2. G. Lincoff and D. H. Mitchel, *Toxic and Hallucinogenic Mushroom Poisoning* (New York: Van Nostrand Reinhold, 1977).

3. D. Arora, *Mushrooms Demystified*, 2nd ed. (Berkeley, CA: Ten Speed Press, 1986).

PART TWO

Mushroom
Poisoning

CHAPTER SEVEN

POISONING NOT CAUSED
BY MUSHROOM TOXINS

IT IS NATURAL to suspect a mushroom toxin when symptoms or signs develop following an accidental ingestion or an intentional meal of mushrooms. However, a variety of other possibilities should be considered, because the outcome and the treatment may be quite different. Adverse reactions to mushrooms fall into the categories shown in the accompanying box.

PANIC REACTIONS

THE FIRST MAJOR mushroom that I learned to pick, cook, and eat was the chanterelle. It is a good beginner's mushroom—easily found in the Northwest, always free of worms and other competitors, moderately flavorful, but best of all it has few "look-alikes." There are some if one is extraordinarily careless, but fortunately these cause little harm apart from abdominal pain, vomiting, and diarrhea—unpleasant enough, but nothing to give one's insurance agent a sleepless night. It is a relatively safe mushroom. For a couple of years I hunted and ate bushels of chanterelles. My confidence grew and each wild mushroom meal desensitized my fear, much like an allergist using small doses to desensitize a patient to bee stings. Not that I was becoming careless or cavalier. I still had a good deal of respect for those chemicals mushrooms had evolved for goodness knows what purpose, only now it seemed more reasoned and rational. But then one day it happened.

It was time to extend my culinary horizons beyond the chanterelle. I knew that I was capable of identifying at least two hundred species with some facility. One weekend I picked a variety of

Adverse Reactions to Mushrooms

Acute adverse reactions not caused by a mushroom toxin
* Panic reactions
* Bacterial food poisoning
    due to spoiled or rotten specimens
    due to improper preparation or storage
* Insecticide, herbicide, or fungicide contamination
* Alcohol intoxication
* Idiosyncratic reaction
* Allergic reaction
    gastrointestinal
    respiratory (usually to spores)
* Intestinal obstruction

Long-term effects of mushroom ingestion
* Heavy-metal poisoning
* Radioisotope contamination
* Cancer

True poisoning by a mushroom toxin

known edible species, just a few of each, determined to break out of the chanterelle rut. The rules of self experimentation were well known to me as each mushroom field guide repeats the same gospel, a liturgy handed down over the generations to would-be gluttons. "The first time eat only a small number." "Never mix species." "Wait a few days before trying another species." "Be absolutely certain of your identification," which is often expressed as "if in doubt, throw it out." All of these seemed logical enough.

I looked over the collection and picked a couple of the firm yellow caps of what I was sure were prime examples of man-on-horseback (*Tricholoma flavoviriens*). The common name made as little sense to me as the scientific one, but it was flavor that I was looking for, not etymology. All the mushroom guides listed it as a choice edible. "Highly esteemed," said a few. With the butter bubbling in the skillet I sliced the caps, sautéing them quickly, unadorned so as to preserve the essential flavor of the mushroom. That may not be the best way to cook a particular mushroom, but it is always a good place to start. I never allow my wife or my children to sample the first time—somehow I would feel cowardly, not to say guilty, if they should take ill. And I already knew that experimenting on mammalian pets was in no way fail-safe, since there are striking differences in species' susceptibilities as well as resistances. The gray squirrel and the rabbit are immune to a meal of *Amanita phalloides*.

As my wife elbowed up to the range waiting for a taste, I popped the last slice into my mouth. She is a trusting soul, more confident of my abilities than am I. I try to hide my fears. "Yes, dear, I am absolutely certain that they are man-on-horseback, but let me check them out first." "Not bad," I thought, "but not very distinctive." I relegated the flavor and the species to a "survival" mushroom—a good mushroom to know if I should ever be lost in the woods, but not one that I would intentionally seek out. Certainly not one that I would die for. But most importantly I had finally come of age—I had sampled a mushroom that I alone had picked and identified. I felt freed and elated. A surge of pride came through me. I had shed the veneer of modern civilization and knew that I had regained some ancient knowledge—knowledge that had allowed our ancestors to feast on what was available. There was the sense that I could now survive, if need be, without a supermarket on the nearest corner, unlike my children who get anxious if we are ever more than a couple of miles from the closest shopping mall. I was more excited about the feeling than the discovery of a new flavor.

An hour or so later, paging somewhat absentmindedly through a field guide in my study, I came across a photograph and description of the species I had so recently consumed. A large skull and crossbones emblazoned the bottom of the page warning the unwary not to confuse this particular edible with some highly toxic, if not deadly, "look-alike." From what I could gather the two were indistinguishable apart from some rather subtle microscopic differences. Within seconds, my pulse soared into the upper hundreds, I broke out in a heavy, cold sweat, turned beet red, and became dizzy and light-headed. Attempts at self-induced emesis were less than successful. It produced a few half-hearted retches, but no mushrooms. I knew that a bottle of syrup of ipecac was hidden upstairs, although there was no possibility that I was going to reveal my plight to my family. The only reason that I didn't call the local poison control center was that I had recently begun consulting with them on cases of mushroom poisoning. I was now in the midst of what medical science calls a "panic reaction," a totally inadequate name for the intense physiological response to the certainty that one is in the process of dying. It was all so real that I even re-read my will—that helped and I decided to survive. Slowly the reaction began to subside. Later I discovered that the deadly "look-alike" does not even grow in North America, let alone in the Pacific Northwest. This episode taught at least two good lessons—never use a field guide written for a different locale, and the great power of suggestion. They were salutary lessons and ones not easily forgotten.

Denis R. Benjamin, personal journal (1984)

A panic reaction is most likely to occur when an amateur begins to experiment with edible or hallucinogenic mushrooms. The precipitating event is a sudden doubt about the identification of the mushroom just consumed. Episodes similar to mine have been recounted to me by a number of mycologists. I have also encountered these reactions at the poison control center. The most memorable involved the president of a nearby mushroom club, who called in a panic when he suddenly realized that he may have eaten an amatoxin-containing *Galerina venenata* rather than a hallucinogenic *Psilocybe*. The combination of the fear and the early effects of the hallucinogen provided a most unpleasant experience.

Panic is not limited to the potential victim. So great is the fear of mushrooms in some individuals that the sight of a small toddler with a mushroom in his or her mouth may invoke a strong reaction in the caregiver. The first task of the medical attendant is to calm the family, most often the mother or the baby-sitter. Reassurance at this stage can be most helpful. Unfortunately, many nurses and physicians are equally ignorant and fearful about mushrooms. They, too, may develop rather inappropriate behavior, even though they may appear more controlled and may disguise their fear behind a professional veil. One of their first panic responses is an exaggerated effort to identify the mushroom. Almost always, the immediate assumption is that the patient is doomed. Either the amateur or the professional mycologist can play a useful role in diffusing the panic. I have often found that merely being available as a resource person is sufficiently reassuring for the treating physician that their inherent tendency to overreact to the situation is abated.

BACTERIAL FOOD POISONING

Mushrooms are highly perishable because of their high water and protein contents and their high metabolic activity. The fruiting bodies have a very short life span. Contamination by microscopic fungi or by bacteria is not uncommon. Decaying mushrooms are an excellent substrate for the growth of many microorganisms.[1] Spoilage is especially likely when old specimens have been picked in the field or when they are inappropriately stored in closed or sealed containers, an environment that accelerates decomposition.

It may seem surprising that people would eat rotten mushrooms, but there are a number of circumstances in which this may occur. Since the 1980s, more and more wild mushrooms have become available in the better food stores and markets in the United States, usually on a very limited and infrequent basis. For the overwhelming majority of American gourmet cooks willing to spend considerable sums of money to try these "new fangles meates," this is their first exposure to wild mushrooms outside a restaurant. They have very little knowledge about what a particular species should look, feel, and smell like. Most often the grocers are equally naive. I continue to be appalled at the mushroom displays in many of the local markets in the Puget Sound area, where I see piles of sometimes mislabeled and partially rotten specimens. These markets are totally dependent on the veracity of the person who sold them the mushrooms. In France, all the pharmacists can identify and classify mushrooms. Many European countries have lists of edible species permitted for commerce, with inspectors who ensure their safety and wholesomeness. And a more sophisticated and knowledgeable clientele are not as likely to tolerate old, decomposed specimens on a vegetable stand.

Some lively debates have taken place in the United States in the past few years regarding the inspection and certification of wild mushrooms for sale in markets or for use in restaurants. In 1986, the U.S. Food and Drug Administration introduced a proposal for controls on the sale of wild mushrooms. This document was not promulgated as a regulation, which is enforceable by law, but as an interpretation regarding food safety. It was and still is up to each local health department to decide whether or not to adopt the guidelines or the proposal. California, as it often does, has been independently pursuing its own set of regulations. But, at the present time, there is no consistent pattern of either regulations or implementation. Of course, should someone get seriously harmed, the situation will no doubt change very rapidly. (Investigative reporters will descend on the victim, regulators will rapidly promulgate complex rules, and a few attorneys will get wealthy.)

Cases of bacterial food poisoning are no different from those associated with any other contaminated food source. Serious poisonings have been reported in a few special circumstances. The most important cases relate to botulism resulting from improper home canning or marinating of mushrooms.[2] In 1988, a New Year's Eve dinner at a prestigious hotel in Vancouver, British Columbia, ended in near tragedy when the chef served chanterelles that he had personally preserved. They were part of a dish called fricassee of mullet and lobster with wild mushrooms. As a consequence of his lack of skill and knowledge, at least a half-dozen cases of botulism developed in the diners, all of whom had spent a considerable sum for a gourmet meal—but got somewhat more than they had anticipated.[3] Three of the New Year's Eve revelers required ventilator therapy. In another case, three members of one family were struck down by this particular scourge. Two family members died, whereas a third had only minor symptoms. Investigation showed home canning to be the culprit; the heat treatment during the canning process had been inadequate.[4]

As illustrated above, home canning or preserving of mushrooms by amateurs carries a distinct risk, because mushrooms, like meat and fish, are a low-acid food and an ideal medium for the growth of *Clostridium botulinum*.[5] Preparing the mushrooms with sufficient vinegar to maintain a low pH greatly reduces that risk.[6] Even commercial food production and canning companies are not immune to an occasional failure. An outbreak of staphylococcal food poisoning erupted a few years back and was traced to contamination of canned mushrooms imported from China.[7] However, although contamination of commercially canned mushrooms is always a possibility,[8] the risk must be very low indeed, considering the millions of cans of mushrooms that are bought and safely consumed each year.

Numerous reports have been made on the isolation of *Clostridium botulinum* from mushrooms kept in airtight plastic packages. The anaerobic conditions created by this environment allow the growth of *Clostridium* species.[9] Since 1987, the U.S Food and Drug Administration has required that fresh mushrooms be packaged in containers with vents to reduce the risk of botulism. The enoki mushroom (*Flammulina velutipes*) is grown commercially and is still frequently packed in airtight plastic bags. Aseptic cultivation and refrigerated storage, however, greatly reduce the risk of bacterial contamination of and growth on this species.[10]

CHEMICAL CONTAMINANTS

WITHIN A FEW years of serious hunting and picking, I had become supremely confident. Dozens of species had crossed my palate and danced on my taste buds. I had long since thrown away the great culinary bible for amateur mycophagists by Charles

McIlvaine, who had by all accounts a most unusual palate, and simply classified my experiments into "merely edible" and "worth hunting for." There was certainly no reason to reexplore those species known to cause physical harm. Like restaurants vying for top honors in the Michelin restaurant guide, very few mushrooms made it into the second, three-star category. One that did was the meadow mushroom, *Agaricus campestris,* the "pink bottom," the one the French call "le champignon." It is the wild cousin to the supermarket mushroom. Vast quantities of this more flavorful but more delicate mushroom are left to rot on lawns as people drive by them on the way to the store to buy mushrooms. I had discovered a wonderful patch in a nearby cemetery that routinely produced abundant crops after the first rains of late summer or early fall. This patch provided many memorable meals.

One Indian summer evening, I stopped by the cemetery on my way home and picked a pound of near perfect meadow mushrooms, their pink gills barely visible. I knew that my wife had already eaten dinner, so I left dozens of prime specimens behind. Once home I prepared a good meal—mushrooms and sliced chicken breasts in a cream-Madeira sauce served over fettucine, the cream and the wine exploiting the full flavor of the mushrooms. Two hours later I thought that I was going to die. Actually I wanted to. I developed severe abdominal cramps and nausea, the pain coming in waves of colic. No position eased the discomfort, although sitting bolt upright was the least unpleasant. I checked my identification again. There was no doubt that they were *Agaricus campestris.* I had eaten them many times before, with no ill effect. I spent the entire night awake, sitting in a chair in our family room or kneeling in agony over a foot stool whenever another wave of pain swept through me. By five o'clock in the morning the cycles of pain and nausea became less frequent and intense. I did not suffer a panic reaction this time—I knew better than that. But considerable curiosity welled up as to what had happened.

The next day I called the cemetery. Obviously I couldn't accuse them of poisoning me, nor did I want to alert them to the fact that they provide a substantial portion of our diet, since there were quite a number of other species that I routinely collected there. "Excuse me, sir," I said, "but could you tell me if the cemetery has been recently sprayed?" "Oh, yes!" answered a voice, with none of the funereal tones expected of an undertaker who had so nearly claimed another client. "Yesterday was a regular monthly treatment by our landscape company. They spray with a mixture of herbicide, fungicide, and insecticide." "Thanks very much," I said and hung up

before he could ask me why I was interested in the subject. So—
the night before I had absorbed the combined toxins of our most
demented chemical minds. Had I been a lawn moth, this book
would never have begun. To this day, cemeteries, rights-of-way,
golf courses, and any other place that humanity has access to with
its poisons have become off-limits to foraging. It is also what we
teach in our beginner's class on mushroom hunting each year. The
message is that one should rather take one's chance with what
nature has produced than with what man is capable of.

Denis R. Benjamin, personal journal (1986)

The application of various herbicides, insecticides, and fungicides is
widespread, often in those areas that favor the fruiting of edible mushrooms. These areas include public places such as parks, golf courses,
cemeteries, school yards, and powerline, railroad, and trail right-of-
ways. Home use of similar agents is common, as Americans attempt to
tame the natural world; and many farmers use various toxic compounds
to control weeds or other unwelcome pests. As a general rule, it is advisable not to pick mushrooms in such places unless you know exactly
what has been sprayed or applied and when. Of course, many people
do collect in public areas and get away with it because the applied compounds lose their acute toxicity fairly rapidly after application.
However, a number of agents are persistent and can potentially accumulate in the environment and, to some extent, in mushrooms. To my
knowledge, this type of chronic toxicity has not yet proved to be a
problem, perhaps because mushroom picking is an intermittent pastime.
    Environmental contamination with potent toxins should be kept in
mind during any form of foraging. Stories similar to mine are recorded
in the medical literature.[11,12]

ALCOHOL INTOXICATION

Differentiating the effects of alcohol from that of mushroom poisoning is sometimes necessary. In addition, the combination of alcohol
with one particular species of mushroom—inky caps (Coprinus atramentarius)—causes a most interesting set of symptoms (see Chapter 15).
    Those adventurous enough to dabble in mycophagy often accom-
pany the meal with wine or spirits. Overindulgence of the latter may
provoke symptoms that can be interpreted by both the participants
and their medical attendants as mushroom poisoning. Sometimes the
revelers wish to indict a mushroom for their state of misery rather than
admitting that they had been a little too deep into their cups. A careful
history usually reveals the truth of the matter.

The combination of wine and mushrooms is time honored and should not be discouraged. There is no substitute for a plate of mushrooms—identified as edible, of course—and a fine bottle of wine, served to celebrate a delicious meal and good fellowship. Furthermore, experimental evidence suggests that for someone unfortunate enough to eat a death cap, alcohol taken at the same time may lessen the effects. Of course, this possibility should not be taken as a justification for drinking excessively. Nor does it give you license to eat *Amanita phalloides*—the death cap could still prove to be deadly.

IDIOSYNCRATIC REACTIONS AND INTOLERANCES

The terms *idiosyncratic reaction* and *intolerance* refer to the myriad adverse reactions to a vast array of products, to any one of which a particular individual is uniquely sensitive but to which the general public is insensitive. In the majority of cases, the mechanism and the cause of the reaction are entirely unknown. Mushrooms contain an amazing assortment of chemical compounds, including unusual carbohydrates, sugars that our guts cannot digest, absorb, or metabolize; large complex polysaccharide molecules for which we lack the necessary degradative enzymes; and unusual polypeptides. Many of these have the potential to cause problems. However, using a term like *idiosyncratic reaction* as an explanation for the symptoms is like hiding behind a huge shield of ignorance, for in reality it explains nothing. It is no more than a convenient wastebasket in which to dump most of what we do not understand.

Perhaps the best studied of the adverse reactions are the intolerances due to the inability of an individual to digest a particular sugar. One of the most common, affecting the majority of humankind, is the progressive loss of the enzyme lactase, normally responsible for our childhood ability to digest lactose, the sugar in milk. Humankind, however, is not designed to drink milk long after having been weaned from the breast. After puberty, many have adverse reactions to milk because of the disappearance of the enzyme from their gastrointestinal tract. This deficiency is not important in regard to mushrooms, because lactose is not present to any degree in fungi, but a number of other sugars are. α-Trehalose is a common sugar in the young fruiting bodies of many mushrooms, including *Agaricus bisporus*. It was proved to be the cause of diarrhea in a patient who developed symptoms every time he ate that mushroom or the sugar alone.[13] α-Trehalose is normally hydrolyzed as the mushroom matures and so disappears from older sporocarps.

However, few people opt for the older specimens at the supermarket stand, preferring the firm, unopened buttons with the pale gills rather than the large, full-flavored specimens with the black gills. A number of patients and even whole families who lack the enzyme trehalase have now been identified; trehalase is responsible for breaking the sugar into more readily absorbable molecules.[14] Although the incidence of the deficiency in the general public is unknown, it appears to be quite rare.

Mushrooms contain compounds that bind to the surfaces of cells or membranes and can interfere with their function. Some of these are proteins or glycoproteins that attach to one of the sugars on the surface of the cell membrane and are known as lectins.[15-18] These lectins may play a role in some patients' adverse reactions to foods.

Some mushrooms contain natural laxatives, the best examples being found in the genus *Suillus*. Others have considerable amounts of vasoactive amines, such as tyramine, which are known to produce a variety of physiological effects in certain individuals. Histamine may possibly be released through a number of different mechanisms, most of which are not immunologically mediated, inducing some of the clinical effects.

In two of the most recent textbooks on the subject of food intolerances—one a tome of over a thousand pages—mushrooms are barely mentioned. Unfortunately, when they are, it is with the misinformation that the effect of *Amanita muscaria* is due to its muscarine content![19,20] One infers from these books that the commonly eaten mushroom *Agaricus bisporus* is either no problem at all or has never been adequately studied. For wild mushrooms, it is safe to assume that absolutely no reliable information is available. Naturally, anecdotes of diverse and sundry problems abound—unsubstantiated claims fill the lay press.

A fascinating and amusing report of the mass "poisoning" of 86 women at a meeting of a doctors' wives committee at a Polynesian restaurant illustrates the difficulties of assigning blame. Fifteen to thirty minutes after eating a light chicken broth to which a couple of pounds of canned sliced "button" mushrooms from Taiwan had been added, 55 of the women developed severe headaches, with bitemporal pressure, and malaise. Some experienced tingling of the lips. A thorough investigation of the incident, initially ascribed to food poisoning, led to the conclusion that the women suffered from muscarine poisoning caused by the "unknown" imported mushrooms. This judgment was based on some rather shaky chromatographic evidence and biological assays of the soup. Interestingly, this diagnosis was made and presented in the medical literature in this form, despite the fact that none of the symptoms vaguely resembled muscarine poisoning.[21] The investigators should have measured the concentration of monosodium glutamate in the soup.

ALLERGIC REACTIONS

Allergic reactions to mushrooms are one of the most problematic areas to evaluate.[22] There was a time when the diagnosis *food allergy* was used to explain almost everything that we did not understand. It was a convenient label, more understandable than the term *idiosyncratic reaction* and therefore more satisfying to the patient. Not much could be done about the patient's symptoms other than recommending that he or she avoid the allergen. In the 1920s and 1930s, the overenthusiastic and uncritical use of this label was widespread. But, eventually, the pendulum swung in the opposite direction, and many physicians avoided the term entirely. They no longer diagnosed many of the purported food allergies, because little was known about them, they were difficult to study, and no easy diagnostic laboratory tests were available. Quackery flourishes in such situations: weird dietary manipulations are prescribed, fanciful pseudoscientific rationalizations are propounded, and self-styled "experts" make a living by preying on the afflicted and gullible public. For example, the idea was promoted that so-called allergic reactions to one's own endogenous fungi, such as the yeast *Candida albicans*, were the cause of everything from pimples to depression.

As is so often the case, the truth lies somewhere between the extremes. Certain susceptible people do develop allergic responses to ingested allergens, but it is probably not nearly as common as some practitioners would like us to believe. Each one of us is bathed on a daily basis in a sea of fungal products, even if we never touch a wild mushroom. The air is filled, both indoors and out, with microscopic spores and other fungal elements, which we inhale with each breath. Our food is laden with the products and by-products of fungal life, from our bread and cheese to miso, soy sauce, and wine. Opportunistic fungi colonize the human skin, intestine, mouth, upper airway, and vagina. From both within and without, we are constantly exposed to fungi. Despite this exposure, most persons have no adverse reactions to these substances.

Our current understanding of the problem of food allergy and intolerance is still remarkably primitive. Because so many people actually believe they have an allergy to particular foods, new eponyms and diagnostic labels have had to be invented for them. Two of the most common are PAR, or pseudoallergic reaction,[23] and FFA, or false food allergy. These labels are not meant to suggest that the symptoms from which these individuals suffer are not real—they merely reflect the fact that neither the cause nor the mechanism of action is understood.

The most dramatic immediate hypersensitivity response occurring after the ingestion of foods such as shellfish or peanuts, which is a typical anaphylactic reaction, has not been reported for the ubiquitous *Agaricus bisporus* (the white supermarket mushroom) or for any of the commonly eaten wild mushrooms. Eczema, chronic digestive problems, acute diarrhea, and "irritable bowel syndrome" have all been associated with the ingestion of the products of the mold fungi; but these conditions do not appear to be the result of the typical immediate hypersensitivity reaction that normally characterizes the symptoms associated with the inhaled allergens. However, a few reports of digestive disturbances have been ascribed to "legitimate" allergic reaction to mushrooms.[24]

The evidence is currently much more compelling for the effects of inhaled spores on the respiratory tract. Bronchial asthma, allergic conjunctivitis, and rhinitis are frequently provoked by the inhalation of fungal allergens. That this has been known for many years is shown by the comment of Sir John Floyer, who wrote in his *Treatise of the Asthma* (1698) about the remarkable case "of an asthmatic who fell into a violent fit, by going into a wine cellar where the must was fermenting."[25] It is not only to the spores of the microscopic fungi that some people react. The large fleshy fungi of the basidiomycetes can also provoke a similar allergic response.[26,27] There is a growing appreciation of the fact that the spores of the basidiomycetes may play a much more significant role in causing asthma than previously believed.[28-30] This is a whole subject in itself and will not be further pursued here.[24,31] Rarely, a more acute pneumonitis results from the inhalation of a massive quantity of spores, as can occur when mature puffballs are disturbed.[32] In this instance, the use of corticosteroids may be warranted. I recently learned of such a case of "lycoperdonosis" in an adolescent who sniffed puffballs in the hope of getting high.

In addition to the problem of bronchial asthma, conditions such as "farmer's lung" and other occupational mold-induced lung diseases are well known. Some of these pathologies were first described—although the cause went unrecognized—in 1770 by Bernadino Ramazzini in the first book entirely devoted to occupational diseases, *De morbis diatriba*. However, some cases of respiratory symptoms related to the inhalation of spores of the large fleshy fungi, both ascomycetes and basidiomycetes, have been reported. Workers who handle mushrooms are particularly at risk.[33-38] As the extent of mushroom cultivation has increased, the problems associated with farming mushrooms have increased accordingly. The organic allergens provoking the disease, which is an extrinsic allergic alveolitis (also known as hypersensitivity pneumonitis), are not always identifiable. Sometimes spores are to blame; they are released in astronomical numbers when mushrooms reach maturity, especially by species such as the oyster mushroom and shiitake.[39,40] Hypersensitivity pneumonitis can be an incapacitating disease. In Japan and China, up to 10% of all mushroom workers have been reported to develop symptoms. No cases in which amateur cultivators are affected have been recorded.

Lung disorders have been linked to the cultivation of *Pholiota nameko* in Japan.[41] In the past these mushrooms were cultivated outdoors, so the risk of spore inhalation was minimized. These mushrooms do not have a veil to protect the gills, and they are harvested at maturity when their spore production is prodigious. Attempts to increase yields and produce a more predictable crop have led the industry to move indoors, where growing conditions can be more carefully controlled. In such an environment, the spores accumulate to high concentrations.

Agaricus bisporus is harvested in the immature button stage, so spore inhalation is not a problem. However, during the spawning process, many other antigens in the organic substrate are released into the environment.[42] These antigens may be the spores of microscopic fungi that live in the compost or bacterial antigens from actinomycetes, which are also resident in the substrate.[43] All these antigens are capable of causing pneumonitis.

The gastrointestinal tract and the respiratory tract are not the only areas affected by the allergenic properties of mushrooms. The skin is occasionally the target organ. Eczema—or more correctly, an allergic contact dermatitis—has been ascribed to the handling of mushrooms.[44] This condition is especially likely to occur in mushroom workers.[45] A variety of species have been implicated.[46] In the United States, various species of *Suillus* have caused the condition in a few professional mycologists. In some instances, individuals have become sensitive to a number of different species because of the presence of cross-reacting antigens. In the usual situation, the person developing such an allergy has often handled mushrooms for a prolonged period, as do pickers and professional mycologists. The usual symptoms include itching, swelling, redness, and vesiculation or blistering. Most cases resolve spontaneously in a week to 10 days once exposure to the offending mushroom is eliminated.

Many other mechanisms may account for an individual's negative reaction to mushrooms. In some patients, a true pharmacological sensitivity to a particular chemical compound may be present, such as has been demonstrated with aspirin. In selected persons, the autonomic nervous system may be triggered by exposure to a certain molecule. Of course, the role of the psyche should never be underestimated. Personal dislikes or true phobias may color a person's response to eat-

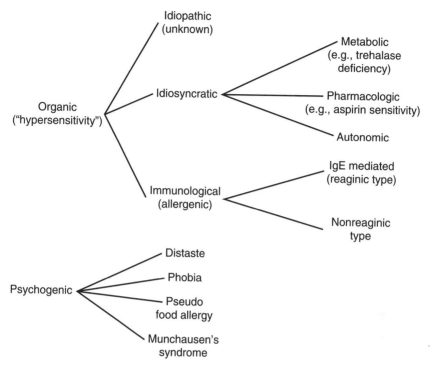

Unusual reactions to foods that are normally nontoxic (modified from Pearson and Rix[23])

ing mushrooms. For some people, this reaction may be assigned the blame for other problems in their lives, being used to explain everything from their lack of energy to impotence. Such false food allergies are emotion-laden subjects. An extreme case is the deceptive use of symptoms to attract attention, garner sympathy, and encourage care and concern. This disorder is known as Munchausen syndrome. I am aware of one case in which a young adult male repeatedly faked the symptoms of abdominal pain by mushroom poisoning in order to obtain the narcotics to which he was addicted.

Whatever the reason, be it an intolerance, an idiosyncratic reaction, or a genuine allergy, many reports from many countries have documented the problems caused by selected mushrooms in certain people. In the United States, the North American Mycological Association, through its poison registry, has accumulated a number of reports relating to this issue. The mushrooms involved include chanterelles, boletes of all varieties, and morels. Indeed, they cover the entire spectrum of edible wild mushrooms. Listing each individual episode would be fruitless. However, Chapters 19 and 20 discuss all those species responsible for more than a single case.

A few people develop no symptoms at all after eating one of the species that normally makes most of humanity ill. It is the flip side or converse of a mushroom sensitivity—a unique resistance to the effect of a specific compound. Resistance to the toxins in raw *Gyromitra* (a very dangerous flavor treat), *Amanita muscaria*, and *Chlorophyllum molybdites* has been documented.

INTESTINAL OBSTRUCTION

Intestinal obstruction is a rare but well-described complication of mushroom gluttony. In part it is due to the relative indigestibility of the chitin in the cell walls. When mushroom cells are intact, they absorb water, thereby resulting in considerable swelling of the mushroom pieces. When the ingested mushrooms are not well chewed, large fragments persist throughout their journey down the gastrointestinal tract. Liberal quantities of these large objects can cause mechanical obstruction. In one patient with an undiagnosed colonic cancer that had narrowed the intestinal lumen, obstruction was complicated by intestinal rupture.[47] Most mushrooms have the potential to cause mechanical difficulties when consumed in excessive amounts.

HEAVY-METAL POISONING

Many species of the higher fungi—from a large number of different genera, including many of the best-known edible varieties—have been shown to concentrate or accumulate trace elements, including some of the toxic metals. This phenomenon was first observed in Europe in the 1960s and is of increasing concern as the environment becomes more and more polluted.[48–53]

Many different trace elements have been detected in mushrooms, although concentrations depended on where the specimens were collected. It is clear that many of the fungi have the ability to preferentially accumulate and concentrate certain elements in their fruiting bodies, even when the soil contains only trace amounts. The reasons for the ability of mushrooms to concentrate these minerals is entirely unknown. Mushrooms may play the role of "sink," removing many of these elements from the environment and reducing their availability to the plant community.

Cultivated mushrooms also concentrate heavy metals, which have been found in the oyster mushroom[54,55] and others.[56] This finding is important because many of the commercial mushrooms are cultivated on waste material, which could potentially be contaminated with heavy metals. Growers must take care to limit the kinds of waste materials used for mushroom cultivation to those without heavy-metal content.

Mercury and cadmium are the elements most often encountered in appreciable levels in species of *Agaricus*. They are not limited to this genus, however. Nor are these elements the only ones accumulated by mushrooms. Vanadium, selenium, arsenic, lead, manganese, bromine, nickel, silver, and gold have been detected. When large quantities are ingested, all pose a health risk, the extent of which is currently unknown. One European guideline recommends that no more than 1 kg—approximately 2.2 pounds—of wild mushrooms be eaten per week, in order to stay below the World Health Organization guidelines for the ingestion of toxic elements.[49] (Many mushroom hunters would be happy to find over 2 pounds of wild mushrooms each week, although at the height of the fruiting season that might not be too difficult.) Other guidelines advise collectors to avoid any mushrooms from known polluted locations, such as the areas beside and around roadways, mines, smelters, incinerators, and disposal sites; to remove all the gills or the pores, because these structures contain the highest concentrations of specific elements; and to avoid too frequent consumption.[58]

Cadmium is strongly accumulated by members of the *Agaricus* genus, especially the species that stain yellow when handled and bruised. Of these, only *Agaricus augustus* and *Agaricus arvensis* are commonly eaten in the United States. The concentration of this element in *Agaricus* appears to be due to a specific cadmium-binding protein.[59,60] However, the accumulation of heavy metals is not limited to the yellow stainers, as *Agaricus campestris* has also been found to contain cadmium.[61] The pattern of mercury accumulation is somewhat similar to that for cadmium (see the table on page 124). Fortunately, only a small proportion of the contaminating mercury is in the more toxic methylmercury form.

Lead is one of the elements not specifically concentrated by mushrooms. For this reason, it is only a problem in areas of high environmental contamination, for example, around lead smelters[63] (and their downwind extension) and along busy roadways, where the lead in the fumes of the old type of gasoline has accumulated in the soil. This latter problem should disappear after the leaded gasolines are finally removed from the market. In a study around a lead smelter in Bohemia, which has been in operation since 1786, mushrooms collected up to 6 km away from the source still contained lead above the safe limit. Around the smelter, the concentrations were over 100 mg/kg of

Concentration of mercury (mg/kg of dried mushroom) in edible mushrooms harvested in Switzerland[62]

Mushroom	Agricultural area	Industrial area	Urban area
Coprinus comatus	0.51	3.59	3.18–5.62
Lycoperdon perlatum	2.87–3.75	3.32	11.0
Agaricus campestris	2.1–4.1	4.37–12.7	9.3–33.6
Amanita rubescens	0.69–2.9		13.2

dry weight, a truly impressive accumulation. Lead concentrations were highest in Lepiota rachodes, the parasol mushroom, and Lepista nuda, the blewitt.[64]

A recent survey of a few edible mushrooms around London, England, showed moderate amounts of lead and cadmium.[61] The investigators concluded that a mushroom weighing 25 g could contain as much as 10 μg of lead and 5 μg of cadmium. In the United Kingdom, the recommended weekly intake of lead is less than 180 μg and of cadmium, less than 140 μg. On the basis of the survey's finding, it would not be too difficult to exceed the proscribed limits with two or three substantial feasts of mushrooms collected from the more highly contaminated areas.[65]

Arsenic has not proved to be a serious problem, although in truth it has not been systematically investigated worldwide. In the common edible mushrooms, it is quite insignificant. However, one edible species, Laccaria amethystina, has been shown to be an accumulator of this element. Although this mushroom is sold in the markets of Switzerland, it is a very minor culinary ingredient elsewhere.[66]

How serious is the heavy-metal problem in the United States? We do not really know. But one area of concern is the increasing use of sewage sludge. Until recently, the commercial market for this product was not large, nor was the average consumer waiting in line to buy some. For the sewage utility companies, sludge was a growth industry. As more and more accumulated in their storage ponds, some way of disposing of it had to be found. A few years ago, it seemed like a good idea to add it back to selected forests in the Northwest as cheap fertilizer. After all, no one eats the trees, or so the reasoning went. The problem with this practice is that the trace metal concentration in sewage sludge is enormous and the metals are readily accumulated by mushrooms.[67] High cadmium levels in the soil have been associated with both sewage sludge and phosphate fertilizers.[68] It would be prudent not to pick mushrooms in an area so kindly irrigated by the sewage utility. Many, but not all, of the sites have been posted with signs informing the unwary mycophagist of the practice.

Unfortunately, sludge is now being used in agricultural settings, so the contamination problem is spreading to our prairies and fields. Saprophytic fungi such as *Agaricus* will accumulate the metals and the hazard will persist—in a different group of organisms. The impact of all of this on animals using mushrooms as part of their diet, such as deer and many small rodents, has not been studied in the United States.

The chemical form of a heavy metal in the mushroom greatly determines its toxicity to humans and animals. For example, inorganic mercury is not nearly as dangerous as the organic forms, such as methylmercury. Conversely, arsenic in the form of arsenobetaine, which is often found in fish, is not absorbed by humans and is harmless. Moreover, the presence of other metals sometimes reduces the effects of otherwise toxic compounds. Selenium, which is present in a variety of edible mushrooms, including *Boletus edulis*, can markedly reduce the biological effects of mercury and cadmium.[53] For these reasons, the mere presence of a heavy metal in a mushroom does not necessarily imply significant human toxicity. We still have a great deal to learn about the interaction of the various compounds, their chemical nature, and their toxic effects in human beings.

RADIOISOTOPE CONTAMINATION

In 1992, *The New York Times* briefly mentioned the interdiction of a tourist returning from the ex-Soviet Union at Kennedy airport customs with a bottle of pickled mushrooms given to him as a gift by a family in the Ukraine. Evidently an astronomical amount of radioactivity was detected. The mushrooms had been picked somewhere in the Chernobyl area. In October 1987, the French insisted that all mushrooms imported from Turkey be tested for radioactivity after the customs inspectors noted that there were no dates on the cans.

Ever since the disaster at Chernobyl on April 26, 1986, when a Soviet nuclear reactor in the Ukraine malfunctioned and released a large amount of radioisotopes into the atmosphere, scientists around the world have noted that a variety of plants and especially fungi have the ability to preferentially concentrate some of the radioactive elements present in the reactor emissions. In the early days following the accident, most of the interest revolved around those radioactive elements with a short half-life, especially iodine-131 and iodine-132. More recently, the long half-life elements, most notably cesium-134 and cesium-137, have been detected in a variety of species from a number of sites.[69–71] (Cesium has a half-life of 30 years, so it will be around for a long time and merits serious consideration.) Great regional variation has been observed and seems to depend largely on the prevailing weather patterns at the time of the radioisotope release. However, many factors determine the content of cesium in any particular fruiting body.[72,73]

In 1988, the meat of domesticated animals grazing in the forests and the mountains of Norway, a country directly downwind of Chernobyl, was noted to have much higher levels of radioisotopes than expected. The vascular plants and the local soil had all shown very stable concentrations of radioactive cesium throughout the growing season. Suspicion fell on the mushrooms, because farm animals are known to consume them avidly if given the chance.[74,75] An environmental survey, including multiple samples of the soil, plants, and fungi within small plots widely scattered in the mountains clearly demonstrated the ability of some fungi to preferentially accumulate many of the trace elements, cesium among them. The levels of radioactive cesium in the fungi were much higher—10 and 150 times higher—than that in the adjacent plants, but the actual amount was very dependent on the species. In one particular *Clitocybe*, the cesium level was 270 times higher than in the surrounding soil or plants. Not only was species variation observed, but within an individual fruiting body, striking differences could be measured. Radioactive cesium was one to two times higher in the gills and the spores than in the flesh of the cap or the stem.

Many of the other macro- and microelements were measured in the same survey, and again species variation was demonstrated. In general, species of *Cortinarius* accumulated large amounts of many of the elements, species of *Russula* and *Lactarius* had intermediate amounts, and boletes tended to have the least. Wildlife in these regions of contamination showed seasonal increases in radioisotope levels coincident with foraging on the more highly contaminated plants, such as bilberry, aquatic plants, and mushrooms.[76]

The accumulation of radioisotopes in areas of contamination or downwind fallout has been documented in many countries.[77] It has even been suggested that mushrooms such as *Ganoderma lucidum*, which concentrate radioactive elements from their woody substrates and directly from the atmosphere, could be used as a biological monitoring tool for measuring radioisotope pollution.[78] In Russia, radioisotope contamination is taken so seriously that efforts are underway to salvage valuable food sources such as mushrooms and fish by developing procedures that will remove the radioactive elements from them.[79]

In Japan, the consumption of mushrooms can range as high as 3.5 kg per person per year. Fortunately, there is no evidence of radioisotope contamination of any of their commonly cultivated mushrooms.[80] And a recent study confirms that edible mushrooms from selected sites in North America also are free of dangerous levels of radioisotopes.[81]

CANCER

It would be a great disappointment to everyone if mushrooms were not in some way associated with the development of cancer, because every other foodstuff that has been vigorously investigated has been shown to contain at least one carcinogen or mutagen. So, why not mushrooms? Many mushrooms contain compounds with structures similar to well-known cancer-producing agents, and this fact has prompted many scientists to investigate the threat that mushrooms pose to the health of humankind.

Agaricus bisporus

One of the best-studied compounds in the common cultivated mushroom, Agaricus bisporus, has the trivial name of agaritine; it is present in rather sizable concentrations. Its chemical name, β-N-[γ-L(+)-glutamyl]-4-hydroxymethylphenylhydrazine, gives a clue to its potential threat, namely, the hydrazine portion of the molecule. Compounds like this one with a N-N bond are metabolized into both hydrazines and diazonium ions in the body. Both these metabolites have known, well-demonstrated carcinogenic properties. One derivative had been shown to produce tumors of the lung and blood vessels in mice. However, this derivative is so unstable that the investigators had to modify the molecule to be able to test it in mice.[82] Another derivative produced tumors in mice when injected subcutaneously, and a third (an actual metabolite from the mushroom) caused gastric tumors in mice.[83] When agaritine itself was injected under the skin, no tumors developed.[84]

Of course, such experiments bear no relationship to the way humans ingest mushrooms. While it is true that the rare case of psilocybe "mainlining" has been recorded, no one would think of subcutaneous injections at an average suburban dinner party. It is just not the best way to savor the taste. This is not meant to disparage the dedicated work of these investigators, because studies of the potential health risks posed by these mushrooms are important. For example, more recent studies have suggested that raw mushrooms fed orally to mice induce tumor production.[85] However, we need to be very careful how these results are extrapolated to the human situation.

Surveys of cultivated mushrooms in the United States and the United Kingdom have shown a variable agaritine concentration—from 80 to 250 mg/kg, depending on the strain, the age of the mushroom, and the time of harvesting.[86,87] The content of agaritine is reduced by blanching, freezing, or canning.

Not all of the experimental situations have produced positive results. The incidence of tumors in a group of rats fed 30% mushroom powder for a year and a half was no higher than that in the control group. However, the amount of agaritine ingested was not stated.[88] Even the role of agaritine in producing mutagenic effects has been seriously challenged.[89] Recent work has suggested that many of the previous observations on the mutagenicity of *Agaricus bisporus* may have been the result of an artifact in the experimental systems used at the time. Newer studies failed to confirm the previous observations.[90] At this time, the data is much too limited to provide any idea of the risk to humans—if there is indeed any at all—associated with the consumption of *Agaricus bisporus*.

Gyromitra esculenta and Relatives

Gyromitra esculenta poses more of a threat than *Agaricus bisporus*, although worldwide *Gyromitra* is eaten by many fewer people than *Agaricus*. However, the false morel contains at least 11 different hydrazines, and many of them have been shown to be actively carcinogenic and highly mutagenic in the standard assays.[91-97] These compounds routinely produce tumors in experimental animals, both in the fresh state and as purified agents. In addition, the tumors are produced regardless of the route of administration. It is difficult to know precisely how many people around the world eat these mushrooms. Although *Gyromitra* is a rare delicacy in spring, large quantities are dried and sold for later use, especially in eastern Europe. But it would be prudent for individuals either to cease eating *Gyromitra* altogether or to limit their intake on an annual basis.

The health risk of eating *Gyromitra* seems to depend on each person's unique metabolism. The most recent studies have shown gyromitrin itself to be a weak methylating agent with rather low carcinogenic potential. It has been suggested that the risk may be greater in those people with a decreased detoxification rate in the liver, in whom larger amounts of one of the hydrazine derivatives (monomethylhydrazine) will be formed.[98] The most recent experiments, in which raw mushroom was fed to mice three days out of every seven during their entire life, showed a striking increase in the number of tumors. Although their life span was not significantly reduced and many of the tumors

were benign, the metabolites in this mushroom pose a clearly demonstrated hazard.[91]

Other Mushrooms

Mutagenic agents have been described in a number of other mushrooms, some of which are not commonly eaten. A powerfully mutagenic
compound has been isolated from one of the milky caps, *Lactarius
necator*.[99–102] However, it is not alone in its ability to damage DNA
or influence the genetic code in deleterious ways.[103] In the current assay
systems, many of the common edible mushrooms, including the chanterelle, demonstrate some mutagenic activity. In one study, 13 of 35
species tested with the bacterial mutagenesis assay were positive.[104] One
investigation of the mutagenicity of canned wild mushrooms showed
no problems at all, an outcome illustrating the variability of the results
depending on how the mushrooms are processed and handled.[105]

The method by which the compounds are extracted and used in
the experimental animals can affect the results. This influence was
illustrated in the investigation of the desert mushroom *Tiramania pinoyi*
(al-faga): a water-based extract showed no mutagenicity; one made
with chloroform was actively mutagenic; and one made with alcohol
was antimutagenic. In other words, the alcohol extract protected the
test bacteria against compounds normally producing mutations.[106]
Recently, extracts of the rim-roll mushroom, *Paxillus involutus*, have
directly produced breaks in the chromosomes of dividing cells. The
compound responsible has not been further identified, except for the
fact that it is thermostable.[107]

Coda on the Cancer Risk

The report of a cancer-causing chemical in one of our favorite foodstuffs or beverages is always heralded with much fanfare by the media.
Most often, the risk is exaggerated. Almost never do the long-term
epidemiological studies validate the initial concern. Human nutrition
is an extraordinarily complex subject and causal relationships are difficult to prove. The induction and progression of cancer is equally complex, with subtle interactions between the genetics of an individual and
the environment.

A recent evaluation of the potential risk to humans of compounds
shown to be carcinogenic in rodents helps to put the preceding comments into perspective.[108] The risk from wine or beer is considerably
greater than that from a mushroom. When one considers the average
daily human exposure to various foodstuffs, coffee, lettuce, and orange

juice rank far ahead of mushrooms in their potential to cause trouble, always assuming that we behave vaguely like rats and mice.

Furthermore, there is a flip side to the cancer risk—mushrooms may actually be useful in treating or preventing cancer, or may mitigate the effects of other mutagenic agents. In other words, some mushroom compounds are antimutagenic and may have antitumor activity. These substances are discussed in greater detail in Chapter 5.

REFERENCES

1. S. Lindroth, E. Strandberg, A. Pessa, and M. J. Pellinen, "A Study on the Growth Potential of *Staphylococcus aureus* in *Boletus edulis*, a Wild Edible Mushroom, Prompted by a Food Poisoning Outbreak," *Journal of Food Science* (1983) 48(1):282–283.

2. E. Todd, P. C. Chang, A. Hauschild, A. Sharpe, C. Park, and H. Pivnick," Botulism from Marinated Mushrooms," *IV International Congress of Food Science and Technology* (1974), pp. 4–6.

3. A. H. Hauschild, B. J. Aris, and R. Hilshseimer, "*Clostridium botulinum* in Marinated Products," *Canadian Institute of Food Science and Technology* (1975) 8(2):84–87.

4. H. E. McLean, S. Peck, R. G. Mathias, W. A. Black, and G. B. Morgan, "Restaurant-associated Botulism from Mushrooms Bottled in House—Vancouver, British Columbia," *Morbidity and Mortality Weekly Report (CDC)* (1987) 36(7):103.

5. A. I. Baranova and M. S. Torgovitskaya, [Case of botulism caused by type A causative agents and associated with ingestion of home-made preserved mushrooms] (Russian), *Zhurnal Mikrobiologii, Epidemiologii i Immunobiologii* (1972) 49(2):139–140.

6. M. A. Horwitz, J. S. Marr, M. H. Merson, V. R. Dowell, and J. M. Ellis, "A Continuing Common-source Outbreak of Botulism in a Family," *Lancet* (1975) 2:861–863.

7. P. Hardt-English, G. York, R. Stier, and P. Cocotas, "Staphylococcal Food Poisoning Outbreaks Caused by Canned Mushrooms from China," *Food Technology* (1990) 44(12):76–77.

8. K. G. Brunner and A. C. L. Wong, "*Staphylococcus aureus* Growth and Enterotoxin Production in Mushrooms," *Journal of Food Science* (1992) 57(3):700–703.

9. D. A. Kautter, T. Lilly, and R. Lynt, "Evaluation of the Botulism Risk in Fresh Mushrooms Wrapped in Commercial Polyvinylchloride Film," *Journal of Food Protection* (1978) 41:120–121.

10. C. J. Malizio and E. A. Johnson, "Evaluation of the Botulism Hazard from Vacuum-packed Enoki Mushrooms (*Flammulina velutipes*)," *Journal of Food Protection* (1991) 54(1):20–21.

11. H. C. Goldberg, "Is It Organophosphate Poisoning or Mushroom Poisoning? A Case History with Treatment Recommendations," *Journal of the Florida Medical Association* (1975) 62(12):26–28.

12. G. Becker, "Unexpected Poisoning by Mushrooms Probably Due to Use of an Herbicide," *Bulletin Mensuel de la Société Linnéenne de Lyon* (1975) 44(10):342–343.

13. R. Bergoz, "Trehalose Malabsorption Causing Intolerance to Mushrooms," *Gastroenterology* (1971) 60:909.

14. J. Madrazarovova-Nohejlova, "Trehalase Deficiency in a Family," *Gastroenterology* (1973) 65:130–133.

15. N. Kochibe and K. L. Matta, "Purification and Properties of an *N*-Acetylglucos-amine-specific Lectin from *Psathyrella velutina* Mushroom," *Journal of Biological Chemistry* (1989) 264(1):173–177.

16. Y. Nagata et al., "Crystallization and Characterization of a Lectin Obtained from a Mushroom, *Aleuria aurantia*," *Biochimica et Biophysica Acta* (1991) 1076(2):187–190.

17. E. R. Gold and P. Baling, *Receptor Specific Proteins: Plant and Animal Lectins* (Amsterdam: Exerpta Medica, 1975), pp. 77–150.

18. J. Guillot, M. Giollant, M. Damez, and M. Dusser, "Isolation and Characterization of a Lectin from the Mushroom, *Lactarius deliciosus*," *Journal of Biochemistry (Tokyo)* (1991) 109(6):840–845.

19. W. J. Rockwell, "Reactions to Mold in Foods." In *Food Allergy: A Practical Approach to Diagnosis and Management*, ed. L. T. Chiaramonte, A. T. Schnieder, and F. Lifschitz (New York: Marcel Dekker, 1988).

20. J. Brostoff and S. J. Challacombe, *Food Allergy and Intolerance* (Eastbourne, Sussex: Bailliere Tindall [W. B. Saunders], 1987).

21. E. K. Rose and P. Rieders, "An Episode of Food Poisoning Attributed to Imported Mushrooms," *Annals of Internal Medicine* (1966) 64(2):372–377.

22. R. Burrell, "Allergy to Mushrooms: Fact or Fancy?" *McIlvainea* (1978) 3(2):11–14.

23. D. J. Pearson and K. J. B. Rix, "Allergomimetic Reactions to Food and Pseudo Food Allergy." In *PAR Pseudoallergic Reactions: Involvement of Drugs and Chemicals*, Vol. 4 (Basel: Karger, 1985), pp. 59–105.

24. A. Koivikko and J. Savolainen, "Mushroom Allergy," *Allergy* (1988) 43(1):1–10.

25. Quoted in G. C. Ainsworth, *Introduction to the History of Medical and Veterinary Mycology* (Cambridge: Cambridge University Press, 1986).

26. A. De Zubiria, W. E. Horner, and S. B. Lehrer, "Evidence for Cross-reactive Allergens Among the Basidiomycetes: Immunoprint-inhibition Studies," *Journal of Allergy and Clinical Immunology* (1990) 86:26–33.

27. C. E. O'Neill, W. E. Horner, M. A. Reed, M. Lopez, and S. B. Lehrer, "Evaluation of Basidiomycete and Deuteromycete (Fungi Imperfecti) Extracts for Shared Allergenic Determinants," *Clinical and Experimental Allergy* (1990) 20:533–538.

28. J. Salvaggio and L. Aukrust, "Mold-induced Asthma," *Journal of Allergy and Clinical Immunology* (1981) 68:327–346.

29. W. A. Check, "Common Mushroom Spores May Cause Asthma and Hay Fever in Fall" [News], *Journal of the American Medical Association* (1982) 247(15):2071.

30. A. Michilis, P. De-Vuyst, N. Nolard, G. Servais, J. Duchateau, and J. C. Yernault, "Occupational Asthma to Spores of *Pleurotus cornucopiae*," *European Respiratory Journal* (1991) 4(9):1143–1147.

31. P. I. Gumowski, J.-P. Latge, and S. Paris, "Fungal Allergy." In *Handbook of Applied Mycology*, Vol. 2, ed. D. K. Arora (New York: Marcel Dekker, 1991), pp. 163–204.

32. R. D. Strand, E. B. D. Neuhauser, and C. F. Sornberger, "Lycoperdonosis," *New England Journal of Medicine* (1967) 277:88–89.

33. E. Jackson and K. M. E. Welch, "Mushroom Worker's Lung," *Thorax* (1970) 25:25–30.

34. A. Sakula, "Mushroom Worker's Lung," *British Medical Journal* (1967) 3:708–710.

35. R. Lockey, "Mushroom Worker's Pneumonitis," *Annals of Allergy* (1974) 33:282–286.

36. J. L. Stolz and P. M. Arger, "Respiratory Disease Due to Mushrooms." In *Mushroom Poisoning*, ed. B. H. Rumack and E. Salzman (West Palm Beach, FL: CRC, 1978).

37. J. Sastre, M. D. Ibanez, M. Lopez, and S. B. Lehrer, "Respiratory and Immunological Reactions Among Shiitake (*Lentinus edodes*) Mushroom Workers," *Clinical and Experimental Allergy* (1990) 20(1):13–19.

38. Y. J. Kamm, H. T. Folgering, H. G. van den Bogart, and A. Cox, "Provocation Tests in Extrinsic Allergic Alveolitis in Mushroom Workers," *Netherlands Journal of Medicine* (1991) 38(1–2):59–64.

39. W. E. Horner, M. D. Ibanez, V. Liengswangwong, J. E. Salvaggio, and S. B. Lehrer, "Characterization of Allergens from Spores of the Oyster Mushroom, *Pleurotus ostreatus*," *Journal of Allergy and Clinical Immunology* (1988) 82(6):978–986.

40. A. Michils, V. P. De, N. Nolard, G. Servais, J. Duchateau, and J. C. Yernault, "Occupational Asthma to Spores of *Pleurotous cornucopiae*," *European Respiratory Journal* (1991) 4(9):1143–1147.

41. T. Nakazawa and T. Tochigi, "Hypersensitivity Pneumonitis Due to Mushroom (*Pholiota nameko*) Spores," *Chest* (1989) 95(5):1149–1151.

42. J. G. Kleyn and T. F. Wetzler, "The Microbiology of Spent Mushroom Compost and Its Dust," *Microbiology* (1981) 27:748–753.

43. K. Yoskida et al., "Hypersensitivity Pneumonitis of a Mushroom Worker Due to *Aspergillus glaucus*," *Archives of Environmental Health* (1990) 45(4):245–247.

44. M. J. Korstanje and W. J. van-de-Staak, "A Case of Hand Eczema Due to Mushroom," *Contact Dermatitis* (1990) 22:115–116.

45. K. Tarvainen, P. J. Salonen, L. Kanerva, T. Estlander, H. Keskinen, and T. Rantanen, "Allergy and Toxiderma from Shitake Mushrooms," *Journal of the American Academy of Dermatology* (1991) 24(1):64–66.

46. J. N. Bruhn and M. D. Soderberg, "Mushrooms Can Cause Allergic Contact Dermatitis: A First-hand Report with a Review," *McIlvainea* (1992) 10(2):46–50.

47. A. E. Alder, "Die Pilzvergiftungen in der Schweiz in den Jahren 1946 und 1947," *Schweizerische Zeitschrift für Pilzkunde* (1948) 26:85–92.

48. R. Seeger, "Toxische Schermetalle in Pilzen," *Deutsche Apotheker Zeitung* (1982) 122:1835–1844.

49. T. Stijve, "Enige potentieel giftige Elementen in Paddestoelen," *Coolia* (1980) 23:28–108.

50. R. O. Allen and E. Steiness, "Concentrations of Some Potentially Toxic Metals and Other Trace Elements in Wild Mushrooms from Norway," *Chemosphere* (1978) 4:317–378.

51. H. Wetzel, "Toxikologisch bedeutsame Schwermetalle in Pilzen," *Mycologisches Mitteilungsblatt* (1979) 23:1–14.

52. C. H. Gast, E. Jansen, J. Bierling, and L. Haanstra, "Heavy Metals in Mushrooms and Their Relationship with Soil Characteristics," *Chemosphere* (1988) 17(4):789–800.

53. T. Stijve, "Mushrooms Do Accumulate Heavy Metals," *Mushroom* (1992–3) 11(1):9–19.

54. G. Bressa, L. Cima, and P. Costa, "Bioaccumulation of Mercury in the Mushroom *Pleurotus ostreatus*," *Ecotoxicology and Environmental Safety* (1988) 16(2):85–89.

55. A. Yasui, C. Tsutsumi, M. Takasaki, and T. Mori, "Absorption of Elements from Heavy Metals Containing Culture Media by Oyster Mushroom, *Pleurotus ostreatus*," *Journal of the Japanese Society of Food Science and Technology* (1988) 35(3):160–165.

56. A. R. Byrne et al., "Studies on the Uptake and Binding of Trace Metals in Fungi. Part 1. Accumulation and Characterization of Mercury and Silver in the Cultivated Mushroom, *Agaricus bisporus*," *Applied Organometallic Chemistry* (1990) 4:43–48.

57. C. G. Zurera, L. F. Rincon, R. R. Moreno, E. J. Salmeron, and L. R. Pozo, "Mercury Content in Different Species of Mushrooms Grown in Spain," *Journal of Food Protection* (1988) 51(3):205–207.

58. A. Bresinsky and H. Besl, *A Colour Atlas of Poisonous Fungi* (London: Wolfe Publishing, 1990).

59. H.-U. Meisch, I. Beckmann, and J. A. Schmitt, "A New Cadmium-binding Phosphoglycoprotein, Cadmium-mycophosphatin, from the Mushroom, *Agaricus macrosporus*," *Biochimica et Biophysica Acta* (1983) 745:259–266.

60. G. A. Jackl, G. Reidel, and W. E. Kollmer, "Identification of the Cadmium-binding Compounds in *Agaricus arvensis* Hyphae Using Cadmium-109," *Applied Radiation and Isotopes* (1987) 38(6):431–436.

61. G. G. Zurera, L. F. Rincon, and L. R. Pozo, "Lead and Cadmium Content of Some Edible Mushrooms," *Journal of Food Quality* (1988) 10(5):311–318.

62. J. P. Quiche, A. Bolay, and V. Dvorak, "La pollution mercurielle de diverses espèces de champignons," *Revue Suisse d'Agriculture* (1976) 8(5):143–148.

63. A. Lepsova and R. Kral, "Lead and Cadmium in Fruiting Bodies of Macrofungi in the Vicinity of a Lead Smelter," *Science of the Total Environment* (1988) 76(2–3):129–138.

64. P. Kalac, J. Burda, and I. Staskova, "Concentrations of Lead, Cadmium, Mercury and Copper in Mushrooms in the Vicinity of a Lead Smelter," *Science of the Total Environment* (1991) 105:109–120.

65. K. Thomas, "Heavy Metals in Urban Fungi," *The Mycologist* (1992) 6(4):195–196.

66. T. Stijve, E. C. Vellinga, and A. Herrmann, "Arsenic Accumulation in Some Higher Fungi," *Persoonia* (1990) 15(2):161–166.

67. D. Zabowski, R. J. Zasoski, W. Littke, and J. Ammirati, "Metal Content of Fungal Sporocarps from Urban, Rural and Sludge-treated Sites," *Journal of Environmental Quality* (1990) 19(3):372–377.

68. R. L. Chaney and P. M. Gioddano, "Microelements as Related Plant Deficiencies and Toxicities." In *Soils for Management of Organic Wastes and Waste Waters*, ed. L. F. Elliot (Madison, WI: American Society for Agronomy and Crop Science, 1977), pp. 235–279.

69. R. Borio et al., "Uptake of Radiocesium by Mushrooms," *Science of the Total Environment* (1991) 106(3):183–190.

70. J. K. Korky and L. Kowalski, "Radioactive Cesium in Edible Mushrooms," *Journal of Agricultural and Food Chemistry* (1989) 37(2):568–569.

71. G. Rueckert and J. F. Diehl, [Increased levels of cesium-137 and cesium-134 in 34 species of wild mushrooms following the Chernobyl disaster] (German), *Zeitschrift für Lebensmittel Untersuchung Forschung* (1987) 185(2):91–97.

72. G. T. Oolbekkink and T. W. Kuyper, "Radioactive Cesium from Chernobyl in Fungi," *The Mycologist* (1989) 3(1):3–6.

73. G. Heinrich, "Uptake and Transfer Factors of ^{137}Cs by Mushrooms," *Radiation and Environmental Biophysics* (1992) 31(1):39–49.

74. K. Hove and P. Strand, "Fungi: A Major Source of Radiocesium of Grazing Ruminants in Norway," *Health Physics* (1990) 59:189–192.

75. K. Hove and P. Strand, "Predictions for the Duration of the Chernobyl Radiocesium Problem in the Noncultivated Areas Based on an Assessment of the Behaviour of Fallout from Nuclear Weapons Tests." In *Environmental Contamination Following a Major*

Nuclear Accident, Vol. 1, ed. S. Flitton and E. W. Katz (Vienna: International Atomic Energy Agency, 1990), pp. 212–220.

76. R. T. Palo, P. Nelin, T. Nyyen, and G. Wickman, "Radiocesium Levels in Swedish Moose in Relation to Disposition, Diet and Age," *Journal of Environmental Quality* (1991) 20(3):690–695.

77. H. Bem et al., "Accumulation of Cesium-137 by Mushrooms in the Rozozno Area of Poland Over the Period 1984–1988," *Journal of Radioanalytical and Nuclear Chemistry* (1990) 145:39–46.

78. L. Tran Van and T. Le Duy, "Linchi mushrooms as biological Monitors for Cesium-137 Pollution," *Journal of Radioanalytical and Nuclear Chemistry* (1991) 155(6):451–458.

79. A. I. Il'enko and T. P. Krapivko, [Results of the studies of the removal of cesium-134 and cesium-137 from the fruiting bodies of edible fungi and freshwater fish] (Russian), *Doklady Akademii Nauk SSSR* (1990) 311(4):1012–1014.

80. Y. Muramatsu, S. Yoshida, and M. Sumiya, "Concentrations of Radiocesium and Potassium in Basidiomycetes Collected in Japan," *Science of the Total Environment* (1991) 105(29):29–39.

81. M. L. Smith, H. W. Taylor, and H. D. Sharma, "Comparison of the Post-Chernobyl [137]Cs Contamination of Mushrooms from Eastern Europe, Sweden and North America," *Applied and Environmental Microbiology* (1993) 59(1):134–139.

82. B. Toth, D. Nagel, K. Patil, J. Erickson, and K. Antonson, "Tumor Induction with the N-Acetyl Derivative of 4-Hydroxymethylphenylhydrazine," *Cancer Research* (1978) 38:177–180.

83. B. Toth, D. Nagel, and A. Ross, "Gastric Tumorigenesis by a Single Dose of 4-(Hydroxymethyl)benzenediazonium Ion of *Agaricus bisporus*," *British Journal of Cancer* (1982) 46(3):417–422.

84. B. Toth and H. Sornson, "Lack of Carcinogenicity of Agaritine by Subcutaneous Administration in Mice," *Mycopathologia* (1984) 85:75–79.

85. B. Toth and J. Erickson, "Cancer Induction in Mice by Feeding of the Uncooked Cultivated Mushroom of Commerce *Agaricus bisporus*," *Cancer Research* (1986) 46(8):4007–4011.

86. M. Sharman, A. L. Patey, and J. Gilbert, "A Survey of the Occurrence of Agaritine in U.K. Cultivated Mushrooms and Processed Mushroom Products," *Food Additives and Contaminants* (1990) 7(5):649–656.

87. J. Speroni, R. B. Beelman, and L. C. Schisler, "Factors Influencing the Agaritine Content in Cultivated Mushrooms, *Agaricus bisporus*," *Journal of Food Protection* (1983) 46(6):506–509.

88. K. Matsumoto, M. Ito, S. Tagyu, H. Ogino, and I Hirono, "Carcinogenicity Examination of *Agaricus bisporus*, an Edible Mushroom, in Rats," *Cancer Letters* (1991) 58(1–2):87–90.

89. C. Papaparaskeva, C. Ioannides, and R. Walker, "Agaritine Does Not Mediate the Mutageniciity of the Edible Mushroom *Agaricus bisporus*," *Mutagenesis* (1991) 6(3):213–218.

90. B. L. Pool-Kobel et al., "Mutagenic and Genotoxic Activities of Extracts Derived from the Cooked and Raw Edible Mushroom *Agaricus bisporus*," *Journal of Cancer Research and Clinical Oncology* (1990) 116(5):475–279.

91. B. Toth, K. Patil, H. Pyysalo, C. Stessman, and P. Gannett, "Cancer Induction in Mice by Feeding the Raw False Morel Mushroom *Gyromitra esculenta*," *Cancer Research* (1992) 52(8):2279–2284.

92. B. Toth, "Carcinogenic Fungal Hydrazines," *In Vivo* (1991) 5(2):95–100.

93. B. Toth and P. Gannett, "Carcinogenesis Study in Mice by 3-Methylbutanal Methylformylhydrazone of *Gyromitra esculenta*," *In Vivo* (1990) 4(5):283–288.

94. B. Toth, J. Taylor, and P. Gannett, "Tumor Induction with Hexanal Methyl-formylhydrazone of *Gyromitra esculenta*," *Mycopathologia* (1991) 115(2):65–71.

95. B. Toth and C. R. Raha, "Carcinogenesis by Pentanal Methylformylhydrazone of *Gyromitra esculenta* in Mice," *Mycopathologia* (1987) 98(2):83–89.

96. B. Toth and K. Patil, "Tumorigenic Action of Repeated Subcutaneous Admin-istration of N-Methyl-N-formylhydrazine in Mice," *Neoplasma* (1983) 30(4):437–441.

97. B. Toth and K. Patil, "Tumorigenicity of Minute Dose Levels of N-Methyl-N-*formylhydrazine of* Gyromitra esculenta," Mycopathologia (1982) 78(1):11–16.

98. K. Bergman and K. E. Hellenas, "Methylation of Rat and Mouse DNA by the Mushroom Poison Gyromitrin and Its Metabolite Monomethylhydrazine," *Cancer Letters* (1992) 61(2):165–170.

99. T. Suortti, "Stability of Necatorin, a Highly Mutagenic Compound from *Lactarius nector* Mushroom," *Food and Chemical Toxicology* (1984) 22(7):579–581.

100. O. Sterner, R. Bergman, E. Franzén, E. Kesler, and L. Nilsson, "The Mutagenicity of Commercial Pickled *Lactarius necator* in the Salmonella Assay," *Mutation Research* (1982) 104:233–237.

101. T. Suortti, A. von Wright, and A. Koskinen, "Necatorin, a Highly Mutagenic Compound from *Lactarius necator*," *Phytochemistry* (1983) 22:2873–2874.

102. T. Suortti and A. von Wright, "Isolation of a Mutagenic Fraction from Aqueous Extracts of the Wild Edible Mushroom *Lactarius necator*" (Preliminary Note), *Journal of Chromatography* (1983) 255:529–532.

103. J. Knuutinen and A. von Wright, "The Mutagenicity of *Lactarius* Mushrooms," *Mutation Research* (1982) 103:115–118.

104. A. Grüter, U. Friederich, and F. E. Wurgler, "The Mutagenicity of Edible Mushrooms in a Histidine-independent Bacterial Test System," *Food and Chemical Toxicology* (1991) 29(3):159–166.

105. P. Morales, E. Bermudez, B. Sanz, and P. E. Hernandez, "A Study of the Muta-genicity of Some Commercially Canned Spanish Mushrooms," *Food and Chemical Toxicology* (1990) 28(9):607–611.

106. M. A. Hannan, D. A. Al, E. H. Aboul, and O. A. Al, "Mutagenic and Antimuta-genic Factor(s) Extracted from a Desert Mushroom Using Different Solvents," *Muta-genesis* (1989) 4(2):111–114.

107. J. Gilot-Delhalle and J. Moutschen, "Chromosome-breaking Activity of Extracts of the Mushroom *Paxillus involutus* (Fries ex Batsch)," *Experienta* (1991) 47(3):282–284.

108. L. Swirsky Gold, T. H. Slone, B. Stern, N. B. Manley, and B. N. Ames, "Rodent Carcinogens: Setting Priorities," *Science* (1992) 258:261–265.

CHAPTER EIGHT

SPECTRUM OF POISONOUS MUSHROOMS

THE NUMBER OF poisonous mushrooms recorded in the literature is immense, although a large portion of the reports of adverse effects are not fully substantiated. The difficulties involved in designating a species as poisonous are fourfold. Mushroom taxonomy is currently in flux, so the species of mushroom involved in a reported event is often uncertain. Similarly, the criteria for diagnosing "poisoning" are not firm. Although allergic, idiosyncratic, and intolerance reactions to any one of a host of compounds present in mushrooms are common, these reactions are in fact no different from any other of the many types of "food reactions" that are not called "poisoning." Adverse reactions to common foods, such as strawberries or broccoli, do not cause the responsible foodstuff to be labeled "poisonous." The individuals experiencing these reactions simply learn to avoid the offending food. Third, the chemical compounds that cause distress in certain individuals vary in concentration within and between species, as has already been noted in Chapter 6. Consequently, species are often only intermittently inedible. How to classify these species in field guides and how to know when they are safe to eat are unresolved questions in mycology. Finally, all the toxic compounds are not well characterized chemically and pharmacologically, a deficiency contributing to a lack of understanding of mechanisms of action and treatments and to identification problems.

The North American Mycological Association Poison Registry in the last decade has listed a variety of edible species of mushrooms as reaction provoking, often on the basis of a very small collection of reports—usually only one. The number of well-documented toxicological studies is limited and cannot support the published claims for over 400 poisonous species of mushrooms. In an effort to avoid further possibly false attributions, I have limited the number of mushrooms discussed in Part Three to only those with proven toxicity.

EDIBILITY CHARACTERISTICS IN THE FIELD GUIDES

A difficulty confronting the potential mycophagist or the physician trying to decide whether a particular specimen is edible or poisonous is the disparity in the descriptions given by the various field guides. A particular species may be called edible in one source and poisonous in another, or more subtle variations in descriptions of edibility may leave the reader with considerable doubt. This situation occurs most frequently with species that are not commonly eaten or are rarely found. In such situations, despite the helpful guidance, collectors are on their own and should not rely on any one particular source. Unfortunately, many of the guides merely repeat the recommendations of past books, without verifying the original source or retesting a particular species. Some texts, however, have wonderfully detailed, personal descriptions of the edibility, taste, or flavor. Few are as colorful as those of Charles McIlvaine in his opus *One Thousand American Fungi*. He disparaged only the rare mushroom, liking almost all, and announcing that many were "second to none." He seems to have had a most remarkable digestive tract.

A second difficulty is the language authors use to describe the edibility and taste of mushrooms. Unlike the world of enology, in which the critics, connoisseurs, and merchants have developed a sophisticated lexicon to describe the characteristics of wine, the vocabulary for the culinary attributes of mushrooms is rather limited. Words such as *choice, esteemed, highly esteemed,* and *edible* do little to convey more than the barest essentials. In fact, there are a number of questionable terms quite widely used in mycological writings, ambiguous phrases such as

- ◆ Edible with caution
- ◆ Not recommended
- ◆ Harmless, but best avoided
- ◆ Tasty, but perhaps poisonous

For most of us, books are repositories of knowledge, and we rightly rely on them for much of our information. Generally, the information is reliable, but it is always worthwhile to maintain a healthy amount of skepticism. Even the authorities can be dead wrong. An edition of the famous cookbook *Larousse Gastronomique*, a culinary bible translated into many languages, went to press with an egregious error. On the day it was released in the 1980s, someone astutely pointed out that the picture of *Amanita phalloides* was labeled "edible." Fortunately, the entire stock was recalled and reprinted. So, when mushroom field guides disagree on the edibility characteristics of a particular mushroom, it is a good idea to experiment with some care—and be willing to withstand a few hours of misery.

Some mushrooms are harmless to some people and toxic to others or are only edible under specific circumstances, or in small quantities, or if properly prepared. These species have been termed "partials." Chapters 19 and 20 are full of the names of such mushrooms. Encapsulating this duality in a simple phrase is not easy, especially in a phrase that can be used in a field guide. It is for mushrooms like these that terms such as *edible with caution* are entirely appropriate.

Cultural taste preferences further confuse field guide readers. Gastronomic differences between countries are often profound. Certain peppery or acrid species of *Lactarius* and *Russula* are regarded with great suspicion by most English-language guides, whereas in eastern Europe they are eaten with relish. *Russula brevipes,* the most widely kicked or stomped-on mushroom in the United States, is listed as edible, but tasteless; conversely, it is one of the most frequently used mushrooms in Russia—along with *Lactarius*—for salting and pickling. *Paxillus involutus* is favored by eastern Europeans. Many in the United States, however, find the smell of this mushroom enough to dissuade them from even trying it.

DISTRIBUTION OF MUSHROOMS IN NORTH AMERICA

One of the points endlessly reiterated in this book is the importance of local knowledge for safe picking and eating of wild mushrooms and for the diagnosis of poisoning, because mycofloras vary so much across the country. Alexander Smith[1] has divided the continent into five large zones of distinctive flora. Obviously there is some overlap of species at the edges of these zones, and certain mushrooms are ubiquitous, but it is a helpful way of looking at the variation in mushroom flora and of keeping track of the movement of mushrooms between regions.

For the amateur mycologist and emerging mycophagist, learning the common edible and poisonous mushrooms in one's home region is the key to a long and healthy life. In addition, the physician called to treat a possible poisoning benefits from knowing what common poisonous mushrooms grow in the area and their likely fruiting periods. In most places, the list of possibilities is quite short. For example, the Gulf Coast area in the Southeast has one of the richest *Amanita* floras in the world, relative to that of the West, where this genus is poorly represented. I say poorly represented with some hesitation, because the representatives growing in the West happen to be some of the worst in the group, including the death cap (*Amanita phalloides*), the panther (*Amanita pantherina*), and the fly agaric (*Amanita muscaria*). The Great Plains region has the smallest number of species of gilled fungi and of poisonous mushrooms in general, but it has a very diverse *Agaricus* flora. The Great Lakes area has an amazing diversity of fungi and probably the largest collection of poisonous species, but this fact may simply be an artifact of observation: Michigan has always been a center for mycological research.

The collector should always remember that our knowledge of the mushroom flora is incomplete, despite the work of many dedicated scientists. There is a veritable dearth of professional mycologists currently engaged in traditional taxonomic field studies. In fact, amateurs can make important contributions in this area, because they are out in the field each weekend. The announcement that a mushroom has been "discovered" in a new region may not mean that it has invaded a new territory or that a species is spreading; it may merely signify that no one has noticed the species before. A prime example of this ignorance involves the distribution of truffles in North America. When the finds of truffles are plotted on a map of the United States, the epicenter is Corvallis, Oregon, with decreasing quantities as one moves away from it. The reason for this regional abundance should not be attributed to natural factors. The real reason is that Jim Trappe at Oregon State University has attracted an enthusiastic group of truffle hunters around him.

VARIATION IN TOXICITY

Each one of us has a unique metabolism and an individual response to a particular food. Considerable study has demonstrated that the concentration, even the presence, of a particular toxin in any single mushroom is highly variable. It is not clear whether this variation is due to fundamental genetic differences or to differences in local growing conditions, that is, differences in substrate, weather, host plants, elevation, and so on (see the box on page 140.) It is the old argument of nature versus nurture. Currently, environmental factors are considered to be most important in influencing the levels of a particular toxin, although many studies still need to be done to clarify this issue.

Factors Leading to Variation in Toxicity

• Genetics
  strain differences
  hybridization with closely related poisonous species (no proven examples)
• Environment
  effect of different substrates
  effect of temperature, light, elevation
• Age of the mushroom
• Uneven toxin distribution within a single mushroom

Support for the environmental influence argument is available in other areas of agriculture and nutrition. For example, cassava, a widely used starchy root, has to be specially handled to rid it of hydrocyanic acid. Normally it is grated, washed, pressed, and drained prior to consumption. However, there is a form called sweet cassava that can be eaten without the usual preliminary treatment. This variety has proved to be genetically identical to the common toxic variety, but it grows under different environmental conditions—and therefore fails to produce hydrocyanic acid.

The many studies of the concentrations of toxins in similar or identical species have shown these concentrations to vary widely.[2] Perhaps most striking is the toxicity of *Gyromitra esculenta*. It is poisonous east of the Rocky Mountains; on the west side, it is said to be widely eaten. (I personally know of very few individuals who still eat this mushroom.) The toxin gyromitrin is present in collections on the west side, but in subthreshold concentrations. Because few individuals ever eat enough of it over a prolonged period, episodes of poisoning have rarely been recorded in the West. The few individuals sickened by this mushroom in states such as Idaho all ate prodigious quantities or committed the unpardonable sin of eating the mushroom raw. Despite this seemingly fortunate circumstance of low concentrations of toxin, *Gyromitra esculenta* cannot be recommended as edible, even west of the Rockies. No one has done the appropriate study: growing a western strain on the east side of the mountains to demonstrate whether the difference in toxin concentrations is an environmental effect or a chemotaxonomic (genetic) difference attributable to distinct varieties.

Specimens of the *Armillaria mellea* complex (honey mushrooms) trouble 10–20% of people who eat it. The toxin causes mild to moderate gastrointestinal distress, although it has not been characterized chemically. The mythology in the Pacific Northwest is that toxicity is greater after the first frost or in specimens growing on conifers rather than on deciduous trees. This folklore has not been confirmed in any systematic way. Similarly, it is believed elsewhere that the honey mushroom should not be eaten if found growing on eucalyptus. An additional problem with figuring out the toxicity of this group is that its taxonomy is an example of chaos theory—its classification in North America is constantly changing.

A third example further reinforces the point about variability. *Phaelepiota aurea*, an absolutely gorgeous, large, golden mushroom, has a dubious reputation. The field guides are not very consistent, but many use the phrase "edible with caution." A Washingtonian couple recently reported their experiences in the newsletter of the Northwest Mushroomers' Association, located in Bellingham, Washington. A decade ago, this couple had eaten a number of large meals of this mushroom near Anchorage, Alaska, without any ill effects. In 1991 they chanced upon a wonderful collection during a hike in Washington. Wanting to reexperience the gastronomic pleasure, they cooked the mushrooms in a seasoned cream sauce. Two hours later, the gastrointestinal symptoms began. The wife became so ill with vomiting and diarrhea that she lost seven pounds, and it took her two days to get back to normal. It seems that the guidebooks are right to advise caution when eating this mushroom, even though the mechanism for its toxicity is not known.

The relationship of toxicity to the growth of the mushroom on the particular substrate is a long-held belief. Collections of *Amanita muscaria* lacking appreciable concentrations of ibotenic acid or muscimol have been described.[3] People in a number of communities, especially in the southwestern portion of Washington State, claim to have eaten *Amanita muscaria* for many years without any ill effects. Great variation in toxicity is the rule, although the factors influencing the concentration of any particular toxin are entirely unknown.

It is no surprise that the relationship between the quantity of mushrooms eaten and clinical manifestations is not always close. For this reason, any estimates of how much of a particular mushroom will produce symptoms should be regarded with some skepticism and should only be used as the roughest of guidelines. These measurements do have value in indicating the minimum amount of a particular species that may be toxic, but there is never a guarantee that symptoms will not develop.

Factors in the environment other than substrate may be important. Temperature may play a role in favoring or retarding the accumulation of a particular toxin, as may the amount of light or the elevation at which the mushroom grows. Hybridization between various closely related species is a possibility, but the extent to which this occurs in the higher fungi in field conditions is not known. Some mycologists believe that hybridization between *Amanita pantherina* and *Amanita gemmata* creates a toxic variety of the latter; and other examples have been proposed. For the most part, however, there is little scientific evidence for such occurrences.

In addition to the variations in toxicity induced by the environment and genetics, the age and the portion of the mushroom eaten have to be considered. As a mushroom matures and ages, its constituents change. This variation is reflected in the color of the mushroom as the concentrations of pigments change. The volatile compounds responsible for flavor undergo similar fluctuations. The simple sugar α-trehalose is present in young specimens but not in old ones. So, too, do the concentrations of toxins change throughout a mushroom's life span.

Some evidence shows that the concentration of toxin ingested depends on the portion of the mushroom eaten. The quantity of toxin in the gills may be different from that in the flesh of the cap or the stem. These anatomical variations in toxin levels are seldom significant from a clinical standpoint. In a few instances, however, they may influence the measurements of a particular compound. Usually mentioned in this regard is *Amanita muscaria*, in which the toxin is claimed to be present only in the cuticle. This assertion is patently false, because the toxins can be detected in high concentrations throughout the flesh and the gills of the mushroom. It is true that one of the toxins, ibotenic acid, not only is present throughout the mushroom but also is bound (conjugated) to one of the yellow pigments of the mushroom. Peeling the mushroom will not ensure its safety, however. It will merely get rid of the color and a little of the poison.

QUANTITY OF TOXINS IN EDIBLE SPECIES

People are always surprised to learn that many of the common edible mushrooms contain small quantities of the same toxic compounds found in the more lethal varieties. However, in this respect mushrooms are no different from a variety of other foods we eat on a daily basis, which, either in their wild state or after improper storage and preparation, can contain potent toxins. It is not widely known by the public that many, if not most, of our common foods contain a variety of secondary compounds, some of which are toxic in certain situations. Potatoes, for example, develop glycosides in and beneath their skin if stored at the wrong temperature (too warm) or if exposed to light for a long time. This change can usually be easily detected by noting a green color of the skin. (Contrary to popular lore, the skin may not be the best part of a potato to eat.) Rhubarb, which some enjoy because of the tart taste, has very toxic leaves containing high concentrations of oxalic acid.

Beans possess a number of interesting sugars that we are not able to digest easily and are partly responsible for the embarrassing consequences of bean eating. In addition, beans contain a variety of potent lectins, compounds binding to the surface of cells and interfering with a number of cellular processes. Fortunately, lectins are heat labile. It is interesting to read the older cookbooks, which used to include instructions for boiling the beans for at least 10 minutes prior to the long slow simmer required for thorough cooking. Such treatment destroys the lectins' activity.

Numerous other examples could be cited, including cassava, lima beans, bamboo shoots (cyanogens), nutmeg (the hallucinogen, myristicin), and celery, which contains furanocoumarins responsible for photosensitizing the skin of people who pick it. Over the centuries, either varieties of these foods have been bred with levels of poison that we can tolerate or we have developed preparation and cooking techniques that eliminate or destroy the toxin. A good example of the second approach is the washing of cassava prior to its consumption. We have also learned to discard the toxic parts of the plant, like the leaves of rhubarb.

There is an immense literature on food toxins, because they are of great interest to the farmer, the botanist, the chemist, and most of all the consumer. (One recent review will get the interested reader into the subject.[4]) It is becoming more and more apparent that these secondary compounds are important in determining the relationships and interactions between adjacent organisms, inhibiting the growth of some and modulating the behavior of others. Seen in this perspective, mushrooms do not differ in any fundamental way from plants in the manner in which secondary compounds are used to modify and control their immediate environment.

Many toxins are either volatile, water soluble, or heat labile. The most notable ones are found in *Gyromitra, Verpa,* and *Morchella* species. Representatives of the genera *Gyromitra* and *Verpa* are traditionally parboiled, and the cooking water is discarded prior to final preparation. The morels cause notorious problems when they are eaten raw; they should always be well cooked. Unfortunately, this knowledge is not always possessed by young chefs who are in the process of rediscovering the flavor of wild mushrooms but lack the lore that should accompany this experimentation. This was best illustrated at a banquet in Vancouver, British Columbia, in the spring of 1992. The feast included a number of city dignitaries, including the heads of the police and the health departments. An exotic salad, liberally sprinkled with pieces of raw morels, was served. Before the end of the meal, many of the participants became ill, and a number required visits to the hospital. Fortunately, no serious health consequences ensued. What happened to the license of the caterer is unknown.

The fairy ring mushroom, *Marasmius oreades*, has a delightful odor of almonds. The odor is quite pronounced when the mushrooms are dried in an enclosed space and is caused by hydrocyanic acid, a potentially deadly toxin. However, the concentrations of toxin are so low and the compound so volatile that this mushroom poses no risk for human health.[5]

The variable toxicity of mushrooms and of an individual's response or reaction to the compounds have been well recognized for years and occasionally commented on. The phrase "harmless to some, poisonous to others" is commonly used to describe the edibility of appropriate mushrooms in field guides or lists of fungi. The problem for the reader is knowing whether he or she belongs to the "some" or to the "other" group. In the early 1900s this problem fascinated John Dearness, who wrote:

The question naturally arises why such opposite effects as nutrition and poison from so many species of one group of plants. Inquiry into the causes may be regarded as practical in view of the fact that there are easily ten times as many people interested in mycophagy as in scientific mycology.

By way of drawing attention to the subject, rather than of throwing light upon it, I beg to cite three types of instances of alleged mushroom poisoning that I have had the opportunity to investigate:

A. A fellow citizen dined on a quart of the common mushroom [*Agaricus campestris*] that he had purchased at a fruit stall. Within twelve hours he was ill enough to have a physician called, who pronounced his case as one of toadstool poisoning. His recovery was complete in two or three days.

B. Near a neighboring town a man collected "a basketful" of supposed mushrooms. His wife was suspicious of them, with the consequence that the collector cooked and ate them without assistance. Before morning he was "sick enough to die," but the promptitude of the doctor "saved his life."

C. A week of wet, warm weather early in May had brought up in a thinly wooded pasture an abundant crop of helvellas [*Gyromitra esculenta*]. Two or three families in the neighborhood collected them for food. One of the families, on a Tuesday evening, ate about two quarts of them, the method of preparation being frying in butter. On the next day at noon a smaller quantity—about a quart and a half—was similarly disposed of.

That night every member of the family was taken ill, and on the Friday, one of them, in spite of the effort of two physicians, passed into a comatose condition which terminated in death. The two others recovered without medical treatment.

The explanation given out in case A was that there had to be a "toadstool among the mushrooms." It is not improbable that wholesome fungi have been blamed for the faults of bad company. In this instance, however, examination of specimens from the basket out of which the quart had been taken revealed a thorough infestation of larvae. Half the quantity of as wormy mutton might have produced worse effects. The limit of edibility of a fungus is reached by the time its "worminess" shows tunnels that can be detected with the unaided eye. [This statement may be contested by "boletovores," who have been known to eat specimens of *Boletus edulis* that have to be shot before being cooked, just to stop the movement of the mushroom across the kitchen.]

In case B the offender proved to be *Lepiota naucinoides*. The victim assured me that he had admitted no other kind to his basket; indeed, that he had been "very particular to gather only clean, fresh specimens." While I suspect that this beautiful and highly commended toadstool is slightly poisonous, I believe that had the consumer under notice made its acquaintance more gradually he might have brought himself to eat "three platefuls of it" with safety. When we consider that, properly remorseful for scouting his wife's advice, he imagined that he had eaten a potful of deadly toadstools, we cannot wonder that his overloaded stomach made him feel sick enough to die.

C. Both terms of the binomial *Helvella esculenta* suggests eating—wholesale and wholesome. Some European mycophagists have written commendations on this fungus. Berkeley's reference to its edibility is tempered with the caution that it is unsafe for some persons, a circumstance that is more dependent on a peculiarity of the person's constitution than upon any deleterious quality of the plant.

John Dearness, *Mycologia* (1911)

As more and more of our foodstuffs come under the scrutiny of chemists, many new compounds are being identified, some of which have rather undesirable effects, at least under certain specific conditions. For example, the enoki mushroom, *Flammulina velutipes*, contains a red cell-destroying toxin. Its action is similar to that of phallolysin, which reacts with the surface proteins to produce a change in the ion channels within the membrane of the cells, thereby causing the cells to explode by osmotic lysis.[6] Fortunately, humans do not absorb enough—if any—of this toxin to produce clinical effects.

The North American Mycological Association's 10-year collection of adverse reactions to mushrooms has documented a small number of cases attributable to commonly edible mushrooms. Probably most of these are allergic or idiosyncratic reactions, because only one or two instances for each species have been recorded. In a few cases, however, more people were affected, a situation suggesting the presence of a toxin that produces a reaction in at least a portion of those who eat it. Some mushrooms that fall into this category have been known for many years and have been termed "partials," a name implying that only some individuals react negatively to these mushrooms, for example, *Armillaria mellea* complex (honey mushrooms), *Verpa bohemica* (false morel), and the *Morchella* group (morels). Other partials have been added to the list in recent years: *Laetiporus sulphureus* (the sulfur shelf), *Pholiota squarrosa*, *Leucoagaricus naucinus*, and *Coprinus comatus*. Whether these reactions are due to a true chemical toxin to which only a portion of the population is susceptible or whether there is some other basis for the sensitivity remains to be determined.

DANGERS OF RAW FUNGI

Throughout this book, the importance of cooking mushrooms well is emphasized. There is no reason to eat mushrooms raw and a lot of reasons why one should not. Reactions after eating raw mushrooms have been well documented.[7–9] Many heat-labile toxins are present in otherwise perfectly edible mushrooms, including comestibles such as morels (*Morchella* sp.), the honey mushrooms (*Armillaria mellea* complex), and the blewitt (*Lepista nuda*). Certain species tend to be mildly toxic for some individuals even after cooking, for example, the milky caps (*Lactarius*), the chicken-of-the-woods (*Laetiporus sulphureus*), and *Russula* sp. These mushrooms, of course, are more predictably toxic when eaten raw. And in a few, additional problems such as hemolysis may surface. The nature of these thermolabile toxins is not known, although the sesquiterpenes are the suspect in some species.

In 1990 four physicians dining at a San Francisco Italian restaurant developed gastrointestinal distress 30 minutes after eating a dish gastronomically described as "raw lobster mushroom in an oil and vinegar dressing." Not only was the mushroom served raw, but it was even misrepresented, because on closer inspection by a mycologist, it proved to be *Laetiporus sulphureus* (chicken-of-the-woods) and not *Hypomyces lactifluorum* (the lobster mushroom).[10]

The habit of using mushrooms raw in salads should be discouraged. Not only are the store-bought, cultivated *Agaricus bisporus* commonly used raw, but also cookbooks and chefs have promoted the use of uncooked thin slices of the cèpe (*Boletus edulis*), the jelly tongue fungus (*Pseudohydnum gelatinosum*), and the orange peel fungus (*Aleuria aurantica*).

Admittedly, an occasional meal of such delicacies will do no harm at all, but it is better not to encourage exceptions. It is more important to get across the message that all mushrooms should be well cooked. The word *well* should be stressed, because the new, young chefs have developed the habit of applying the flame for the briefest of times, barely warming ingredients to release their volatiles and flavors. Although this technique may be fine for herbs, it can lead to problems when mushrooms are in the pan.

BIOLOGICAL ROLE OF MUSHROOM TOXINS

The biological significance and role of these secondary compounds in the life of mushrooms are virtually unknown. However, it is clear that toxic chemicals are present in many of the large fleshy fungi, reaching clinically significant concentrations in only a small minority of specimens. Amanitins have been detected in the cèpe, the chanterelle, and others, usually in such minute quantities that detection by very sensitive techniques, such as radioimmunoassay, is required. Studies have shown that 1 kg (2.2 pounds) of *Amanita rubescens* (red blusher) contains only 0.1% of the presumed lethal dose for humans.[11] In other words, edible mushrooms in which only trace amounts have been detected are perfectly safe to eat, at least those that contain trace amounts of amanitin.

The taxonomic position of a mushroom is no indication of the toxin it may contain. The amatoxins, for example, are found in various species of four unrelated genera, other representatives of which are perfectly edible. In certain cases, however, most of the species within a given genus contain toxins. The best example of this is the presence of muscarine in the overwhelming majority of *Inocybe* species.

Why toxins are present in certain species is unknown. Only the most anthropocentric person would consider the presence of toxins to be a defensive strategy by the mushroom to avoid human consumption. The role of most of these so-called secondary compounds in the plant and the fungal kingdoms has been grist for the speculation mill. These substances do not appear to be crucial for growth, development, or reproduction. However, they may play important roles in the relationship of the organism to surrounding plants, trees, animals, and insects. Some investigators have suggested that certain compounds have evolved to discourage consumption by herbivores.[12] While plant–herbivore interactions have been well studied, little is known about fungal–herbivore relationships. These compounds almost certainly fulfill important roles in the regulation of metabolism within the organism and may well have evolved useful functions for mediating both positive and negative relationships with other individuals of the same species or with all the adjacent organisms with which it has to interact. In the natural situation, the invertebrates, rodents, and mammals dining on mushrooms have coevolved in a way to ensure that certain species of mushrooms, potentially dangerous to life and livelihood, are seldom eaten. Other animals have adapted to the high levels of toxins in certain species; the classic example of this adaptation is the ability of the gray squirrel to eat *Amanita phalloides*. Of course, the gray squirrel may dine on *A. phalloides* very infrequently, in which case the adaptation may be purely fortuitous. It has been shown in the laboratory, however, that some genetic variants of the fruit fly *Drosophila melanogaster* have specifically adapted to high levels of amanitin and can use this trait to their advantage.

The hypogeous (underground) fungi like the truffle have evolved an amazing array of volatile compounds to attract the animals they rely on for spore dispersal. For some animals, like the flying squirrel and the California red-backed vole, these mushrooms form a major component of their diet. Other wildlife, such as deer, are also decidedly mycophilic.[13,14]

DOMESTIC ANIMALS AND MUSHROOMS

The situation changes rather dramatically for humans and domesticated animals. Many domesticated herbivores—a.k.a. cattle—love fungi just as much as deer, reindeer, and elk do. Poisoning has been recorded in a number of instances.[15–17] Sheep have been poisoned by the toxic species of *Cortinarius*,[18] and many pets have become the victims of poisonous fungi.[19] Scattered throughout this book are references to domesticated animals who have become intoxicated on mushrooms, from the horse who got high (a high horse?) on *Psilocybe*[20] to a herd of cattle who mixed the common inky cap with alcohol (but not without the help of human beings). Dogs appear more vulnerable than cats (a fact that comes as no surprise to cat owners, because the latter obligate carnivores are very particular about their food.)[21] Liver failure has been reported in dogs.[22,23] Even exotic pets, like the miniature Chinese pot-bellied pig, do not always "know" what food is safe.[24] Poison centers around the country are quite accustomed to calls from very anxious pet owners about a possible mushroom ingestion. These calls should generally be handled in the same manner as one for a human ingestion. (Unless the physician is dealing with a particularly wealthy and attached dog owner, liver transplantation oversteps the bounds of reason—unless it is part of a university research program!) Unfortunately, the susceptibility of domesticated animals to mushrooms is even less understood than is that of humans.

MUSHROOM POISONING OF CHILDREN

Some readers may be offended by the order of topics—a section on animal poisoning followed by one on poisoning in children. As we all know, however, the resemblances can be quite remarkable. There are two outstanding similarities between animal and child mushroom poisonings: The mushrooms are usually eaten raw, and the child may be as unable as the animal to give a verbal history of mushroom ingestion. In the toddler age group, it is always wise to assume that anything on the ground might be placed in the child's mouth. Perhaps this assumption should be more forcefully expressed: Sooner or later, everything on the ground will be placed in the child's mouth.

Mushrooms are generally eaten raw by children, so the chance for significant toxicity increases and can be complicated by the presence of heat-labile toxins. A child's metabolic system, although very similar to that of an adult, does undergo continual developmental changes with growth. The pharmacokinetics of a particular toxin is usually slightly different in a child and can be strikingly so in some cases. For example, the length of time a toxin is active in a child may be longer or shorter than in an adult, depending on the particular circumstances of metabolism and excretion. Because many mushroom toxins show a dose–response relationship, the low body weight of a child generally means that the serum concentrations reached by the poisons are relatively higher. Children are also far more vulnerable than adults to the loss of fluids, either through vomiting or diarrhea. They become dehydrated more rapidly and show earlier cardiovascular instability than most adults. Their nervous systems are in the process of maturation, and seizures as well as other central nervous system manifestations are often more prominent in children than in their elders. All these differences conspire to create the need for more careful monitoring of fluid and electrolyte imbalances and an anticipation that the symptoms may be a little more severe than in an adult. To their advantage, children have remarkable powers of recuperation and a relatively short memory for such unpleasant events.

The special problems of mushroom poisoning in children have been reviewed by Lampe.[25]

REFERENCES

1. A. H. Smith, "Poisonous Mushrooms: Their Habitats, Geographic Distribution and Physiological Variation Within Species." In *Mushroom Poisoning: Diagnosis and Treatment*, ed. B. H. Rumack and E. Salzman (West Palm Beach, FL: CRC Press, 1978), pp. 59–65.

2. V. E. J. Tyler, R. G. Benedict, L. R. Brady, and J. E. Robbers, "Occurrence of Amanita Toxins in American Collections of Deadly Amanitas," *Journal of Pharmacological Science* (1966) 55:590–593.

3. R. Catalfomo and C. H. Eugster, "*Amanita muscaria*: Present Understanding of Its Chemistry," *Bulletin of Narcotics* (1970) 22(4):33–41.

4. R. C. Beier, "Natural Pesticides and Bioactive Components in Foods," *Reviews of Environmental Contamination and Toxicology* (1990) 113:47–137.

5. L. Göttl, "Blausäurebildende Basidiomyzeten: Hat Cyanogenese einen taxonomischen Wert?" *Zeitschrift für Pilzkunde* (1976) 48:185–194.

6. A. W. Bernheimer and J. D. Oppenheim, "Some Properties of Flammutoxin from the Edible Mushroom *Flammulina velutipes*," *Toxicon* (1987) 25(11):1145–1152.

7. A. Bresinsky and H. Besl, *A Colour Atlas of Poisonous Fungi* (London: Wolfe Publishing, 1990).

8. E. Pieschel, "Die Rohgiftigkeit einiger Lebensmittel und Pilze," *Mycologisches Mitteilungsblatt* (1964) 8(3):69–77.

9. A. E. Adler, "Pilzvergiftungen in der Schweiz während 40 Jahren," *Schweizerische Zeitschrift für Pilzkunde* (1960) 38:65–73.

10. Anonymous, "Mushroom Mixup at San Francisco Restaurant." In *Mycena News* (Mycological Society of San Francisco, 1990).

11. H. Faulstich and M. Cochet-Meilhac, "Amatoxins in Edible Mushrooms," *FEBS Letters* (1976) 64:73–75.

12. S. Camazine and A. T. Lupo, "Labile Toxic Compounds of the *Lactarii*: The Role of the Lactiferous Hyphae as a Storage Depot for Precursors of Pungent Dialdehydes," *Mycologia* (1984) 76(2):355–358.

13. R. Fogel and J. M. Trappe, "Fungus Consumption (Mycophagy) by Small Animals," *Northwest Science* (1978) 52:1–31.

14. H. A. Miller and L. K. Hall, "Fleshy Fungi Commonly Eaten by Southern Wildlife." In Research Paper SO-49 (Washington, DC: USDA-FS, 1969).

15. R. L. Ridgway, "Mushroom (*Amanita pantherina*) Poisoning," *Journal of the American Veterinary Medical Association* (1978) 172:681–682.

16. P. L. Piercy, G. Hargis, and C. A. Brown, "Mushroom Poisoning in Cattle," *Journal of the American Veterinary Medical Association* (1944) 105:206–208.

17. M. N. Dos Santos, S. S. De Barros, and C. S. Lombardo De Baros, "Cattle Poisoning by the Mushroom *Ramaria flavo-brunenescens*," *Pesquisa Agropecuaria Brasileira, Serie Veterinaria* (1975) 10(8):105–109.

18. J. Överas, M. J. Ulvund, S. Bakkevig, and R. Eiken, "Poisoning in Sheep Induced by the Mushroom *Cortinarius speciocissimus*," *Acta Veterinaria Scandinavica* (1979) 20:148–150.

19. R. S. Hunt and A. Funk, "A Mushroom Fatal to Dogs," *Mycologia* (1977) 69:432–433.

20. J. Jones, "'Magic Mushroom' Poisoning in a Colt," *Veterinary Record* (1990) 127(22):554.

21. C. H. Mullenax and P. B. Mullenax, "Mushroom Poisoning in Cats: Two Possible Cases," *Modern Veterinary Practice* (1962) 43(1):61.

22. A. D. Liggett and R. Weiss, "Liver Necrosis Caused by Mushroom Poisoning in Dogs," *Journal of Veterinary Diagnosis and Investigation* (1989) 1(3):267–269.

23. F. M. Cole, "A Puppy Death and *Amanita phalloides*," *Australian Veterinary Journal* (1993) 70(7):271–272.

24. F. D. Galey, J. J. Rutherford, and K. Wells, "A Case of *Scleroderma citrinum* Poisoning in a Miniature Chinese Pot-bellied Pig," *Veterinary and Human Toxicology* (1990) 32:329–330.

25. K. Lampe, "Mushroom Poisoning in Children Updated," *Paediatrician* (1977) 6:289–299.

CHAPTER NINE

INCIDENCE OF
MUSHROOOM POISONING

MUSHROOM POISONING is not a major public health problem. It is hardly a minor footnote in the pantheon of human ills. The reason for the relative rarity of poisoning, even in societies in which wild mushrooms are regularly gathered and sold, is that only a handful of truly dangerous mushrooms fruit with any frequency, and all of these are normally easily identifiable (see the figure). Of course, to the rare individual unfortunate enough to ingest an amatoxin-containing mushroom, the result is catastrophic. The impact of even a minor poisoning on the media is entirely out of proportion to its significance, perhaps because the report validates the mycophobic attitude of the journalist and the editor, and satisfies their need to perform a public service by warning the population of the potential danger. It is rare for a serious poisoning to go unreported. Conversely, ingestion of wild mushrooms followed by minor symptoms is actually quite common, but this affliction involves a small segment of the population, is usually mild, occurs in a group relatively knowledgeable about mushrooms, and therefore is of little medical consequence.

In other regions and times, mushroom poisoning has been more of a problem. A report from the Vosges area of northeastern France in the late 1800s mentioned 60 cases, and the annual death toll in that part of the country was said to be about 100 souls; in Japan, at the beginning of the twentieth century, 480 deaths occurred in eight years.[1] A booklet by Sartory in 1912 gives the details of 61 episodes of poisoning in France involving 241 victims, 89 of whom perished. The majority of the deaths (51) were attributed to *Amanita phalloides*. In Switzerland during the years 1943 and 1944, 93 poisoning episodes were recorded, involving 356 people.[2] Two fatalities were due to *Amanita phalloides* and two to *Inocybe patouillardi*, a muscarine-containing European species rarely seen

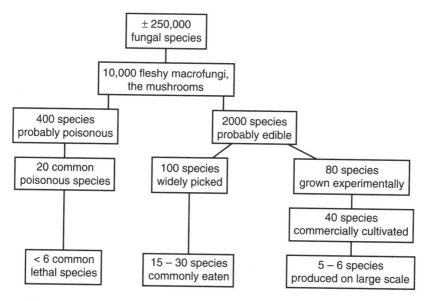

Culinary classification of the fungal kingdom

in the United States. In this series of patients, most of the problems were caused by *Tricholoma pardinum* and *Entoloma lividum*, neither of which cause life-threatening illness, although, admittedly, the gastrointestinal problems provoked by these mushrooms can be quite unpleasant.

During the 30-year period from 1920 to 1950, 39 fatalities were recorded in England and Wales.[3] In the United Kingdom, 32,892 calls were received by the British National Poisons Information Centers in 1977 alone. Of these calls, less that 10% involved either insect bites or plants, only a small fraction of which were mushrooms. For Scotland, there were only 16 inquiries, all in small children, none of whom had symptoms.[4,5] In the following years, these numbers increased quite noticeably, especially in young adults. Most of this increase can be explained by the more frequent use of the hallucinogenic species and better reporting. Even so, the number of persons poisoned in Great Britain is vanishingly small.[6] As expected, the incidence of poisoning in countries like Canada and Australia is similarly low. Up to the 1970s, Canada averaged around 150 reports a year, three-quarters of which were in the province of Ontario and 80% in children under nine years old. Very few required hospitalization.[7]

Probably the best data from Europe comes from the Munich area, collected over 10 years (1975–1985) by the Clinical Center of the Technological University. During that period, 516 patients were acutely poisoned, 105 of whom had amatoxin poisoning. Fourteen patients died of this form of poisoning, a mortality rate for the most severely poisoned group of only 13%. This is an excellent record in terms of the success of management and outcome. Fungi accounted for 2.1% of all cases of poisoning and 2.7% of all the fatalities due to a toxin.[8] The incidence of mushroom poisoning in this area is significantly different from that in the United States, where mushroom poisoning accounts for less than 1% of all poisonings.

A major survey recently published in Japan covered the years from 1959 to 1988.[9] During these three decades, there were 2096 separate episodes, involving 10,924 patients and 72 deaths. The number of cases reported each year has been gradually declining over the years. The species responsible for most of the problems in Japan are very different from those in either Europe or North America, the patterns reflecting the different mycoflora and the cultural preferences for different species of wild mushrooms. For Japan, the three most frequently cited species were *Lampteromyces japonicus*, *Entoloma rhodopodium* (*Rhodophilus rhodopolium* or *sinuatus*), and *Tricholoma ustale*. The first is a pleurotoid fungus growing on logs and is likely to be confused with the oyster mushroom (*Pleurotus ostreatus*).

INCIDENCE OF MUSHROOM POISONING IN THE UNITED STATES

The recorded history of poisoning in the United States is short. At the end of the nineteenth century, Palmer collected 33 cases with 21 deaths from the eastern seaboard around Boston; from Charlestown, 44 cases with 14 fatalities were recorded.[10] Ford, who in 1906 attempted to produce an antitoxin against *Amanita phalloides*, comments on 12 to 15 deaths in one year due to mushroom poisoning.[11] Although the mushrooms were never specifically identified, 22 deaths occurred in the vicinity of New York in a 10-day period in September 1911. The unique vulnerability of the immigrant was noted at that time:

> On the other hand, this country has seen, during the past decade, an great influx of peasants from Italy, Hungary and Bohemia where even the children know the difference between poisonous and harmless mushrooms. As a result, many of the edible species of fungi which grow in the woods are gathered by this foreign-born population either for themselves or for sale in the markets. In consequence of this greater interest in the subject, mushroom poisoning has become somewhat more common in America despite the warnings issued from time to time, both to native Americans who are ignorant of the first principles of mycology, and to our foreign-born citizens who are misled by variations in color and other properties which fungi exhibit in other countries.[1]

For the twentieth century, we have a sound estimate of the number of fatal cases, because these have invariably been reported in the press or to local mycological clubs. Seventy deaths were reasonably attributed to mushrooms up to 1972, an average of a death a year for that time period.[12] In just over half of these, amanitin-containing species were the culprits, with *Gyromitra* accounting for the second-largest group. A sprinkling of other mushroom poisonings, each responsible for a single fatality here or there, are recorded; and in at least 18 cases the species was never identified. Our information about the number of potential exposures and less serious cases is sadly deficient. However, this situation has improved somewhat in the last few years because of the cooperation of the regional poison control centers. Each year, nationwide statistics are published and provide a simplified picture of the problem.

A number of years ago, in an effort to gather more detailed information about all poisonings, the North American Mycological Association (NAMA), a national amateur mycology group, organized a toxicology committee to collect these data. Unfortunately, only a small fraction of the cases reported to poison control centers are also reported to NAMA.[13] In the reports NAMA does receive, the symptoms and the identification of the offending mushroom are accepted as given, without any attempt to validate them. Even so, merely cataloguing these cases has proved to be valuable. It gives a glimpse of the (slightly biased) way in which the cases are accrued and reported (see the table). The most recent data about and an analysis of all the poisonings reported during 1989[14] and 1991[15] have been presented by

Species responsible for poisonings reported to NAMA, 1984–1987

Species	Number of reported poisonings
Unknown	135
Mixed	54
Chlorophyllum molybdites	35
Amanita virosa	22
Amanita pantherina	20
Armillaria mellea complex	17
Amanita phalloides	16
Omphalotus olearius	16
Amanita muscaria	14
Gyromitra esculenta	11
Pholiota squarrosa	9

National Poison Control Center data on mushroom exposures, 1991

Toxin type	Number of exposures (total = 9482)	Percentage of total
Unknown	8572	90
Hallucinogens	242	2.4
GI irritants	225	2.4
Nontoxic mushrooms	189	2.0
Monomethylhydrazine	131	1.4
Cyclopeptides (amatoxin)	47	0.5
Toxic mushrooms	42	0.4
Ibotenic acid/muscimol	16	0.2
Coprine	13	0.1
Muscarine	3	0.03
(Orellanine	2	0.02)*

*Not definitely confirmed; see Chapter 13.

the chairman of the toxicology committee of NAMA, John Trestrail, and are available from that organization.

It also is informative to examine the reports of the many regional poison control centers in the United States. In 1987, for example, the following data were accumulated from around the country:[16]

◆ 7023 cases of mushroom "exposure" (individuals calling their local poison control center about a possible ingestion)
◆ 5807 (of 7023, 82%) involved children under 6 years
◆ 1462 patients received some form of medical care
◆ 12 were considered to have major toxicity and required hospitalization
◆ 0 deaths

In 1991, the total number of exposures was 9482. An analysis of the types of toxins involved shows that most of them were unidentified (see the table). One wonders how many of these reactions were panic reactions or false food allergies.

GROUP OR MASS POISONINGS

Eating is a collective social activity. It is therefore not uncommon for mushroom poisoning to involve a large number of people. Often families or small groups are affected. But sometimes, a sensational poisoning occurs at a banquet, at a large communal gathering, or in a public institution. In September 1918, 31 Westphalian children died within five days after eating a dish made with *Amanita phalloides;* the poisonings occurred at Bierzglin, near Wrzesnie, not far from Poznań in Poland. Many of incidents from eastern Europe, especially prior to the disintegration of the Iron Curtain, went unheralded in the Western press. We heard nothing of the "epidemic" around Zagreb in the autumn of 1988, in which about 140 people were poisoned. The afflicted group included 18 children, 4 of whom died.[17]

The most recent example of a mass poisoning occurred in Russia and the Ukraine in the autumn of 1992. Newspaper headlines around the world announced the poisoning and deaths of hundreds of people.

Facts. In September 1992, news reports from Russia and the Ukraine claimed that at least 400 people had been poisoned by mushrooms, with as many as 100 deaths. The Russian communication media issued repeated warnings about the potential hazards—this to a public for whom wild mushroom gathering is a national passion and pastime. According to the scanty details accompanying these reports, symptoms appeared a number of hours to days after the mushroom meals.

Unknowns.

◆ The types of mushrooms involved and their precise identification
◆ Carefully documented clinical features
◆ Accurate number of people poisoned, number of fatalities, and location of each episode

Speculations. Because one of the areas implicated was relatively near the Chernobyl region, the site of the nuclear power plant disaster in 1986, the immediate speculation in the press was "radioactive fungi kills hundreds." When it was pointed out that no form of radioactivity kills that rapidly, except for a thermonuclear explosion, the story was switched to "previously safe mushrooms undergo mutation." Few journalists or the public could resist the temptation to blame the previous environmental tragedy for the problem. When it was next pointed out that many of the incidents had occurred in areas with no known evidence of radioisotope contamination (always a problematic statement in the lands of the superpowers), the story gradually faded into obscurity.

To this day, we do not know the true extent of the problem or its cause. Numerous efforts to get information from people in these areas have proved fruitless. Although the reports are all compatible with amanitin poisoning, the precise identification of the mushrooms involved is not known. It is interesting to hear the comments the Russians themselves made as this episode was unfolding. These remarks were reported to me by a passionate amateur mycologist, Dennis Bowman, who has led mushroom collecting trips to the old

Soviet Union and was in St. Petersburg during the "mushroom crisis." The reactions of the Russians were, "This is nothing new, it happens every year." "We just hear about it now because we have a more open press. As in the West, they like the sensationalism." "It's due to mutant mushrooms because of the radiation or the pollution." "It's to scare others away, so that there will be more for the people in the media." "Many more folks are picking this year because of the price of food and widespread hunger." The reactions ranged from cynicism to anger and bitterness. But what was the true story?

Many other episodes of mass poisonings have occurred in the past. These have not always been well reported or well documented, but we know of their occurrence from news reports. A reasonable possibility is that they are due to nothing more than simple mushroom misidentification. Mushroom identification is difficult because of the ephemeral nature of the fruiting bodies and the relatively unpredictable fruiting patterns of certain species. Many species fruit every year, regardless of the weather conditions. The only thing that varies is their relative abundance. In this respect, there are good years and bad years, depending mainly on temperature and moisture. Other species are much more contrary. For reasons not well understood, the fungus may remain happily in its vegetative state for long periods, seemingly unwilling to expend the energy required for sexual reproduction. Then, when the conditions are just right, it explodes in a riotous display of sexuality. *Amanita phalloides* did this in California in the early 1980s, causing a number of deaths as a result of misidentification with the expected *Amanita calyptroderma*, an edible and highly desirable species among those people who comfortably experiment with the genus *Amanita*.

The problem of unpredictable fruiting patterns is compounded by the practice of passing foraging lore from one generation to the next. If during one's formative years the extremely poisonous varieties are not common, then the knowledge about their characteristics may never be passed on. Folklore is widely believed (or hoped) to be accurate and reliable. A taxonomy of sorts develops for the edible species, but, unfortunately, knowledge of the inedible or the poisonous taxa is usually woefully inadequate. In fact, the inedible mushrooms frequently are lumped into two big groups: "poisonous" and "we don't eat those."

One could predict that within a closed and stable community, episodes of poisoning will occur every 15 to 25 years—in other words, once in each generation. The data about mass mushroom poisonings are not sufficient to prove or disprove this hypothesis, but it seems more reasonable than an explanation like "radioactive" mushrooms. Of course, the possibility of a mutation cannot be completely excluded.

Because many of the edible mushrooms already have the capacity to produce a number of toxins and do so in minute concentrations, a genetic alteration in the production or the metabolism could allow a toxin to accumulate. But such an unlikely event would be a very local phenomenon, limited to one specimen at one site. The odds of it happening across a wide region is so remote as to be readily discounted. Other mycologists investigating the Russian epidemic have similarly concluded that misidentification is still the most credible explanation.[18] Perhaps folk wisdom is not as wise as some would like to believe!

MUSHROOM POISONING AMONG AMATEUR MYCOLOGISTS

One of the myths, perpetuated by the media, is that mushroom identification is so difficult and so dangerous that even the experts make fatal mistakes. There is no question that professional mycologists have differences of opinion—merely attending a mycology conference provides convincing evidence that little unanimity exists about some of the classifications and nomenclature—but the number of mycologists who have been poisoned approaches zero.

What about amateur mycologists—the weekend "myco-hackers"—who plunge through the woods and meadows in search of their favorite mushrooms or a new gastronomic experience? This group is difficult to define, because they come from widely divergent backgrounds with variable amounts of biological knowledge and a mycological education based on anything from some self-study to a course or two at a community college or lectures by the local mycological society or mushroom club. In other words, I shall define this group as anyone who has attempted to get some modern scientific knowledge and whose identification, picking, and eating is not based purely on folklore. It is a heterogeneous group, sometimes disparagingly termed "pot-hunters."

Surveys of poisonings among amateur mycologists are available, one by the toxicology committee of the Mycological Society of San Francisco[19] and one performed in 1992 with the members of the Puget Sound Mycological Society.[20] The results of the two surveys were remarkably similar, although some differences in the relative popularity of the edible species were noted. The discrepancies probably reflect differences in availability, because the two habitats are not identical. The average length of time that people had been collecting and eating wild mushrooms—in other words, their exposure risk—as well as their experience, was about 12 years. These clubs are both large clubs; Puget

Sound Mycological Society sometimes has as many as 600 members. No mushroom-related deaths have been recorded in the history of either organization, or, as far as I am aware, in any mushroom society in the history of the United States. But in both groups, between 35 and 40% of all respondents reported some adverse effects on at least one occasion with one or more species. The overwhelming majority of the reactions consisted of mild gastrointestinal symptoms. In many of these, the victim was not certain of the role or the identification of the mushroom. Very few serious incidents were noted; in most of the cases, recovery was rapid and complete within a few hours.

The spectra of mushrooms causing the reactions were similar for both groups and are discussed more fully in the chapters on the gastrointestinal toxins (Chapter 19) and the miscellaneous group of toxins (Chapter 20). It is likely that many of the problems reported by the respondents were the result of individual "allergic" or idiosyncratic reactions to particular mushrooms.

REFERENCES

1. W. H. Ford and E. D. Clark, "A Consideration of the Properties of Poisonous Fungi," *Mycologia* (1914) 6:167–191.

2. C. M. Christensen, *Molds, Mushrooms, and Mycotoxins* (Minneapolis: University of Minnesota Press, 1975).

3. J. Ramsbottom, *Mushrooms and Toadstools* (London: Collins, 1953).

4. P. A. J.-L. Margot, "Identification Programme for Poisonous and Hallucinogenic Mushrooms of Interest to Forensic Science," Doctoral thesis, University of Strathclyde, Glasgow, Scotland, 1980.

5. G. M. Ola'h, "Le genre *Panaeolus*," *Revue de Mycologie* (1969) Mémoire Hors série 10, Paris.

6. J. N. Edwards and J. A. Henry, "Medical Problems of Mushroom Ingestion," *Mycologist* (1989) 3(1):13–15.

7. J. Lough and D. G. Kinnear, "Mushroom Poisoning in Canada: Report of a Fatal Case," *Canadian Medical Association Journal* (1970) 102(8):858–860.

8. A. Bresinsky and H. Besl, *A Colour Atlas of Poisonous Fungi* (London: Wolfe Publishing, 1990).

9. Y. Ishihara and Y. Yamaura, "Descriptive Epidemiology of Mushroom Poisoning in Japan," *Nippon Eiseigaku Zasshi* (1992) 46(6):1071–1078.

10. Forster, "Mushrooms and Mushroom Poisoning," *Boston Medicine and Surgery Journal* (1890) 123:267–272.

11. W. H. Ford, "The Toxins and Anti-toxins of Poisonous Mushrooms," *Journal of Infectious Diseases* (1906) 3(April):191–224.

12. R. W. Buck, "Acute Encephalopathy in Children after Eating Wild Mushrooms." In *Mushroom Poisoning: Diagnosis and Treatment*, ed. B. Rumack and E. Salzman (West Palm Beach, FL: CRC Press, 1978). pp. 191–197.

13. J. H. Trestrail and K. F. Lampe, "Mushroom Toxicology Resources Utilized by Certified Regional Poison Control Centers in the United States," *Clinical Toxicology* (1990) 28(2):169–176.

14. J. H. Trestrail, "Mushroom Poisoning in the United States—An Analysis of 1989 United States Poison Center Data," *Journal of Toxicology and Clinical Toxicology* (1991) 29(4):459–465.

15. J. H. Trestrail, "Mushroom Poisoning Case Registry: North American Mycological Association Report—1991," *McIlvainea* (1992) 10(2):51–55.

16. T. L. Litovitz, B. F. Schutz, N. Matyunas, and T. G. Martin, "1987 Annual Report of the American Association of Poison Control Centers National Data Collection System," *American Journal of Emergency Medicine* (1988) 6:479–515.

17. M. Kacíc, M. Dujsin, S. Puretíc, and J. Slavicek, [Mycetismus in children—report of an epidemic of poisoning], *Lijec Vjesn* (1990) 112(11–12):369–373.

18. G. Fourré, "Hecatombes de mycophages dans l'ex-URSS: Les hypothèses plausibles . . . et les autres," *Documents Mycologiques* (1993) 23(89):57–60.

19. P. Vergeer, D. Orr, and P. Duffy, Report of Questionnaire Survey, Toxicology Committee, Mycological Society of San Francisco, Inc., May 1975 (unpublished).

20. D. Benjamin, Survey of Mushroom Poisoning in Members of the Puget Sound Mycological Society, 1992 (personal observations, unpublished).

CHAPTER TEN

SOCIOLOGY OF MUSHROOM POISONING

"**E**XPERIENCED MUSHROOM HUNTER dies of poisoning," the newspaper headlines declare. "Mushroom expert makes fatal mistake," blares another. The press bears full responsibility for perpetuating this mistaken view of many cases of mushroom poisoning. It is in fact almost never the case that an experienced mycophagist dies from mushroom poisoning, and these headlines merely enhance the mythology surrounding the dreaded wild mushroom. Even the most superficial investigation of these stories usually reveals that the victim was far from experienced. To date I am unaware of a single fatal poisoning in any member of an amateur mycological society, although, as has been discussed, a sizable proportion have had minor adverse encounters with mushrooms. These are seldom of much consequence and certainly do not merit newspaper headlines. One professional mycologist is known to have perished many years ago. In 1944, a renowned German mycologist, Dr. Schaeffer, died after eating a dish of *Paxillus involutus*. He generally did not eat mushrooms, because he suffered from a variety of allergic or idiosyncratic reactions to them. In this instance, he developed vomiting and diarrhea, which compromised an underlying kidney condition. His problem was not one of misidentification but rather a most unfortunate individual reaction to a mushroom, a reaction that exacerbated a preexisting disease.

If neither the amateur nor the professional mycologists are the ones contributing to the news, then who is making the headlines? The accompanying box lists the most common groups involved. In the United States, a fairly standard and predictable pattern of wild mushroom poisoning has developed. Although the details may differ in other countries, the underlying principles appear to be the same. In essence, the nature of the victims can be quite simply stated—vulnerable or

Sociology of Mushroom Poisoning

- Children
  - toddlers: grazing
  - young girls: playing house
  - young boys: dares
  - teenagers: experimentation
- Chemically dependent individuals
  - seeking hallucinogenic experience
- Immigrant peoples
  - using wrong field guide
  - hunting in an unfamiliar habitat
  - picking an unfamiliar species
  - misidentifying nonnative mushrooms
  - confused by unusual fruiting season
- Elderly individuals
  - misidentification
- Mycologically naive individuals
  - relying on folklore knowledge only
  - foraging before learning basic identification skills
- Psychologically disturbed individuals
  - attempting suicide

reckless. The vulnerable are the young, the elderly, and immigrants; the reckless are those whose knowledge is based purely on folk wisdom and those seeking chemical recreation. I do not wish to disparage folk knowledge, which has great value. Nevertheless, it has equally great limitations.

CHILDREN

By far the most frequent calls to any poison control center concern toddlers who have been observed chewing, swallowing, or spitting up fragments of mushroom. This "grazing" behavior, in which everything on the ground is transferred to the mouth, is standard operating procedure for any normal child between the ages of two and four years. Over 80% of the calls involve toddlers.

Outside the toddler group, accidental poisoning is quite rare, perhaps because the intense fear of mushrooms harbored by most parents is readily transferred to the children. In the English-speaking countries, the coining of the word *toadstool* did as much to prevent mushroom poisoning as the introduction of childproof packaging did for drug intoxication. Unfortunately, it also promulgated a most unhealthy and negative attitude toward the fungal kingdom. Nevertheless, mushroom poisoning has occurred in older children in a number of unusual situations. One such scenario involved a group of young girls. In the process of playing "house," the "hostess" served her friends tea and mushrooms, the latter of unsavory reputation. Another common scenario involves young boys, who, during the competition of puberty, are known to dare one another to do a variety of dangerous things. Sampling potentially lethal mushrooms qualifies as such an adventure.

CHEMICALLY DEPENDENT INDIVIDUALS

The number of individuals labeled "chemically dependent" reached its zenith during the 1960s and 1970s, when hallucinogenic mushrooms played a significant role in the counterculture of the time. Since then, the "magic" mushroom has suffered serious decline in favor of other, more available drugs. The fad is likely to return, however, and a hard-core group of devotees has kept the tradition alive. They can be seen each fall, crawling on the well-manicured and manured lawns of parks and college campuses. Pastures also are a favorite haunt. (Indictments for trespassing are often easier to uphold than those for possession of a controlled substance.)

In the teenage years, the desire to experiment with illicit chemical compounds appears. *Psilocybe* use may occur at this time. In 1992, a 16-year-old female from Tacoma, Washington, was handed a bag of dried mushrooms by a male friend and told to eat three or four caps for an interesting experience. The young woman, firmly believing that more is better, ate approximately 50 mushrooms. As promised, she did indeed have a mind-altering experience. About 12 hours later, she developed abdominal pains and vomiting and by 36 hours was mildly jaundiced, with markedly elevated levels of liver enzymes. A number of caps of the amatoxin-containing *Galerina venenata* were discovered in the collection. This species is not uncommon on lawns in the Pacific Northwest and has been reported on a number of occasions to have been confused with a *Psilocybe* species. Fortunately, the thrill-seeking teenager recovered from her brush with mortality.

As illustrated by the preceding story, the biggest danger for the *Psilocybe* hunter is the inadvertent collection of toxic species that also grow in the targeted habitat. The majority of the *Psilocybe* species fall into that disparaged category of mushrooms called LBMs (little brown mushrooms). Despite the bluing reaction on the stem of the mushroom, which develops with handling or bruising, many of them are easily confused with a host of other similar mushrooms of dubious character and identification. *Galerina* species pose the greatest threat, for although they grow primarily on wood rather than grass, fragments of buried and rotting wood are very common in lawns. One ex-college student I heard about has been poisoned twice by *Galerina* while picking mushrooms on a university campus. (He was offered a course in mushroom identification for his own safety but declined. He has not been heard from since 1992.)

Some of the *Psilocybe* species, being saprophytic, are easily cultivated, so an underground of growers exists. These mushrooms are grown largely for home consumption but occasionally reach the streets. Fortunately, in the 1990s the market for these hallucinogens is not large.

Other lawn-inhabiting species can also be picked by mistake; members of the muscarine-containing *Inocybe* group are a particular hazard. *Conocybe* is another potential threat, especially where wood chips are common.

Perhaps the most interesting variation on the use of mushrooms for chemical recreation is illustrated by the case of a young man who presented himself at the emergency rooms of small rural hospitals in both Oregon and Washington, claiming to be suffering from mushroom poisoning. He complained of severe abdominal pain, for which he highly recommended a narcotic as the treatment of choice. He was convincing enough to persuade a number of physicians to inadvertently minister to his drug habit. The ruse was finally uncovered when the regional poison control center received three identical calls in a single week.

IMMIGRANTS

Mr. L. of upstate New York had eaten mushrooms all his life. When some small *Lepiotas* appeared under the pines in his front yard, he gathered them and prepared them for lunch. He identified them as *Lepiota excoriata*, a species with which he was familiar years ago in his native Finland. The mushrooms were not *Lepiota excoriata* (but probably *L. josserandi*, an amatoxin-containing species) and five days later he died from progressive liver damage.[1]

After toddlers, immigrants are the most vulnerable group. In many respects, the history of mushroom poisoning in both the United States and the United Kingdom reflects the waves of immigrants who have come ashore over the years. Most of these groups are from countries where wild mushroom eating is an accepted part of the culinary tradition. Unfortunately, their knowledge about mushrooms is based only on folklore and tradition. It is handed down from one generation to the next by simple demonstration in the field and forest. They are unprepared for the different flora encountered in a new land, and many suffer the consequences. From the beginning of the twentieth century to the end of the Second World War, most cases of inadvertent poisoning involved eastern Europeans or Russians. Since the 1970s, the majority of poisonings, especially in the United States, have occurred in immigrants from Southeast Asia.[2] The following newspaper headlines illustrate the point:

16 Laotians Suffer from Mushroom Poisoning (*San Francisco Chronicle*)

Japanese-American Family Pick Wrong Mushroom (*Seattle Times*)

New species can be encountered in many ways. A subtle change in flora sometimes occurs, which may produce a huge abundance of a particular species. This situation occurred in California in the 1980s, when mass fruitings of *Amanita phalloides* led to confusion between it and the usually abundant, edible *Amanita calyptroderma*. The latter mushroom is avidly hunted by Italian immigrants and their offspring, perhaps because of its slight resemblance to *Amanita caesarea*.

The most desperate immigrants are illegal aliens. In 1985, four men came across the border from Mexico and lived in a shack in San Diego County. After their food and money ran out, they ate some mushrooms, presumably *Amanita phalloides*. The symptoms began between 24 and 48 hours later. They had each eaten from one to six fried mushrooms. Three died two days after admission to the hospital; the fourth perished on the tenth day following the lethal meal.[3]

So far, the focus has been on immigrants from abroad. But everyone becomes an immigrant in a mycological sense when he or she moves from one part of the country to another. Although many species are quite widespread, each habitat has its own particular flora. Hunting in an unfamiliar area poses a significant danger unless a local field guide is used or someone with local knowledge is consulted. As mentioned earlier, a few mushrooms, such as those in the *Gyromitra* genus, are overtly poisonous in one region of the country and not in others.

Even experienced collectors may be naive under certain circumstances. New species of mushrooms are constantly being imported into new habitats, most frequently as mychorrizae on the roots of trees. The widespread planting of imported trees in place of the native varieties also has the potential of altering the flora with which the general public is familiar. The appearance of introduced species (immigrant mushrooms!) renders folk wisdom less than wise and has been responsible for a number of poisonings. The most dramatic example of this phenomenon has been the introduction of *Amanita phalloides* at many places across the United States. Whether this species is spreading by itself or continually being introduced on the roots of foreign trees is still open to investigation and debate. This problem is discussed in detail in Chapter 12.

The world's insatiable desire for wood products continues to alter the woodland and forest habitats and their mycoflora. Foreign trees have been introduced, bringing with them mushrooms that are unexpected in these new habitats. Biodiversity has been altered in some regions. A recent import to Scotland is *Chlorophyllum molybdites*, which has fruited in Edinburgh, a place where this species was almost unknown just a few years ago. Investigation fingered a tree imported from Florida as the vector for this new species.[4] Because *C. molybdites* is not a mycorrhizal fungus, one assumes that the mycelium was present in the soil in which the tree was transported. Also in Scotland, the use of Sitka spruce from North America is significantly changing the spectrum of mushrooms. As these "alien" forests mature, certain toxic species, such as *Cortinarius speciosissimus*, are becoming much more frequent. Fruitings in unexpected places and at unexpected times are recognized phenomena.[5]

Records of mushrooms gathered at forays are valuable sources of information regarding the distribution of particular species. A few scientists have suggested that changes in the world's temperature and climate patterns can be detected by the spread of usually tropical species into more temperate areas. However, this intriguing suggestion—that mushrooms (at least, selected species of mushrooms) can be used as indicator species for detecting long-term weather changes such as global warming—has to be regarded as speculative at best. While it is true that fungi can rapidly adapt to any new ecological niche made available to them, many factors determine the distribution of a particular fungus, not the least of which are the human-made changes in the habitat. This relationship has been beautifully documented in Holland, where profound changes in the mycoflora have occurred. The effect of changing climatic conditions on the distribution and fruiting of the macrofungi deserves and requires much more detailed study.

The use of the wrong field guide is a subtle variation on the theme of hunting in an unfamiliar habitat. An Englishman and his Chinese girlfriend were poisoned in California after they picked a mushroom they thought was "the prince" (*Agaricus augustus*). They based their identification on a picture in a British publication. Unfortunately, the illustration was less than optimal, and the couple ate a meal of *Amanita phalloides*. In this particular case, the fault may have been due to the illustration, but, in general, local field guides are preferable to more generic, countrywide ones, which of necessity only illustrate a small fraction of the mycoflora and do not give the readers any idea of the prevalence of a particular mushroom in the habitat in which they are hunting. Appendix 2 contains a list of regional field guides most useful in the United States and Canada.

ELDERLY

The cases of mushroom poisoning in the elderly that I know about were generally due to simple misidentification. Declining vision, forgetfulness about a particular mushroom feature, and the loss of observational skills in general—all conditions that may accompany aging—contribute to the problem. I recall an 80-year-old woman who picked a dozen *Amanita muscaria* buttons in the mistaken belief that they were puffballs. *Amanita pantherina* specimens were confused with shaggy manes in another instance.

Many elderly individuals, however, safely hunt and pick mushrooms well into their later years. Indeed, some of the most experienced and astute members of the amateur mycological fraternity are the most senior ones. Mushroom hunting is a passion and pastime that can be pursued throughout one's lifetime.

MYCOLOGICALLY NAIVE

A large number of potential poison victims can be conveniently grouped under the category of the mycologically naive. This group includes all those individuals whose knowledge is based on common folklore. Although knowledge learned from one's parents or peer group is extremely valuable, it is beneficial only when used in the context and habitat in which it was garnered. Within a stable population and an unchanged environment, this knowledge is both accurate and reliable. However, greater movement of people around the globe has rendered this folklore meaningless and even dangerous, as illustrated by the poisoning accidents among migrant populations.

Moreover, a change in fruiting patterns and the appearance of unexpected mushrooms triggered by shifts in the weather or habitat may tax local knowledge to its limits and beyond. Perhaps the most vulnerable are the back-to-nature enthusiasts of the last few decades. Firmly believing in the good providence of Mother Nature, these harvesters learned of a few edible wild mushrooms from colleagues and associates, without much instruction about the hazards.

Another group of victims fitting conveniently under the rubric of mycologically naive are the neophytes, those individuals recently introduced to the field of mycophagy. It takes time, patience, and some facility to learn the characteristics of the common edible and poisonous mushrooms. Unfortunately, these attributes are not always found together in the same individual. In the last few years, the North American Mycological Association has attempted to keep track of the mistakes individuals have made in identification. (Obviously, many more mistakes are made than those reported, but because most collected mushrooms are not eaten, they do not lead to either embarrassing or fatal consequences.) Chapter 6, a catechism for safe foraging, was specifically included for the neophyte. By following the guidelines given there—and by exercising a modicum of common sense—we should be able to eat wild mushrooms without jeopardizing our health.

PSYCHOLOGICALLY DISTURBED

Of all the available methods for killing oneself, mushroom poisoning is perhaps the most unreliable, the slowest, and the most miserable. It has almost nothing to recommend it. It is little wonder then that only a few cases are documented in which poisonous mushrooms were intentionally eaten in an attempt to commit suicide. The usual choice has been *Amanita phalloides*, but at least one patient admitted to eating *Cortinarius orellanus*.[6,7] *Amanita muscaria* was selected by another woman, with decidedly disappointing results.[8] In a few circumstances, mushrooms have merely been one item in a potpourri of potential poisons ingested by the desperate. The reason so few mushrooms have been used for self-dispatch, despite being widely believed to be rapidly lethal, is that few people have the slightest notion of how to identify a really toxic species. Moreover, most lethal mushrooms are uncommon and ephemeral. Seldom are they likely to be around when wanted. A most inconvenient poison indeed!

REFERENCES

1. J. H. Haines, "A Tragic Case of *Lepiota* Poisoning," *McIlvainea* (1984) 6(2):21–23.

2. Anonymous, "Mushroom Poisoning Among Laotian Refugees—1981," *Morbidity and Mortality Weekly Report* (CDC) (1982) 31(21):287–288.

3. J. L. McClain, D. W. Hause, and M. A. Clark, "*Amanita phalloides* Poisoning: A Cluster of Four Fatalities," *Journal of Forensic Science* (1989) 34(1):83–87.

4. R. Watling, "A Striking Addition to the British Mycoflora," *The Mycologist* (1991) 5(1):23.

5. E. Hazani, U. Taitelman, and S. M. Shasha, "*Amanita verna* Poisoning in Israel— Report of a Rare Case Out of Time and Place," *Archives of Toxicology, Supplement* (1983) 6:186–189.

6. N. Delpech, S. Rapior, P. Donnadieu, A. P. Cozette, J. P. Ortiz, and G. Huchard, [Voluntary poisoning by *Cortinarius orellanus*: Usefulness of an original early treatment after determination of orellanine in the biological fluids and tissues] (French), *Néphrologie* (1991) 12(2):63–66.

7. N. Delpech et al., [Outcome of acute renal failure caused by voluntary ingestion of *Cortinarius orellanus*] (French), *Presse Médicale* (1990) 19(3):122–124.

8. G. Donalies and G. Völz, "Ein Selbstmordversuch mit Fliegenpilz," *Nervenarzt* (1960) 31:182.

CHAPTER ELEVEN

DIAGNOSIS AND MANAGEMENT OF MUSHROOM POISONING

There is probably no area of clinical toxicology more steeped in folklore, misunderstood, or mismanaged than plant and mushroom poisoning.[1]

K. Lampe (1974)

SPRING 1991, rural King County, about 50 miles southeast of Seattle. Adam, a five-year-old, and a couple of his buddies are playing out on the lawn. He samples a mushroom growing under a tall Douglas fir. His friends decline the offer. Forty-five minutes later, he vomits and staggers toward the house. He trips and falls down three rather small steps. He tells his parents what he has done and proceeds to fall into a deep sleep, from which he cannot be aroused. This state is conventionally termed coma by the medical world. Someone dials 911, and within minutes a group of paramedics arrives. Because of the story of the fall, he is placed on a board with a neck collar and rushed, lights flashing, sirens screaming, to a local emergency department. At the hospital, Adam is found to be breathing normally, to be nice and pink, but impossible to arouse. He is taken off to radiology, where he has X-rays of his neck and chest as well as a CT scan of his head. Back in the emergency room, he has a generalized seizure for which he is immediately given two types of intravenous medication. Fifteen blood tests later, he is still comatose, even more so now because of the medication. A tube is inserted into his stomach, which is washed out with a few liters of water, and some charcoal is put in to hopefully adsorb any remaining toxin. But this is a mushroom ingestion! It's time to send him to the major medical center. Promptly, too.

The airlift helicopter is called in, and because Adam is going to be transported, a tube has to be inserted into his airway. Twenty minutes later, he is in the intensive care unit on a ventilator. He shows no evidence of respiratory depression. He has blood tests, intravenous drips, monitors, a tube in his bladder, and all those other therapies that are part of the ritual of an intensive care unit. An amateur mycologist is called in to identify the mushroom. *Amanita pantherina*. There is no antidote. The only treatment is time.

About six hours later, Adam awakes, a trifle worse for wear, but with a smile on his face. No one really knows what a five-year-old child hallucinates about, but thank goodness he did and probably won't recall what was done to him. A few hours later, he is back on the front lawn. The only differences are that a lawn mower has flattened the beautiful ring of mushrooms under the fir tree and the parents have to figure out how they are going to pay the bill, a mere $6000. That is the price of fear and the cost of not knowing what to expect.

Denis R. Benjamin, personal journal (1991)

This account illustrates the management of a serious mushroom poisoning in the 1990s. Throughout the course of the treatment, a great sense of urgency, sometimes bordering on panic, drove the health care providers. And once the therapeutic ball got rolling, it was very difficult to stop it. Only after the child reached the intensive care unit did anyone try to identify the possible toxin involved. As this case illustrates, the tendency is to treat first and ask questions later, because, in the average emergency department, neither a mycologist nor the expertise needed for toxin identification is available. A sample of the offending mushroom seldom accompanies the patient, but when it does, it is frequently fragmented, decomposed, cooked, or partially digested.

Much of the literature and advice on the subject of mushroom poisoning has been written by mycologists or toxicologists, for whom identification of the mushroom or the biochemistry of the toxin is of paramount importance, and yet the treating physician needs to manage the situation with what he or she has at hand, most commonly an ill patient. Nevertheless, armed only with the knowledge of the basic patterns of poisoning (presented on pages 178–183), the attending physician can reasonably and effectively manage the vast majority of cases, without knowing much mycology at all. It is certainly comforting to have the assurance that one is treating appropriately by having a confident identification of the presumed mushroom, but this should never be the overriding concern. This was perhaps best expressed by Lincoff, who said, **"Treat the patient, not the mushroom!"**

A second fundamental guideline is to avoid overtreatment. Many therapies, if vigorously applied, can cause more problems than the mushrooms themselves do.

Third, treat only in response to symptoms. The reflexive administration of drugs like atropine should be avoided.

I do not mean to disparage or discourage attempts to obtain a satisfactory identification of the ingested specimen. Indeed, I encourage physicians to do so. Collecting this data is the only way to increase our knowledge and understanding of this group of poisons. Part of our present state of ignorance stems from the fact that in many cases of mushroom poisoning a satisfactory identification was never made. A variety of resources can be utilized to identify specimens, such as the local universities, community colleges, mushroom societies (Appendix 3) or amateur mycologists. When available, mushroom specimens should be treated with care. They can be dried and held for later identification. Fresh mushrooms can be refrigerated in paper towels or paper bags for up to 24 hours. Fresh specimens should not be kept for any length of time in plastic bags, or enclosed specimen cups, especially at room temperature, as they rapidly decompose into unidentifiable, often foul smelling, slimy messes. Details for handling and identifying specimens are given at the end of the chapter.

TREATMENT OF COMMON CLINICAL PRESENTATIONS

Medical personnel see three common clinical scenarios involving mushroom poisoning:

♦ A toddler who has been discovered with a piece of mushroom in his or her mouth but is asymptomatic
♦ A symptomatic patient known to have ingested mushrooms
♦ An unconscious patient, for whom no history is available

Without some index of suspicion, cases of mushroom poisoning can go undetected, as probably occurred with orellanine poisoning for many years. The patient may not connect the symptoms to a mushroom meal, either because of a long latent period or because the patient has been eating that particular mushroom for many years and has no reason to doubt its edibility. Unexpected renal or hepatic failure are the two clinical syndromes in which mushroom poisoning should be considered if their cause is unknown. A third, very rare possibility in the United States is the immune hemolytic anemia due to *Paxillus involutus*. This mushroom is almost exclusively eaten by the older generation of eastern European immigrants. Should one encounter a patient of this particular heritage with an undiagnosed hemolytic anemia, questions relating to mushroom ingestion are appropriate.

The following approach to management has been adapted from the work of Lincoff, Mitchel, Rumack, Lampe, and McCann.[2-4] It has been subtly modified and extended by a few others including myself, as our experiences with mushroom poisoning have grown. This general approach is based on the clinical signs and symptoms of mushroom poisoning and provides a general approach to the initial treatment of almost any case in which mushrooms are suspected.

When samples of the ingested mushroom are not available, or not immediately identifiable, the clinical situation is then similar to almost every other case of poisoning in the emergency department in which the causative agent is unknown. However, with a history of a mushroom ingestion, the physician should be able to successfully manage and treat the vast majority of cases on the basis of the presenting symptoms. The widely available POISINDEX[5] has become the standard toxicological reference for many poison control centers.[6] This publication provides an excellent source of current information and recommendations for therapy. The combination of the POISINDEX and this book should meet the needs of physicians treating mushroom intoxication over the next decade. However, the fungal identification system in the POISINDEX has yet to be refined enough to replace a competent mycologist,[7] thereby ensuring some element of job security for this already rare breed.

Asymptomatic Child Suspected of Eating a Mushroom

The clinical situation involving a toddler found with a mushroom fragment in his mouth is common. It accounts for the overwhelming majority of calls for help. In the United States, regional poison centers now successfully manage almost all cases of intoxication and consequently receive the bulk of calls concerning mushroom poisoning.

A potential poisoning in a pediatric patient differs in one main respect from poisoning in adults—the mushrooms are usually eaten raw. Although this is of marginal significance when the mushroom ingested is one of the most poisonous species in which the important toxins are resistant to heat, it does make some difference for a number of other mushrooms that contain gastrointestinal irritants, some of which are heat labile and therefore inactivated by cooking. The child chewing on raw mushrooms is at greater risk for an adverse reaction than an adult who has ingested a cooked specimen.

The time elapsed between the ingestion of the mushroom and contact with medical personnel determines the initial clinical approach. The care giver should call the nearest poison center. If the ingestion has occurred within two hours of the contact and if the parents or care givers feel that the child has actually swallowed more than a small amount (i.e., anything bigger than a dime-sized portion of a specimen), syrup of ipecac and oral fluids should be administered. The recommended dose for infants (6–12 months) is 5–10 mL with 15 mL of clear fluids per kilogram body weight. Children (12 months to 12 years) should receive 15 mL with 240 mL of clear fluid. For individuals over the age of 12 years, up to 30 mL of syrup of ipecac can be administered with 240–450 mL of fluid. In addition, vomiting should be induced in any patient in which the quantity ingested is in doubt. The child should be observed at home for the next 24 hours. There is no indication for hospitalization. Almost all pediatric cases can be managed at home with the support of the poison center. Should any symptoms develop, the child must be reassessed for possible intoxication.

If ingestion occurred more than two hours before the medical contact, one of two actions can be taken: Either observe at home for the development of any symptoms, or induce emesis with syrup of ipecac and then observe. Many poison centers adopt the first option, because there is little evidence for the efficacy of induced vomiting when more than two hours has passed since ingestion. The common mushrooms generally produce symptoms within four hours.

This approach—emptying the stomach if the ingestion is very recent and then watchful expectancy—is based on the assumption that the likelihood of a child eating a significant dose of a potentially lethal amatoxin-containing mushroom in the gardens or parks of the United States is exceedingly small. The two other types of serious poisoning—those due to monomethylhydrazine (gyromitrin syndrome) and orellanine—are virtually unknown within urban areas. If a family had been camping in the woods, the spectrum of possible poisonings increases. Nevertheless, the initial therapy in all cases is to remove the toxin from the stomach.

It is reasonable to expect a better outcome for any poisoning if absorption of the toxin is limited, if the toxin can be rapidly eliminated from the bloodstream by inducing more rapid excretion, and if it can be prevented from binding to the cells. However, no controlled trials have been done to show an improved outcome for amanitin intoxication if therapy, other than emesis, was instituted during the presymptomatic phase.

In many situations, the family does not bring a sample of the mushroom with them, thereby avoiding the difficulty of identification. By playing the odds, which are astronomically on the side of it having been a nonfatal species, the chance of a misadventure is very remote. It is reminiscent of the statement, "Don't ask the question if you don't want the answer!"

If the parents do have a mushroom with them, the problem is compounded. One might take a casual glance, declare it to be no problem at all, and then have a sleepless night worrying about the remote consequences. However, most medical attendants feel obligated to make some attempt at identification. If one has a resident mycologist on site, the question can be readily resolved. The key is not so much to identify the mushroom but to exclude the possibility of an amatoxin containing species. Given a decent specimen, this identification can usually be done by following the guidelines on pages 189–191. Attempting to identify the mushroom over the telephone is fraught with difficulties and should generally be avoided, except when a very experienced and skillful mycologist can be consulted.

Symptomatic Patient Known to Have Ingested Mushrooms

The following guidelines are applicable for the individual or group of individuals who develop symptoms that they attribute to a recent meal of wild mushrooms—ingested either as a tasty treat or an intoxicating substance. These cases usually involve only adults, but children may be included if the mushrooms were served at a family gathering.

The most crucial information is the time of onset of the symptoms. When signs and symptoms develop within two hours of the meal, serious, life-threatening poisoning is unlikely. In other words, the sooner the symptoms occur, the better. If symptoms are delayed and do not appear for four or five hours, or even longer—say, until the next day—the likelihood of serious poisoning is quite high.

This simple diagnostic scheme assumes that only one species was eaten, which is not always the case. Simple syndromes tend to occur mainly in textbooks, seldom in the real world. Some mycophagists favor the ratatouille approach to cooking, mixing many species in the stew. Such cases can be very confusing, because of the overlapping onset of various symptoms. The effect of concomitant alcohol consumption also has to be considered in the initial evaluation.

Regardless of the type of toxin ingested, gastrointestinal symptoms often dominate the early picture and are not too useful in the differential diagnosis. In most patients, these symptoms are the only manifestations of poisoning, because mushrooms with gastrointestinal toxicity are those most frequently ingested. However, for all the other toxins, additional symptoms usually develop and help the physician identify the specific type of toxin and determine the management of the patient.

Unconscious Patient with No History of Mushroom Ingestion

The patient who arrives unconscious in the emergency department always constitutes a major dilemma, because the differential diagnosis is so broad. In terms of mushroom intoxication, the most significant possibility is poisoning with a mushroom containing ibotenic acid and muscimol, for example, *Amanita muscaria* or *Amanita pantherina*. It is most unlikely for any other mushroom toxin to produce such a clinical picture, except for a huge overdose of one of the hallucinogenic mushrooms. Both these possibilities are much more likely to occur in a child than in an adult. Of course, the only way to make the diagnosis is to consider it. Questioning the family or the people who arrive with the patient about the likelihood of a mushroom ingestion is sometimes helpful.

Without a history or a specimen, one is reduced to looking for the spores in the gastric contents or in the stool. Details for this procedure are described on pages 193–195. Until an etiology is uncovered, the patient should be managed according to the protocol standard for someone with undiagnosed coma.

MAJOR CLINICAL SYNDROMES

Eight common clinical syndromes have been described. The great majority of cases treated at medical facilities and through poison centers can be placed into one of the categories shown in the figure.

Although these eight descriptions cover the most common syndromes, there are a number of other symptom complexes that are caused by toxins and do not fall into any of these groups, such as the renal effects of *Amanita smithiana*, the immune hemolytic anemia due to *Paxillus involutus*, or the coagulation disorder caused by *Auricularia auricula*. These are described in Chapter 20. In addition, there is always the possibility that a constellation of symptoms that look like a poisoning are not due to a mushroom toxin but to some other cause. I have termed these syndromes pseudo-mushroom poisonings. The possible causes of these symptom complexes are discussed in Chapter 7.

The following questions may help the medical attendant distinguish between true mushroom poisonings and symptoms due to other mechanisms, as well as assist in identifying the cause.

◆ Where were the mushrooms picked? Were they in public places that could be contaminated with insecticides, fungicides, or other pesticides?

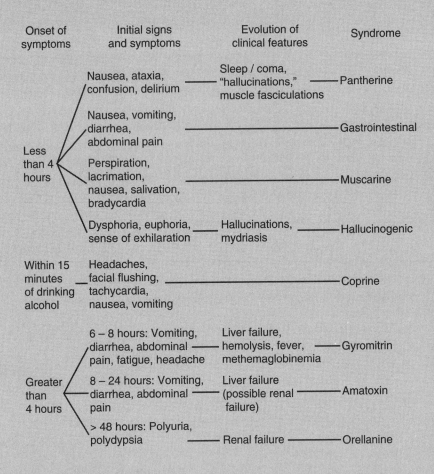

Onset of symptoms	Initial signs and symptoms	Evolution of clinical features	Syndrome
Less than 4 hours	Nausea, ataxia, confusion, delirium	Sleep / coma, "hallucinations," muscle fasciculations	Pantherine
	Nausea, vomiting, diarrhea, abdominal pain		Gastrointestinal
	Perspiration, lacrimation, nausea, salivation, bradycardia		Muscarine
	Dysphoria, euphoria, sense of exhilaration	Hallucinations, mydriasis	Hallucinogenic
Within 15 minutes of drinking alcohol	Headaches, facial flushing, tachycardia, nausea, vomiting		Coprine
Greater than 4 hours	6 – 8 hours: Vomiting, diarrhea, abdominal pain, fatigue, headache	Liver failure, hemolysis, fever, methemaglobinemia	Gyromitrin
	8 – 24 hours: Vomiting, diarrhea, abdominal pain	Liver failure (possible renal failure)	Amatoxin
	> 48 hours: Polyuria, polydypsia	Renal failure	Orellanine

Major clinical syndromes

♦ How were the mushrooms prepared? Where they raw, cooked, pickled, or canned?

♦ What is the state of health of other people who shared the meal?

♦ How much time elapsed between the meal and the onset of symptoms?

♦ How many species of mushrooms were eaten?

♦ What was the condition of the mushrooms and how had they been stored?

♦ Did any individuals who did not eat the mushrooms have similar symptoms?

Poisonings with Delayed Onset: Amatoxin, Gyromitrin, and Orellanine Syndromes

When the onset of the symptoms occurs six hours or more after mushroom ingestion, the attending physician must assume that the patient has one of the four most serious forms of poisoning, those due to amanitin, gyromitrin, orellanine, or the toxin in *Amanita smithiana*. The patient should be admitted to the hospital for immediate evaluation and management. Because these poisonings are life threatening, efforts should be made to obtain a specimen and a positive identification of the mushroom. Many of the other clues to the correct diagnosis will take from hours to days to develop, and it may be helpful to begin therapy as soon as possible. Laboratory tests can assist in the differential diagnosis.

Amatoxin Syndrome   Onset of the gastroenteritis is at least six hours after mushroom ingestion. The gastrointestinal symptoms are followed after a variable time (48–96 hours) with the development of evidence of acute hepatic toxicity, which may rapidly lead to liver failure.

Gyromitrin Syndrome   Onset of the gastroenteritis is six to eight hours after ingestion, accompanied by fever, severe headache, muscle cramps, evidence of hemolysis, and progressive liver failure.

Orellanine Syndrome   Onset of the gastroenteritis is at least 12 hours after ingestion; the delay in onset is often much longer, sometimes many days to even weeks. The gastroenteritis is followed by the development of an increased frequency of urination (polyuria), intense thirst (polydypsia), and other evidence of kidney failure.

Differential Diagnosis   Serum levels of transaminases (ALT, alanine aminotransferase; AST, aspartate aminotransferase) become elevated as a result of the liver toxicity caused by both amanitin and gyromitrin. The cellular damage caused by the toxins leads to leakage of the enzymes from the cells. The serum levels rise markedly once the cells die. Increases are usually apparent within 24–48 hours of the ingestion of amanitin and as early as 12–24 hours after gyromitrin poisoning. These enzyme elevations precede the increase in serum bilirubin by at least 48 hours and are currently the most sensitive tests for detecting the onset of liver damage.

Measurements of methemoglobin, free hemoglobin, and haptoglobin in the serum and hemoglobin in the urine are appropriate, because hemolysis is a specific, but uncommon, characteristic of gyromitrin poisoning. If hemolysis has developed, it is usually evident at the time of initial presentation. The amount of serum haptoglobin decreases, because it binds to the free hemoglobin. Assessment of serum hapto-globin levels is the most sensitive test for intravascular hemolysis, because all the haptoglobin must be bound before free hemoglobin appears in the serum. With severe hemolysis, the free plasma hemoglo-bin increases, and the serum may be frankly pink. The presence of hemoglobin in the urine is also characteristic of intravascular hemoly-sis in the more severe cases. Methemoglobinemia and methemoglo-binuria occurs in a small percentage of cases. The methemoglobin can be measured with a cooximeter.

Renal failure is the hallmark of orellanine and *Amanita smithiana* poi-soning and can complicate the later stages of amanitin toxicity. Measurement of serum creatinine and blood urea nitrogen (BUN) are the most useful tests. The BUN may be elevated in any patient who has lost excessive fluid as a result of vomiting and diarrhea, so it is very important to differentiate the dehydration-related pre-renal azotemia from intrinsic renal failure. Serum creatinine is the more useful mea-surement in this situation.

In the absence of any evidence of hemolysis or renal failure, especially when the diarrhea has been profuse and very watery—with or without the presence of flecks of blood or mucus—the presumptive diagnosis is amanitin poisoning. The detailed discussion of the management of a patient with amanitin poisoning is given in Chapter 12.

When there is evidence of hemolysis and other manifestations of gyromitrin toxicity, such as jaundice, fever, loss of muscle coordina-tion, vertigo, and convulsions, the patient should be managed as out-lined in Chapter 14.

When true renal failure is present, without any signs of liver toxicity or hemolysis, then the diagnosis is most likely poisoning by orellanine or by the toxin in *Amanita smithiana*, and treatment for the acute renal failure can be instituted, as discussed in Chapter 13.

In all cases in which the ingestion of a potentially fatal mushroom toxin is suspected, attempts to identify the culprit should be vigorously pursued. Remnants of the meal may still be available or the family may recall where the mushrooms were collected and hence acquire a fresh specimen. Although identification is not mandatory, because a patient can usually be treated adequately on the basis of a presumptive diagno-sis, it is best to have positive and reliable knowledge about the cause of the symptoms. The guidelines for identifying mushroom specimens are on page 189–191.

Poisonings with Rapid Onset

When the onset of symptoms occurs soon (two hours or less) after the ingestion of one or more mushrooms, constructing a differential diagnosis is still possible, especially if signs and symptoms other than purely gastrointestinal complaints develop within a few hours. What is perhaps more important and reassuring is that a poisoning with rapid onset is seldom lethal. However, a serious poisoning cannot be excluded, because some individuals eat a veritable potpourri of wild mushrooms in a single dish—despite exhortations to avoid mixing species, especially the first time they are tried. Prudence recommends 24 to 48 hours of monitoring for anyone poisoned seriously enough to require specific treatment in the hospital. This monitoring will detect any signs of delayed toxicity due to an amatoxin-containing mushroom.

Muscarine Syndrome   Perhaps the most characteristic syndrome, usually distinguishable in this rapid-onset group, is that caused by muscarine. It frequently develops within 30 minutes of ingestion and features perspiration, salivation, and lacrimation (PSL syndrome). This syndrome is also called SLUDGE syndrome (for salivation, lacrimation, urination, defecation, gastrointestinal distress, emesis). A number of other manifestations may also be associated: blurred vision with constricted pupils (myosis), dizziness due to a fall in blood pressure (hypotension), and a decidedly slow pulse rate (sinus bradycardia) (see Chapter 18). The major danger is the possible confusion between the PSL syndrome and a panic reaction. However, the heart rate is always elevated in a panic reaction, whereas it is normal or even slowed by muscarine. In the panic reaction, the pupils will often be dilated rather than small. In addition, the panic reaction features a cold and clammy skin, unlike the flushed, sweaty appearance of muscarine stimulation. During the initial assessment, superficial similarities between these two syndromes are often evident (see the accompanying table).

Comparison of muscarine poisoning and panic reaction

Sign or symptom	Panic/fear reaction	Muscarine poisoning
Sweating	++	+++
Lacrimation	+ (after vomiting)	++
Salivation	+ (after vomiting)	++
Skin	Cold and clammy	Hot, flushed
Heart rate	Rapid (tachycardia)	Slow (bradycardia)
Pupils	Dilated (mydriasis)	Constricted (miosis)
Urinary urgency	Absent	May be present

Inebriation Syndrome   This intoxication syndrome lacks the obviously cholinergic symptoms of perspiration, salivation, and lacrimation. Onset of an inebriationlike state occurs between 30 and 120 minutes after ingestion of the mushroom. This syndrome is characterized by dizziness, staggering, and incoordination (ataxia). Often there is a waxing and waning of symptoms, with periods of excessive activity alternating with lethargy or a deep, comalike sleep. Hallucinations and other symptoms may occur. For a patient exhibiting these symptoms, the presumptive diagnosis is ibotenic acid/muscimol poisoning. This inebriation, or pantherine, syndrome is detailed in Chapter 16. Little specific needs to be done in these cases except to allow the patient to sleep off the effects of the toxin and to ensure that the airway is protected and can be monitored. In children, however, seizures and other complications are more common and may require specific therapy. The case report that introduced this chapter is an excellent example of this syndrome.

Hallucinogenic Syndrome   Should hallucinations be the most prominent feature—often with a sense of dysphoria, great gaiety, and excessive laughing—the odds are that the patient is experiencing the effects of psilocybin, or one of the other hallucinogens. Individuals having a "bad trip" as a result of eating these mushrooms can be extraordinarily anxious and very fearful, even simulating the panic reaction. Their truly bizarre behavior and complete lack of attachment to reality can usually be detected by the most casual observer. The onset of symptoms is usually within an hour of ingestion. The group of compounds causing these effects is covered in Chapter 17, and the syndrome is also called the magic mushroom syndrome.

Coprine Syndrome   Most uncommon, but very characteristic, is the reaction produced when the compound coprine, found in the species *Coprinus atramentarius*, is combined with alcohol. This is a classic "Antabuse" type of reaction. It occurs very rapidly—within minutes after the alcohol is imbibed. It is not related in onset to when the mushroom was eaten, an event that may have occurred from hours to a few days earlier. The sensation of facial flushing, a pounding headache, nausea, a metallic taste, and a rapid heart rate are the usual symptoms. Other symptoms and signs may be present: dizziness with low blood pressure, nausea, and sweating (see Chapter 15). The diagnosis may be difficult to make because the patient may not give an account of the previous mushroom meal, having no reason to relate the mushroom with the reaction. Furthermore, the effects of the alcohol itself may modify the clinical picture.

Gastrointestinal Syndrome The gastrointestinal syndrome is the result of one of the most frequent adverse reactions to mushrooms. Onset of the gastroenteritis is usually within two hours of ingestion. Fortunately, it is only occasionally severe enough to bring an individual to the hospital. The patient presents with an undifferentiated case of "food poisoning" characterized by abdominal pain (usually cramping) nausea and vomiting, and diarrhea. The clinical picture may be indistinguishable from many of the other causes of acute gastrointestinal distress, and it is only by eliciting a history of mushroom ingestion—if the patient is inclined to mention it or the physician astute enough to inquire—that the cause is found. In some cases, the symptoms may be severe enough to simulate an acute "surgical" abdomen. A host of mushrooms can produce these effects; they are catalogued in Chapter 19.

GENERAL MANAGEMENT OF ANY CASE OF SUSPECTED MUSHROOM POISONING

A variety of general measures should be adopted in most cases of mushroom poisoning. These may need to be modified on the basis of the clinical condition of the patient or any other health problems from which the patient may be suffering. Moreover, once a specific cause has been identified—on the basis of either the characteristic clinical picture or a positive identification of the mushroom—the therapy can be tailored to the diagnosis and the anticipated symptoms.

The general principles of management in all potential poisonings are self-evident:

- Ensure that all vital functions are maintained.
- Reduce or eliminate any further absorption of the toxin.
- Enhance excretion and elimination, if possible.
- Provide supportive measures.

Elimination of Further Toxin Absorption

The initial treatment of all patients that arrive at a medical facility within one to two hours of ingestion and are still likely to have the suspected poisonous mushrooms in the stomach is the induction of emesis—assuming that the patient is awake, alert, oriented, and cooperative. Gastric lavage, with airway protection, is used in those patients who are in coma, convulsing, or unable to protect their own airway. An emetic should not be used in any patient with a questionable or deteriorating mental status. Physicians must exercise clinical judgment in each case when deciding the risks and benefits of inducing emesis or performing gastric lavage.

Emesis is an ancient, time-honored therapy, well documented in the classic literature on mushroom poisoning. It is still the most valuable treatment for the rapid-onset syndromes caused by mushroom toxins. Unfortunately, the most serious poisonings are caused by toxins with such a long delay between the meal and the first symptoms that the mushrooms have long since departed the stomach. In fact, in most delayed-onset poisonings, the remains of the mushrooms will be long past the anus, especially if the patient has had much diarrhea. Producing emesis in these cases will do no more than complicate fluid management. Furthermore, many mushrooms induce vomiting quite well on their own; and it is not uncommon for the patient to have vomited copiously prior to arrival. In these situations, there is no need to further assist the vomiting center.

Emesis is appropriate when the following conditions are met:

◆ Patient is fully conscious.
◆ Patient has not already had much vomiting.
◆ Ingestion occurred less than two hours before admission.

Emesis is most efficacious when performed within the first 30 minutes after ingestion.

The most effective emetic agent is syrup of ipecac (30 mL for adults; 15 mL for children 1–12 years old; 5–10 mL for infants 6–12 months old), followed by oral fluids. Adults should be given 250–500 mL of oral fluids; children require at least 20 mL per kilogram of body weight; and infants need 15 mL oral fluids per kilogram of body weight. For example, an average toddler of 15 kg (approximately 33 pounds) should receive 15 mL of ipecac (0.5 fluid ounce) and at least 300 mL of fluids (10 fluid ounces). It is most effective when given within 30 minutes of mushroom ingestion. If vomiting does not occur within 30 minutes of treatment, the dose can be repeated again. If emesis is still not provoked, the physician should perform gastric lavage.

If the decision is made to perform gastric lavage, the procedure must be done with a wide bore tube (3 to 4 French), because the physician never knows how well the patient has chewed his or her food. Mushrooms are often not well digested, so blockage of the lavage tube by mushroom fragments can be a problem. Lavage is usually performed with normal saline in children (50–100 mL aliquots, up to a total of 2 liters), or tap water or normal saline in adults (200 mL aliquots, up to a total of 2–10 liters). Lavage must be continued until the fluid that is returned is clear.

Gastric lavage is indicated in the following situations:

◆ Patient is in coma, convulsing, or cannot protect his or her airway.
◆ There is no history of repeated vomiting.
◆ Ingestion occurred less than four hours before admission.

Lavage has been performed in conscious patients as a disincentive to repeated use of a particular mushroom. However, this "punitive" approach to intoxication, even for self-induced intoxication, is now deprecated and should never be used. In fact, lavage may exacerbate the symptoms of *Psilocybe* intoxication.

Activated Charcoal   The addition of activated charcoal (which has great adsorptive capacity) at the completion of emesis or lavage has become part of the standard therapy for almost any kind of toxic ingestion.[8] Some practitioners have even suggested that charcoal alone is as effective for many intoxications as the combination of emesis and charcoal is. Normally, 30–100 g of the charcoal is mixed with water and given as a slurry, either orally or through the gastric tube (adults). The volume of water used for the slurry should be 240 mL for every 30 g of charcoal. In children, 15–30 g is the usual dose and can be repeated every three to six hours. In the United States, prepackaged preparations of activated charcoal are routinely available. No specific studies have validated its use in mushroom poisoning per se, but it is so much a part of the current tradition that its use will continue. In other forms of poisoning, any delay in the administration of activated charcoal diminishes its effectiveness. The earlier it is used, the better. No obvious side effects of this therapy have been described, but its efficacy needs to be proved in patients poisoned by the toxins of specific mushrooms.

Recently, recommendations have been made to mix the charcoal with sorbitol, the latter acting as a cathartic. Sorbitol should only be used in the initial dose of charcoal. In seriously ill individuals, the recommended dose for adults is 1 g charcoal in 4.3 mL of 70% sorbitol per kilogram of body weight. In children, 30% sorbitol should be used.[9] It is very important to monitor fluid balance whenever sorbitol is used.

Cathartics (Purgatives)   Many texts recommend the use of cathartics in cases of poisoning. However, diarrhea is a common accompaniment to mushroom poisoning and hardly needs additional assistance. In fact, it is very difficult to justify the use of cathartics in mushroom poisoning. For those to whom tradition is important or who can justify its use in an exceptional case, 250 mg sodium or magnesium sulfate per kilogram of body weight can be used. It probably will not change the course or the outcome of the case. Hypermagnesemia may complicate the repeated use of magnesium-containing cathartics, so multiple doses should be avoided. Rare deaths have been reported after the use of these cathartics in poison management.

Sorbitol has also been recommended in a dose of 1 to 2 g sorbitol/kg body weight up to a maximum of 150 g. In children, the dose should be reduced to 1 to 1.5 g sorbitol/kg body weight up to a maximum of 50 g. Sorbitol is administered as a 30–35% solution. No more than one dose should be given. If a cathartic is used, especially in small children, the patient should be admitted to the hospital for observation, monitoring, and fluid replacement when necessary. Frankly, there is little justification for using a cathartic, except possibly in cases of amatoxin poisoning. Even in this circumstance, the diarrhea produced by the mushroom is usually sufficient.

Enhancement of Excretion

Despite the many recommendations in the literature, there is little carefully documented scientific evidence supporting the use of forced diuresis. Maintaining an adequate urine output by appropriate hydration is probably all that is needed.

Nevertheless, forced diuresis has been widely used in many of the more serious cases of mushroom poisoning, such as amatoxin poisoning, because urinary excretion is a common elimination pathway for toxins and their metabolites. Diuresis was accomplished by administering liberal oral fluids, if possible, or intravenous fluids, if necessary, and a diuretic such as furosemide (1 mg/kg body weight/hour intravenously up to a maximum of 40 mg as a single dose). The goal was to achieve a urine output of between 3 and 6 mL/kg body weight/hour.[10] The input and output of fluids had to be carefully monitored, especially when other ongoing fluid losses, such as continued diarrhea, were present. In the majority of cases, the combination of 0.45% NaCl and 5% dextrose was a reasonable replacement solution.

Charcoal hemoperfusion (hemadsorption) was recommended at one time, but there are probably no clinical situations in which this treatment is applicable. Although it is said to be effective for the removal of ibotenic acid and muscimol, poisoning with these toxins is almost never a life-threatening situation, and the therapy poses a greater risk than the toxins. Hemodialysis alone has no role to play in the treatment of mushroom poisoning, because the amanitins are not removed during this process. Similarly, unless the patient is already in renal failure, there is no indication for peritoneal dialysis. In some countries, hemadsorption is widely employed for many forms of supposed intoxication. It may evoke a satisfying mental image of "scrubbing the blood," but in the United States it has never achieved prominence as a treatment for serious mushroom poisoning.

Maintenance of Vital Functions

Standard practice should be followed to ensure that all vital functions are adequately maintained. An adequate, patent airway must be established and maintained, especially if the patient is in coma or convulsing. In cases of ibotenic acid/muscimol poisoning, respiration is not usually depressed; however, excessive use of anticonvulsants can depress respiratory activity, so this sign should be monitored. Intubation should be used to protect the airway from aspiration if the patient develops seizures or becomes unconscious. Oxygen should be administered as required.

Blood pressure may require maintenance with intravenous fluids, especially in patients who have suffered significant fluid losses. In rare instances, when the central venous pressure is normal, vasopressors may be required (dopamine at a dosage of 2–15 µg/kg body weight/minute or norepinephrine given in a dosage 0.1–0.2 µg/kg body weight/minute) to maintain the blood pressure. Because none of the mushroom toxins have a very significant direct impact on the heart, myocardial function is generally unimpaired.

Supportive and Symptomatic Care

Fluid and Electrolyte Balance   In many of the severe cases of poisoning, water and electrolytes deficits are the most serious problems, especially in young or elderly patients. Careful attention should be paid to the degree of dehydration resulting from ongoing losses and to the patient's daily water and electrolytes requirements. Standard replacement and maintenance therapy should be used.

Pain   With some gastrointestinal irritants, including amanitin, severe, colicky abdominal pain can occur. It seldom lasts for more than a few hours and often is not treated. However, in certain patients, it may be severe and distressing. If symptomatic relief is needed, opiates such as meperidine (0.5–1.0 mg/kg body weight) or morphine (0.05–0.1 mg/kg body weight) can be administered intramuscularly. Other analgesics may also be tried.

Convulsions   Seizures are most likely to occur in children who have ingested mushrooms containing ibotenic acid/muscimol. The anticonvulsant diazepam should be administered as an intravenous bolus. In an adult, 5–10 mg can be used initially and repeated every 15 minutes up to a total of 30 mg. In a child, 0.25 to 0.45 mg/kg body weight/dose is used, up to a maximum of 10 mg in a child over the age of five years, or up to a maximum of 5 mg in a child under that age. If the seizures are not controlled with diazepam, then phenobarbital can be employed. Because this barbiturate may cause respiratory depression, the patient must be closely watched and intubated and ventilated, if required. The recommended dosage of phenobarbital for adults is 10–20 mg/kg body weight diluted in 60 mL of 0.9% saline, administered at a rate of 25–50 mg/minute. Additional doses of 120–240 mg may be given every 20 minutes. Phenobarbital dosage for children is 15–20 mg/kg body weight at a rate of 25–50 mg/minute. Doses of 5–10 mg may be given at intervals of 20–30 minutes. Intubation may be required when the total dose is greater than 20 mg/kg body weight. In addition to watching for respiratory depression, careful attention also needs to be paid to the cardiovascular system, by frequently monitoring the patient for signs of developing hypotension. Measurement of serum levels of the anticonvulsant agent may be performed after the appropriate intervals to ensure therapeutic but nontoxic levels.

Anxiety or Hallucinations   For the majority of anxious or hallucinating patients, reassurance, a safe, quiet environment, and the technique of "talking the patient down" are the only treatments needed. If required, sedatives such as diazepam (5 mg in adults or 0.1 mg/kg body weight in children) or chlorpromazine (2.5 mg/kg body weight) can be used. Recently, specialists have recommended that agents with anticholinergic action, such as chlorpromazine, not be employed in cases of psilocybe intoxication. Another alternative, of more recent vintage, is midazolam (Versed®) in a dose of 0.1 mg/kg body weight, up to a total dose of 0.5 mg/kg body weight. This treatment carries the risk of respiratory depression or arrest, so careful monitoring is required after midazolam is administered.

Nausea and Vomiting   Once the attendant is confident that the stomach has been emptied of all its contents, an antiemetic can be used. Metroclopramide (0.1 mg/kg body weight) is an old standby, whereas ondansetron (Zofran®), a new 5-hydroxytryptamine antagonist, has found favor in the treatment of chemotherapy-induced nausea. Antispasmodic agents, such as glycopyrrolate (1 mg, three times a day, taken orally) may help to alleviate the cramping abdominal pain. If this agent is insufficient, opiates may be required.

Monitoring

The input and output of fluids should be carefully monitored on a continuous basis until the patient is back to normal. In the first 48–72 hours, serum transaminases (ALT, AST), prothrombin time, urinalysis, BUN, and creatinine should be performed to assess the presence of liver or renal dysfunction. Serum electrolytes may be required for fluid management.

HANDLING AND IDENTIFICATION OF MUSHROOM SPECIMENS OR FRAGMENTS

A positive identification of a mushroom is seldom possible in the emergency situation. However, when the patient or family does provide some material, it is worthwhile to attempt an identification. The goal is to exclude the possibility of one of the three most serious types of poisonings. It is usually easy to eliminate one of these, namely, *Gyromitra* species, on the basis of the time of the year and a cursory look at the specimen. *Gyromitra* species fruit in the spring or early summer, often at the edges of receding snow banks or in ground that has recently lost its snow cover. They are convoluted, cerebriform structures, varying from a pale ochre through a dark brown to black. There are no gills, tubes, or spines, because the spore-bearing surface is on the outside of the structure. Showing the patient or the family a photograph (color plate 6) is usually sufficient to get a reasonably reliable response. For the second serious poisoning syndrome (orellanine syndrome), the ingestion usually has occurred so far in the past that mushrooms are no longer available. The renal complications are a distinctive diagnostic feature.

The most important mushrooms to exclude, because their toxin is the most common cause of fatality, are the amatoxin-containing varieties. An examination of well preserved, intact specimens by a careful observer—even a mycologically naive one—can usually exclude the three amatoxin-containing taxa. There are situations, however, in which identification is not an easy task.

Is the Specimen an Amanita?

Examples of the amanitin-containing species of *Amanita* are illustrated in color plates 1–3.

◆ Gills are usually white.

◆ Gills usually do not attach to the stem. There is a small gap between where the gills and the stem come together, so that when the cap is broken upward it is cleanly removed from the stem. In some species, however, the gills may barely attach to the stem.

◆ Spores are white. Note that the spore color cannot be determined from the color of the gills. They must be determined from a spore print. Place a mushroom cap or a portion of the cap, with the gill side down, on a piece of paper and cover with a cup or a drinking glass. In about two to four hours, the spores will have fallen onto the paper and their color will be obvious. The time needed for a decent spore print is quite variable, and in a few mushrooms can be considerably longer than four hours. The specimen should be kept at a cool temperature, because high temperature (for some species, even room temperature) may retard spore release. It is best to observe the color of the spore under natural light, because fluorescent or other artificial lighting can distort the true color. By looking at the paper at an angle or with light coming from the side, the observer can easily see even white spores. Very immature mushrooms may not release sufficient spores to yield a decent spore print. The same is true of very old, very wet, or dehydrated specimens. When viewed under the microscope, *Amanita* spores appear oval (globose) and clear (hyaline), have no pores, are 8–10 μm in size, and stain amyloid (blue) with Melzer's reagent (20 mL H_2O, 1.5 g KI, 0.5 g iodine, 20 g chloral hydrate).

◆ A ring is usually present around the stem (the remnant of the partial veil). A ring is not present in all species, because some do not have a partial veil.

◆ A cup (volva) is present around the swollen, bulbous base of the stem. Note that the volva is not often observed in the specimens brought in for identification, because the cook or the individual mushroom collector usually cuts the mushrooms off at the base, leaving the volva in the ground. Therefore, the absence of a volva must not be a deciding factor unless one is certain that one has the whole mushroom.

Is the Specimen a Galerina?

This genus is illustrated in color plate 4. The habitat and fruiting pattern of this genus are described on pages 205–206. *Galerina* specimens have the following features and habits:

- Cap is usually small (2–6 cm) and brown to yellow
- A small ring is present around the stem
- Spores and gills are brown
- Frequently found growing on decaying wood
- Autumn fruiting

Is the Specimen a *Lepiota*?

An example of an amanitin-containing *Lepiota* is illustrated in color plate 5. These mushrooms have many features in common with those in the genus *Amanita*, but they lack a volva at the base of the stem. Common *Lepiota* features are

- Gills and spores are white
- A ring is present around the stem (but not always)
- No volva is present (assuming that the entire mushroom has been picked and that the base has not been left in the ground)

Testing for Presence of Amatoxin

If the specimen looks like any of the three groups described above, then the next question is, "Does this particular specimen contain amatoxin?" A professional mycologist will be able to identify the species, and such help should be sought when doubt exists about whether or not one is dealing with one of the extremely toxic mushrooms. Appendix 3 lists mycological associations that can be called for help. Local universities and community colleges can be contacted, and many of the poison control centers have mycological consultants available. In addition, a screening test can be done in the office, emergency room, or the laboratory. A very simple and underutilized test was described in 1979 (see next section). It is most useful in answering the question "Does this mushroom contain any of the amatoxins?" This is always the major concern.

The Weiland Newspaper Test, or Meixner Test, for Amatoxins

The Weiland newspaper test is based on the reaction of lignin (found in cheap newsprint) with the amatoxins in the presence of concentrated acid to produce a blue color.[11,12] High-quality, glossy paper has had almost all the lignins removed and should not be used. Filter paper, if available, is ideal. The basis for this test was discovered in the Weilands' laboratory in the late 1940s; the test was validated, revived, and popularized by Meixner 30 years later. In a study of 200 specimens of *Amanita phalloides* from California, 199 reacted positively; extracts of one unreactive mushroom were also negative on thin layer chromatography. A few specimens of *A. bisporigera* and *A. ocreata*, all the *Galerina autumnalis*, and half the specimens of *A. virosa* were positive. A specimen of *A. verna* was negative, as were half the *A. virosa* specimens. Thin layer chromatographic studies showed that these mushrooms were devoid of amatoxins, a result confirming the sensitivity of the assay. Some specimens from the eastern part of the country were also tested and showed a more variable reaction, but all the *A. phalloides* specimens were judged to be positive. A moderate number of false-positives were observed for various other species (approximately 15% of all non–amatoxin-containing species tested).[13]

The test is very sensitive for the presence of amatoxins; it will detect as little as 0.02 mg/mL of toxin. However, it is not that specific, producing a positive reaction with a variety of other mushrooms. Despite this lack of specificity it is still very useful. A negative result provides reasonable assurance that the patient has not eaten an amatoxin-containing species. But it does not ensure that the mushroom was edible, because toxins other than amatoxins may be present. Moreover, one must be sure that the mushroom tested is the same species as the one ingested. The main problem with this simple test is the large number of false-positive results. Mushrooms giving a positive reaction must be immediately identified by a competent mycologist. Moreover, in the Midwest, where *Amanita verna*, *A. virosa*, and *A. bisporigera* are the more common toxic species, further validation of this test's performance is required.

To perform the Weiland/Meixner test, squeeze one drop of juice from the cap of the fresh mushroom onto a piece of newsprint or filter paper (not glossy or treated paper). Circle the drop with a pencil and allow the spot to dry at room temperature or dry gently with a hair dryer (protect from direct sunlight). Apply one drop of 8 to 12 N (concentrated) hydrochloric acid to the dried spot. Add one drop of the same acid to an adjacent blank circle to act as the negative control (some newsprint will turn blue with acid by itself). A blue color appearing within 15–20 minutes (up to an hour) should be regarded as evidence of amatoxin, assuming that the control spot is negative (not blue).

A number of color reactions may develop in response to the presence of many other chemical compounds. These can generally be discounted. A false-positive occurs when the paper is heated above 63°C or is exposed to sunlight. A variety of other species—especially those containing psilocybin, terpenes, and bufotenine—also give a blue color.

Spore Identification

When no mushroom specimen is available, a positive identification is not possible. However, a presumptive diagnosis of mushroom ingestion can sometimes be made by looking for spores in the gastric contents or stools. It is also useful if any uneaten remains of the meal can be found, because the spores are generally resistant to cooking, gastric acid, and the digestive enzymes. One recent study examined the effect of simulated gastric secretion on the morphology of spores from some of the more important edible and toxic species.[14] No significant changes in the light microscopic appearance of the spores were observed. Other structures, however, such as cystidia, which can be very helpful in identification, almost completely disintegrated. It was also found that mounting the sample in 10% aqueous ammonia aided in identification because this reagent removed some of the debris.

In the three most serious forms of poisoning, the gastric contents are usually devoid of any residual mushrooms because the patients usually present more than 12 hours after ingestion of the toxic mushrooms. However, if gastric contents are available, or if some diarrheal stool can be obtained, a drop of the material should be examined under a coverslip at the highest magnification possible (oil immersion, 1000–1600 ×). Mushroom spores come in many different shapes, sizes and ornamentations (see the accompanying figure), but generally they are elliptical and approximately the size of a red blood cell (6–8 μm). For a single species, they are relatively uniform in both size and shape.

As noted earlier, a spore from an *Amanita* species will have the following characteristics:

- Thin walls
- No pore
- Colorless and transparent (hyaline) walls and contents
- Globose (oval) to, at most, slightly ellipsoid in shape
- 8–10 μm in diameter

Numerous mushrooms in addition to *Amanita* have spores that fit this description, so seeing spores with these characteristics does not constitute an exclusive identification of an *Amanita* species.

If no spores are visible in a direct preparation or if so much debris obscures the field that one cannot distinguish spores, then some form of concentration procedure is useful. This method was originally described for the identification of the spores of *Chlorophyllum molybdites*, but it is useful for all mushroom spores.[15] The sample (10 to 50 mL of

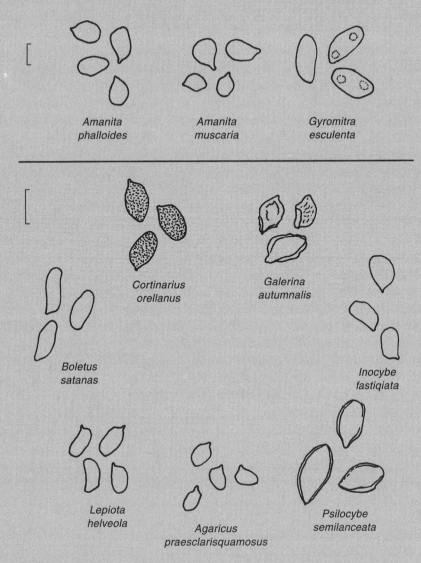

Spores of common toxic mushrooms. Each bracket indicates 10 μm.

fluid) is filtered through three to four thicknesses of cheesecloth to remove the larger particulate matter. The filtrate is then centrifuged at $7000 \times g$ for 10 minutes to sediment the spores. The supernatant fluid is discarded by careful decanting or is withdrawn with a pipette, and the sediment is resuspended in one drop of water or in a drop of 10% aqueous ammonia. The drop is placed on a slide, covered with a coverslip, and examined.

Spores can be made more visible by staining them with 1% acid fuchsin. This solution should be available in any histology laboratory. A drop of sediment is deposited on a slide, then the slide is gently heated or placed on a heating block to evaporate the fluid. A drop of 1% acid fuchsin is added, and the slide is heated again to evaporate the liquid. The slide is gently washed, and a coverslip is placed over the wet slide. The spores should be readily visible as bright red objects. An alternative procedure is to resuspend the sediment in a drop of cotton blue in lactophenol. This stain colors the spores blue. It is usually available in microbiology laboratories.

More elaborate procedures for spore concentration and specific staining reactions have been well described.[16] The possibilities and limitations of spore identification have been reviewed.[17] Results of stool examination are mediocre at best and generally require an additional concentrating method to increase the number of spores and reduce the debris. However, when the victim has watery diarrhea, the latter is not too much of a problem. The largest series—282 cases—of poisonings in which spore diagnosis was attempted, was reported from the Trieste commune in Italy.[18]

ALTERNATIVE DIAGNOSTIC APPROACH

An interesting alternative approach to the differential diagnosis of mushroom poisoning, using a microcomputer-based system, was developed in the 1980s in Scotland. It relied on the identification of rather simple and stable characteristics of the mushroom, microscopy of the spores in stomach contents, the symptoms, habitat in which the mushroom was picked, and a few other criteria. Unfortunately, this algorithm has not reached commercial circulation, nor has it been adequately field tested to validate its effectiveness, but it does have a number of intriguing features.[19]

REFERENCES

1. K. Lampe, "Systemic Plant Poisoning in Children," *Pediatrics* (1974) 54:347–351.

2. D. H. Mitchel and B. H. Rumack, "Symptomatic Diagnosis and Treatment of Mushroom Poisoning." In *Mushroom Poisoning: Diagnosis and Treatment*, ed. B. Rumack and E. Salzman (West Palm Beach, FL: CRC Press, 1978), pp. 171–179.

3. G. Lincoff and D. H. Mitchel, *Toxic and Hallucinogenic Mushroom Poisoning* (New York: Van Nostrand Reinhold, 1977).

4. K. Lampe and M. A. McCann, "Differential Diagnosis of Poisoning by North American Mushrooms with Particular Emphasis on *Amanita phalloides*-like Intoxication," *Annals of Emergency Medicine* (1987) 36(9):956–962.

5. *POISINDEX* (Denver, CO: Micromedex, 1991).

6. J. H. Trestrail and K. F. Lampe, "Mushroom Toxicology Resources Utilized by Certified Regional Poison Control Centers in the United States," *Clinical Toxicology* (1990) 28(2):169–176.

7. C. B. Fischbein, J. W. Lipscomb, G. M. Mueller, and J. B. Leikin, "Field Test of the New Fungal Identification System in Poisindex," *Veterinary and Human Toxicology* (1993) 35:204–206.

8. K. Kulig, "Initial Management of Ingestions of Toxic Substances," *New England Journal of Medicine* (1992) 326(25):1677–1681.

9. A. Minocha, E. P. Krenzelok, and D. A. Spyker, "Dosage Recommendations for Activated Charcoal-Sorbitol Treatment," *Journal of Toxicology and Clinical Toxicology* (1985) 23:579–587.

10. S. Vesconi, M. Lange, and D. Constantino, "Mushroom Poisoning and Forced Diuresis," *Lancet* (1980) 2:854–855.

11. J. A. Beutler and P. Vergeer, "Amatoxins in American Mushrooms: Evaluation of the Meixner Test," *Mycologia* (1980) 72(6):1142–1149.

12. A. Meixner, "Amatoxin-Nachweis in Pilzen," *Zeitschrift für Mykologie* (1979) 45:137–139.

13. P. P. Vergeer, "The Meixner Test Evaluated," *McIlvainea* (1986) 7(2):61–70.

14. S. J. Fender and R. Watling, "Effects of Digestion on Agaric Structures," *Edinburgh Journal of Botany* (1991) 48(1):101–106.

15. F. I. Eilers and B. L. Barnard, "A Rapid Method for the Diagnosis of Poisoning Caused by the Mushroom *Lepiota morgani*," *American Journal of Clinical Pathology* (1973) 60:823–825.

16. A. Bresinsky and H. Besl, *A Colour Atlas of Poisonous Fungi* (London: Wolfe Publishing, 1990).

17. M. Josserand, "L'examen de débris de champignons et celui des fèces d'un intoxiqué," *Cryptogamie et Mycologie* (1983) 4:199–205.

18. L. Tomini, S. Gregorutti, and L. Triolo, [Sporological diagnosis in 282 cases of suspected fungal poisoning in the municipality of Trieste from 1967 to 1984] (Italian), *Minerva Medica* (1987) 78(5):321–327.

19. P. A. J.-L. Margot, "Identification Programme for Poisonous and Hallucinogenic Mushrooms of Interest to Forensic Science," Doctoral thesis, University of Strathclyde, Glasgow, Scotland, 1980.

Mushroom
Poisoning
Syndromes

CHAPTER TWELVE

AMATOXIN SYNDROME
Poisoning by the Amanitins

Amanita phalloides

IN THE FALL of 1989, a Portland family of Korean immigrants drove to the Columbia River gorge to pick chestnuts from a grove known to members of the Asian community. While gathering chestnuts in the county park, they found mushrooms resembling paddy straw mushrooms, *Volvariella volvacea*, which are sold commercially in Korea. Relatively few Asian consumers ever see them growing in their native state, because paddy straw mushrooms are widely cultivated. The chestnut pickers were thrilled—a double harvest, chestnuts and mushrooms. Convinced that the mushrooms were edible, almost two bucketfuls were added to a stir-fry. Each of the participants ate an estimated 10 to 12 caps—a rather prodigious quantity, considering the misidentification. Ten hours after the feast, they all began to have cramping abdominal pain, nausea, vomiting, and diarrhea. For three, the symptoms were severe enough to prompt a visit to the hospital. Two other victims felt they only had a touch of food poisoning and did not seek help at this stage.

At the hospital, the specter of possible mushroom poisoning was raised. One patient was released but readmitted later. Attempts to identify a mushroom signature in the vomit—spores or gill pieces—proved difficult. Despite the lack of a positive identification, it was wisely decided to treat the three patients on the basis of a presumptive diagnosis of amatoxin poisoning.

One of the patients mentioned that uneaten mushrooms were still at home. The amateur mycologist Preston Alexander, who had tentatively identified the mushrooms in the hospital, visited the house, only to find two more victims suffering from the early symptoms of poisoning. The mushrooms in the remains of the meal also proved too difficult to identify with certainty. A visit to the chestnut grove a day or two later—by Jan Lindgren, the chairperson of the toxicology committee of the Oregon Mycological Society—uncovered another 40 specimens of *Amanita phalloides*.

Despite initiation of therapy about 12 hours after the meal, including administration of activated charcoal, intravenous fluids, and oral silymarin, the poisoning symptoms persisted and progressed. (Silymarin, extracted from the milk thistle, has been used in its injectable form in Europe in cases of amatoxin poisoning.) Within 72 hours, four of the five patients had evidence of severe liver dysfunction. The fifth patient was said to have had two helpings of the meal, yet he appeared to be less severely afflicted. His servings, it turned out, had been in small rice bowls, whereas the rest of the group had each eaten a large plateful. Four patients had liver transplants. The fifth recovered after an episode of moderate liver and kidney failure.

As is typical for this type of poisoning, onset of symptoms was delayed for up to 10 hours. The initial symptoms were ignored by some of the victims for a time. Identification of the culprit was very difficult, ultimately requiring a visit to the picking site. From a medical standpoint, the cases were well managed. A reasonable presumptive diagnosis was established early, and the patients all received aggressive treatment. Conservative therapy would have been insufficient in these patients because of the massive doses ingested: liver transplants were required in four of the five victims.

It is impossible to know whether oral administration of silymarin had any benefit. The optimal preparation of that agent, its injectable or parenteral form, is unavailable in the United States. The oral form, which has no proven efficacy, has not been approved for human use in the United States and also is not available through the pharmaceutical houses.

This case history illustrates two common situations that lead to serious poisonings. The victims were immigrants who assumed that their mushroom find was similar to a species found in their native land. Their knowledge was folklore based and limited. The mushrooms also were immigrants, having been imported on the roots of chestnut trees from Europe sometime in the 1920s or 1930s.

INCIDENCE OF AMATOXIN POISONING

California has been the epicenter of amatoxin poisoning in the United States, especially during the 1980s.[1,2] This dubious distinction is due to both the influx of many mycophilic immigrants from Southeast Asia and an increase in the number of toxic mushrooms throughout the state. Recently, the heavily urbanized states along the East Coast have begun to claim a share of the victims.[3]

In Europe, *Amanita phalloides* is responsible for the overwhelming majority of fatal mushroom poisonings—various estimates range as high as 90% of all mushroom-related deaths. This species fruits very widely across the European continent and in Russia, sometimes abundantly. Perhaps the single most tragic episode occurred in 1918 near Poznań in Poland. Thirty-one schoolchildren died after eating a dish containing *A. phalloides* that had been prepared at school.

Although *Amanita phalloides* causes most of the serious mushroom poisonings, a number of other amatoxin-containing species have been the documented culprit. *Amanita virosa*,[4] *A. ocreata*, and *A. verna* are amatoxin-containing mushrooms that have each caused a few serious poisonings.

Amanita virosa

Running a distant second to *Amanita phalloides* in number of amatoxin poisonings are *Galerina autumnalis* and *G. venenata*, which are generally mistaken for *Psilocybe* species. Because these specimens are much smaller than the *Amanita* mushrooms, the absolute quantity of amatoxin consumed is generally small. Consequently, the majority of patients poisoned by *Galerina* species develop mild to moderate signs of liver toxicity, followed by spontaneous recovery. *Galerina venenata* was responsible for one death in Washington in 1981.[5] Altogether only a handful of deaths have been ascribed to these species, but they have caused considerable morbidity.

The small toxic species of *Lepiota* have been responsible for a few poisonings in North America, Chile, Israel, southern Europe,[5,6] and Turkey.[7] All the people involved in recent poisonings in Turkey were migrants; they were seasonal cotton laborers working in a part of the country with which they were unfamiliar. In one incident, a mother and three children died. The husband, who did not like mushrooms and refused to eat the meal, survived. A second incident involved two

Lepiota castanae

families; all seven victims perished. Twenty-seven additional individuals were poisoned in the autumn of 1988, after consuming meals of either *L. helveola* or *L. castanae*. Of the first 11 patients poisoned during this particular "epidemic," 10 died;[8] all 11 failed to seek medical assistance until many hours after the mushroom meal. Subsequent victims presented themselves for therapy much earlier in the course of the intoxication. This behavior was almost certainly in response to television reports of the earlier poisonings, which alerted the public to the danger. Later victims all sought medical attention the morning after ingesting mushrooms the previous evening, and shortly after the onset of vomiting and nausea. In this second group of patients, who were treated early with the usual potpourri of techniques, mortality was 4 out of 18. No evidence of renal involvement was noted in any of the patients.

In the United States in 1986, a Finnish-born engineer misidentified some *Lepiota* specimens in upstate New York and succumbed to the amatoxins in *L. josserandi*.[9] It is most unwise to scorn the potential toxicity of the small *Lepiota*.

DISTRIBUTION AND HABITATS OF THE AMATOXIN-CONTAINING MUSHROOMS

The accompanying species list identifies the mushrooms that have been shown to contain the clinically important amatoxins, namely, the toxins that go under the rubric amanitins.[5,15,16] The distribution of the genera *Amanita*, *Lepiota*, *Galerina*, and *Conocybe* across the North American continent is quite extensive.[17–20]

Amanita Species

Most of the toxic species of *Amanita* are mycorrhizal fungi and are found in association with trees. Although woodlands are their primary habitat, they also appear in parks, golf courses, and other small wooded areas near, but not necessarily directly under, a particular tree. The few *Amanita* species that are not mycorrhizal are of little toxicological interest.

The mushrooms purportedly responsible for the poisonings described in the case history at the beginning of the chapter were specimens of *Amanita phalloides*. The history of this species in the

AMATOXIN-CONTAINING SPECIES AND THEIR COMMON NAMES

Amanita species

A. *bisporigera*   Destroying angel
A. *ocreata*
A. *phalloides*   Death cap, deadly amanita
A. *verna*   Spring destroying angel, spring amanita, white death cap, fool's mushroom
A. *virosa*
A. *tenuifolia*
A. *suballiacea*
A. *hygroscopia*
A. *magnivelaris*

Lepiota species

L. *helveola*
L. *josserandi*   Deadly parasol
L. *heteri*
L. *castanae*
L. *subincarnata*
L. *brunneo-incarnata*
L. *scobinella*

Galerina species

G. *marginata*   Marginate pholiota
G. *autumnalis*   Deadly Galerina[10]
G. *venenata*   Deadly lawn Galerina
G. *unicolor*
G. *beinthii*
G. *badipes*
G. *fasciculata*[11]
G. *sulcipes* (tropics; not United States; extremely toxic)[12,13]

Conocybe species

C. (*=Pholiotina*) *filaris* (United States)[14]
C. *rugosa* and other related species may contain toxin

United States is rather poorly understood. In Europe, *A. phalloides* is quite abundant and well described. There is speculation that it is not native to the United States, but the mechanism of introduction of *A. phalloides* into this country is unknown. Early reports of *A. phalloides* poisonings from around the turn of the century are generally discounted by current mycologists, even though many accounts are clearly compatible with amatoxin poisoning.[21] Most of the evidence points to the importation of the species in the 1920s or 1930s along with European nursery stock. It arrived either in the soil or, more likely, in mycorrhizal association with ornamental trees and shrubs. Now that it is here, keeping track of its migrations will give some indication of its habitat preferences and its adaptability. If it keeps spreading, the risk of serious poisoning is likely to increase.

The first well-authenticated identification of *Amanita phalloides* in the eastern part of the country was made by Harold Burdsall in Maryland in 1967. Perhaps it fruited earlier and more widely, but, if so, it was not recognized. Subsequently, numerous collections were made in the region around Rochester, New York, in Pennsylvania, and in a variety of other eastern habitats.[17] Sightings of *Amanita phalloides* in Delaware, New Jersey, Virginia, Rhode Island, Massachusetts, and Maryland have now been reported.[22] Despite the number of states listed, the mushroom is neither widespread nor common. The regions of the Southeast, the Midwest, and the Mountain states seem to be free of it. It has fruited in California for a number of years and appears to be gradually moving up the coast through Oregon and into Washington. However, it is extremely rare north of central Oregon. Confirmed identification was made in Washington in 1965. Ben Woo, the founding president of the Puget Sound Mycological Society, followed the annual fruiting of *A. phalloides* specimens near a Seattle home until 1986, when the mushrooms disappeared after the owners relandscaped the yard. A possible relationship to one of the surrounding plants was not established in this case, because a dogwood, a red oak, and a juniper, all of unknown parentage, were nearby (B. Woo, personal communication).

In California, *A. phalloides* tends to associate with oak trees. In Europe, it grows in deciduous or mixed deciduous woods of oak, beech, and hornbeam. Any conifers in the vicinity are fortuitous. On the eastern seaboard, from Maryland to Vermont and throughout New York and Pennsylvania, it is associated with various hardwoods and conifers.

The range of *A. phalloides* appears to be expanding gradually, and collectors in neighboring states could well see specimens in the next few years. All amateur mycologists should watch for the appearance of this mushroom, so that its progress can be tracked. Its fruiting season is variable, depending on the climatic conditions. In California, it has been reported from August to January; in the East, it is rarely seen after November.

Amanita ocreata has an even more restricted range than *A. phalloides* does, being found primarily along the Pacific Coast in California.[19] It has not been well studied, however, and may have a wider distribution. It associates primarily with oak trees. It fruits in the spring, which in California is as early as January. This mushroom has caused a number of poisonings.

Some of the other toxic amanitas are more widespread. *Amanita verna* fruits in the spring, summer, and early fall and associates with oaks and chestnuts. *Amanita virosa* is found in hardwood and mixed forests. It is most common throughout the southeastern part of Canada and the eastern seaboard of the United States. It is sometimes quite common along the shores of Lake Ontario. *Amanita bisporigera* occurs across the northern tier of states and throughout Canada, favoring the eastern regions and becoming quite rare in the West. It is found in hardwood forests or mixed woodlands, most often associating with oaks or, in the West, with birches and aspens. The remaining amatoxin-containing amanitas occur largely in the warmer southern and southeastern portions of the United States. These regions of the country have the largest variety of *Amanita* species.

Galerina Species

The ecological niches inhabited by members of the genus *Galerina* are quite different from those of most amanitas. *Galerina marginata* and *G. autumnalis* are saprophytic species that grow on dead and decaying wood, so they are found on logs and stumps or pieces of woody debris, such as wood chips and sawdust. The dead wood may be buried, thereby giving the emerging mushrooms the appearance of terrestrial species. This false appearance deceives the *Psilocybe* picker, who generally frequents grassy areas where the hallucinogenic mushrooms are found. Other mushrooms of the genus *Galerina*, for example, *G. venenata*, a lawn-inhabiting species, are also lying in wait for the careless collector. *Galerina venenata* is found mainly in the West.

Members of the genus *Galerina* fruit primarily in the late summer and fall. However, *Galerina autumnalis*, despite its moniker, is known to fruit in the spring in some years. This habit has been noted especially in the eastern United States, where it may fruit during the morel season.

Galerina species

Lepiota and *Conocybe* Species

The small mushrooms belonging to the genus *Lepiota* are also typically fall-fruiting mushrooms, but they may be seen at other seasons, including the summer months. The various species of this genus are widely distributed throughout the United States. The habitats occupied are quite variable. *Lepiota* specimens can be found at the edges of woods, on roadsides, on lawns, and in meadows, parks, and gardens—a rather catholic genus, in terms of its habitat preferences.

Only North American specimens of *Conocybe filaris* have been shown to contain amatoxins. This mushroom has not been associated with any documented human poisonings, but it is a lawn-inhabiting species and could be misidentified by the *Psilocybe* seeker. It is especially partial to habitats richly sprinkled with wood chips or sawdust. Its distribution across the country is not well studied, but it has been reported in the Pacific Northwest, where it fruits during the warm, rainy months. It can also occur early in the year in well-watered yards.

METABOLISM AND MECHANISMS OF ACTION OF AMATOXINS

The history of chemical investigations of the toxins in poisonous *Amanita* species is a 200-year saga in which persistence, false leads, and extraordinary, single-minded devotion played large roles. Although many individuals were involved in the research, the Weiland family can be credited for much of our current knowledge. That story and a complete summary of present-day knowledge of amatoxin toxicology is well recounted by T. Weiland.[15] Not only did the research lead to a full description of the toxins, but, in the process, a number of powerful, molecular biological tools were developed.

The prototypical mushroom, the one responsible for the largest percentage of mushroom-related fatalities, is *Amanita phalloides*, the death cap. This mushroom produces two families of extraordinarily potent peptides, phallotoxins and amanitins.[15,23–25] These compounds, constituting a large family of related molecules, are all bicyclic peptides. Amanitins contain eight amino acids (they are octapeptides) and have a molecular weight of 900. Phallotoxins have one less amino acid than the amanitins and are all heptapeptides. This family of toxins is finding increasing use in biological research.[26]

Phallotoxins

When phallotoxins are given parenterally to an animal or when isolated liver cells are exposed to them, the phallotoxins form strong complexes with F-actin (filamentous actin, a critical cytoskeletal protein), impair the cortical web associated with the plasma membrane of the cell, and cause rapid cell death. But, despite these dramatic parenteral or in vitro effects, phallotoxins have no role in damaging the liver because they are not absorbed from the gastrointestinal tract in a biologically active form. This fact has been confirmed in experiments with isolated intestinal cells.[27] The older idea that phallotoxins are destroyed by enzymes in the gut now seems invalid.

Phallotoxins cannot even be responsible for the initial gastrointestinal phase of amatoxin poisoning, which develops 10 to 12 hours after ingestion. It has been shown that phallotoxins act very rapidly on susceptible cells. Therefore, if phallotoxins were active in the gut, any symptoms produced by them would develop very soon after ingestion.

Instead, there is no evidence of early gastrointestinal mucosal cell sensitivity to this group of toxins.

Amanitins

Amanitins are clearly the cause of the symptoms of amatoxin poisoning. They have a very high specificity for RNA polymerase II, completely inhibiting its activity. Because this enzyme is responsible for the transcription of DNA to messenger RNA, all protein synthesis ceases when it is inhibited, and cell death ensues. Because of their mechanism of action, amanitins most severely affect cells that have a rapid protein turnover.

Amanitins inhibit the transcription process in all eukaryotic cells, although sensitivity to the toxin varies from one cell type to another. Not surprisingly, the cells of those mushrooms that normally accumulate amanitins have the lowest sensitivity; these cells are approximately 10,000-fold less sensitive than mammalian cells are. However, some organisms, for example, a mutant strain of the fruit fly *Drosophila melanogaster* (an organism so beloved by geneticists and developmental biologists), actually flourish in the presence of amanitin, because they possess a mutated gene for RNA polymerase II.

The liver is not unusually sensitive to amanitins; it just happens to be the first organ, after the gastrointestinal tract, exposed to high concentrations of the toxin. And because the liver is responsible for a major portion of protein synthesis in the body, its failure produces symptoms that overshadow the symptoms produced by other poisoned organs. The fact that all the cells in the body are affected by amanitins may become more significant as "artificial liver" techniques improve. If such machines are used to support future amatoxin-poisoned patients, then it is likely that other failing organs will become more important, especially when a large dose of toxin has been ingested.

Other Amatoxins

Poisons other than phallotoxins and amanitins have been described in *Amanita* mushrooms, but their role in human poisonings has not been elucidated. Virotoxins are monocyclic heptapeptides. A virotoxin and a compound called amaninamide have been identified in *Amanita virosa*.[28,29] Phallolysins constitute a family of compounds with hemolytic properties. These hemolysins are heat labile and sensitive to the low pH in the stomach, so it is assumed that they play no role in human poisoning.

Absorption, Excretion, and Transport

What percentage of the toxin in a mushroom is actually absorbed in the human gut is unknown, but experiments in other animal species have shown that the amount of absorption varies from one animal species to another. Rodents such as the mouse and the rat cannot be poisoned orally. Dogs do absorb amatoxins from the gut, but, to produce comparable effects, oral doses must be five times larger than intravenous doses; for cats, the oral dose must be 10 times larger than the intravenous dose.[15] Because of the great variation in toxicity between various animal species, it is not a good idea to use the "pet" test to determine whether a mushroom is safe to eat—nor is it a good idea to provoke the ire of animal rights groups.

Amatoxins are also taken up by the epithelial cells lining the gastrointestinal tract, where they immediately begin to inhibit protein synthesis. These mucosal cells rapidly undergo degenerative alterations.[30]

Once in the bloodstream, amatoxins circulate to the liver, where they are taken up by hepatocytes.[31] An active transport system, similar or identical to the multispecific transport mechanism mediating the uptake of bile salts, seems to be the likely mechanism of toxin uptake in liver cells. A recent study proposed that Kupffer cells (hepatic macrophages) are involved in the uptake of amatoxins.[32]

In one series of experiments, toxin transport into liver cells was blocked by competing compounds, which included bile salts (taurocholate), prednisolone, antanamide, and silibinin. Antanamide is one

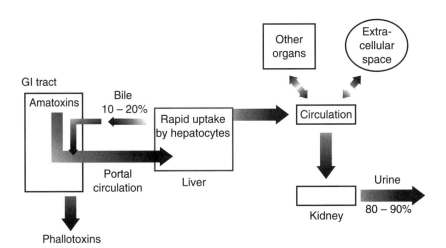

Uptake and excretion of amatoxins

of the nontoxic amatoxins present in small concentrations in many toxic amatoxin-containing species. Unfortunately, it also inhibits reexcretion of the amatoxins into bile, so it appears to have no value as a therapeutic agent. Antanamide and the other substances appear to compete specifically with the amatoxins. Interestingly, neither penicillin G nor thiotic acid demonstrated this effect.[33]

A large fraction of the circulating amatoxins undergoes rapid renal excretion. In a dog model, over 90% of an ingested dose was excreted by the end of six hours.[34] In human victims, renal excretion is equally striking. The concentration of amatoxins in the urine is 100- to 150-fold higher than that in the blood. A 15-year-old boy attempted suicide by eating an *Amanita phalloides*. Amanitin was detected in his urine two hours after ingestion.[35] The importance of renal excretion in the elimination of toxin has been amply demonstrated.[36] These studies indicate that efforts to enhance renal excretion should be instituted as early as possible.[37]

Urinary excretion can continue for more than 72 hours because of the persistence of mushrooms in the gastrointestinal tract or because of the recirculation of toxins within the enterohepatic system.[38,39] A fraction of the amatoxin taken up by the liver is excreted in the bile. This route of elimination begins immediately after the amatoxins enter the hepatocytes. Unfortunately, enterohepatic circulation allows the amatoxins excreted into the bile to be reabsorbed in the gut and again by the liver cells, thereby prolonging the liver's exposure to the toxin.[38,40] In experimental situations, excretion into the bile can be inhibited by penicillin and by silymarin or silibinin.[41] The importance of enterohepatic circulation in perpetuating and enhancing the hepatic toxicity of amatoxins was demonstrated in dogs, in which the creation of bile fistulae (to circumvent the reabsorption of toxin) proved to be life saving.[42]

There is no evidence that amatoxins cross the placenta in the third trimester, so a fetus should not be affected when the mother eats an amatoxin-containing mushroom and survives. A 21-year-old woman who was eight months pregnant was moderately poisoned with amatoxin. At admission, the amount of α-amanitin in her serum was 18.5 ng/mL (measured by high-pressure liquid chromatography); there was no detectable toxin in the amniotic fluid. She subsequently delivered a full-term, healthy baby with no evidence of hepatic damage.[43] Similarly, a 13-week fetus survived the poisoning of its mother, with no complications at birth or in the postnatal period.[44] A hapless 19-year-old woman and her seven-month fetus died as a result of the ingestion of *Amanita phalloides*. The mother's liver showed typical massive necrosis, whereas the fetus's liver was entirely normal.[45]

The behavior of amatoxins in the first few weeks of pregnancy is still unknown. There is one report of a nine-week fetus that was therapeutically terminated three weeks after the mother ingested amatoxin; the fetus presumably was damaged.[46]

Amatoxins do get into breast milk. In one unusual circumstance, a nursing infant was affected.[47]

TOXICITY OF AMATOXINS: HOW MUCH WILL KILL YOU?

α-Amanitin is extraordinarily toxic for humans. The LD_{50}, the dose that will kill half the people ingesting it—if such an experiment could be done—has been estimated to be about 0.1 mg/kg body weight.[25] This amount is equivalent to 6–7 mg for the average person, easily the quantity of toxin in one decent-sized mushroom cap. In quantitative studies, the concentration of α-amanitin varied from 0.5 to 1.5 mg/g of mushroom,[48] and the average Amanita phalloides cap weighs about 60 g.

Chilton compiled a list of the concentrations of all the various toxins from a number of different collections of amatoxin-containing mushrooms.[49] The toxins were found in all portions of the mushroom, but not in equal amounts. The gills contained the largest amount (about 46% of the total amount), whereas the stem and the cap each had about 23%. The remainder of the toxin was in the volva.[23] Touching a mushroom is not dangerous—it has never proved to have adverse consequences.

Numerous studies have evaluated the presence and concentration of the amatoxins in mushrooms collected in different countries and the variation of toxin concentrations among specimens in a single collection.[22,48,50–52] Concentration varies considerably from one specimen to the next, even among specimens collected at the same time. This variability is quite striking for Amanita virosa and A. verna. Some collections contained little amanitin, whereas others had potentially lethal amounts.

Concentration differences between species were greater than those within a species. And even though clinically significant concentrations of amatoxins were found only in the small group of mushrooms listed earlier, very sensitive techniques have detected minute quantities in a number of edible species.[53]

When amatoxin is artificially bound to protein, its toxicity appears to increase.[54] This serendipitous observation occurred during experimental attempts to raise antibodies against the toxins. Because the poisons are such small peptides, they are not immunogenic and must be bound to albumin or a similar protein before they will stimulate antibody production. During the immunization process with such a conjugate, most of the inoculated animals died. Whether this lethal response was due to decreased renal elimination of the protein-bound toxin,[54] thereby allowing it to prolong its action, or to increased uptake by hepatocytes is unknown. A more recent attempt to use a monoclonal antibody raised against α-amanitin produced the same disastrous results. The authors concluded their abstract with a supreme piece of understatement: "To our knowledge this is the first reported case where immunoglobulins or their fragments enhance rather than decrease the activity of a toxin. Accordingly, immunotherapy of *Amanita* mushroom poisoning in humans does not appear promising."[55] Fortunately, amatoxins generally do not bind to human serum proteins.[56]

The concentration of amatoxins in *Galerina* generally is less than that in a toxic *Amanita*. The lethal dose has been estimated at 100–150 g of fresh mushrooms. Because *Galerina* specimens are small, 10–20 mushroom caps constitute a fatal meal. Coincidentally, and sometimes unfortunately, this number is similar to the average dose of *Psilocybe* ingested by individuals seeking a mushroom high.

The amatoxins are heat stable and cannot be inactivated by freezing or drying.

HISTORY OF THE TREATMENT OF AMATOXIN POISONING

The prevalence and severity of amatoxin poisonings in Europe have spawned numerous attempts to find specific antidotes or treatments to improve the survival of poisoned victims. It is a fascinating history, full of strange twists and turns, wrong leads, poorly designed experiments, and occasional flashes of insight.

In the early nineteenth century, at least two toxins were believed to be important in mushroom poisonings, a hepatotoxin responsible for the degeneration of the liver and kidneys and damage to the intestines and a neurotoxin thought to cause paralysis and other damage to the central nervous system. We know today that the latter effects are largely due to metabolic derangements associated with liver failure. The amatoxin-containing mushrooms probably contain no neurotoxin.

The traditional therapies, such as cathartics, purgatives, cautery, and bleeding, frequently failed to benefit the patient, so it was only natural to look for more satisfactory alternatives. One of the first appeared in the early 1800s. This therapy was based on the observation that rabbits could eat poisonous mushrooms without ill effects. It was assumed that some agent or principle in the stomach of rabbits must be able to neutralize the poisons. The "antidote" formulated on the basis of valid observations—but flawed logic—was a concoction consisting of the stomachs of three and the brains of seven rabbits, all finely minced or ground, fashioned into pellets, and served—or rather, administered—with sugar or jam. It is doubtful whether the sugar coating improved the flavor or the palatability much, especially for the poor patient who was already nauseated or vomiting. Of course, some patients survived both the mushroom and this treatment, because the mortality rate at that time—without therapy—was not much more than 50%. These "successes" were accepted as evidence that the treatment was effective.

This initial effort at devising a therapy for amatoxin poisoning was followed near the end of the nineteenth century by the introduction of an antiserum from the Pasteur Institute in Paris, a gift from the burgeoning new science of immunology. Antibodies were produced by inoculating animals with unheated extracts of mushrooms. The antibodies produced were largely directed against the hemolysins (phallolysins), which play no role in human poisoning (see page 207). Once again, poisoning victims survived, and the mythology of a successful antidote was perpetuated. The Institute continued to produce and distribute the antiserum until the 1960s.

Another Frenchman, Dr. P. Bastien, developed a very personal and unique approach.[57,58] He carried out clinical "trials" from 1957 to 1969 on 15 amatoxin-poisoned patients, all of whom reportedly recovered. His method reached the public's attention in the 1970s. It consisted of a "cocktail" comprising multiple intravenous injections of ascorbic acid (vitamin C), administered as 1-g doses morning and evening or up to 3 g/day by intravenous infusion; nifuroazide, two tablets three times a day (1200 mg/day); and abiocine (dihydrostreptomycin), given in the form of two tablets three times a day (1500 mg/day). A carrot soup broth was the sole source of nutrition during the three days of therapy.

In 1974, to prove and promote the effectiveness of his therapy, Dr. Bastien ate a specimen of *A. phalloides* (65 g), treated himself, and recovered. He gave a repeat performance in Geneva in 1981, when he ate 70 g of *A. phalloides* and recovered after receiving his own treatment. A number of published articles quote a variety of European experiences that appear to confirm the efficacy of his treatment.[57,59] However, this treatment is not the standard of practice in France. Until a well-described series is published, with the details needed for evaluation, this form of therapy remains an intriguing enigma.

The continuous and unabated search for an antidote that will reverse the effects of amatoxin is a testament to human optimism. The best agent would be a compound with the ability to disrupt the interaction

Historical highlights of the treatment of amatoxin poisoning

Date	Investigator	Treatment
Early 1800s	H. Limousin, Paris	Ground rabbit stomach and brains
1821	Joseph Roques	Large quantities of sugar water
1897	Albert Calmette	Antiserum "sérum antiphallonique," Pasteur Institute
1936	Le Calve	Half-hourly administration of one teaspoon of salt in a glass of water
1968	J. Kubicka	Thiotic acid[125]
1970	P. Bastien	Vitamin C and antibiotics[58]
1983	K. Hruby	Silybin and silibinin[106]
1985	E. Woddle	Liver transplantation[115]

between amanitin and RNA polymerase II, thereby allowing the enzyme to function again, and to change the released amanitin into a harmless and excretable molecule. Barring that miracle, investigators continue to search for agents that limit the damage. This approach is reasonable, because not all the amanitin is immediately bound to the liver cells. The principle of limiting the damage underlies duodenal suction and drainage, which is a treatment that decreases the entero-hepatic recirculation of the toxin in the bile, thereby interrupting the cycle and limiting further damage to the liver. The same principle motivates the use of many of the "hepatoprotective" agents that have found favor at various times. Most of these are thought to inhibit binding or uptake of toxin by liver cells.

During the search for antidotes, inappropriate experimental animal models have been used. The profound species differences in absorption of and sensitivity to amatoxins have led to several invalid interpretations, and these errors have been compounded by the use of alternate routes of administration of the toxin—many experiments bypass the gastrointestinal tract completely. On rare occasions, even the sex of the experimental animals has proved to make a difference in the outcome. One experiment convincingly demonstrated that cytochrome c protected female mice from intoxication.[60] Unfortunately, this agent had no effect in male mice; and if females were oophorectomized and given androgens, they no longer benefited from the treatment.[61,62]

Experiments that do not reproduce the poisoning situation in human beings are not reliable measures of the therapeutic effectiveness of a particular agent. For example, it has been reported that mice pretreated with kutkin, an iridoid glycoside extracted from *Picrorhiza kurroa*, are protected against an intraperitoneal injection of *A. phalloides*.[63] There was even a smidgen of evidence that delaying the dose of kutkin for a couple of hours still provided some protection, although many of the animals died. However, intraperitoneal injection of a mushroom extract introduces many additional toxins, especially the phallotoxins, that are not normally important in human poisoning by the oral route. And few individuals are astute enough to know that they are about to be poisoned by a mushroom, so prophylactic therapy is useless. Of course, it is intriguing that a compound used in native American folk medicine does indeed have a "hepatoprotective" effect. But it is a very long way from this anecdotal observation to a therapeutic application.

Historically, the outcome for many amatoxin poisonings was miserable. In the early part of the twentieth century, mortality rates of 60–70% were common. However, the introduction before the World War II of good fluid replacement therapy and better supportive care lowered the mortality rate to approximately 30%. With further refinements in intensive care, overall mortality is now less than 30%. Of course, it varies from one location to another, depending on the level of sophistication and experience of the treating institution. In Floersheim's 1982 report of a series of 205 patients, the overall mortality rate was 22.4%, although mortality was considerably higher for the children treated (over 50%) and very much lower in the adults.[64] In other collaborative studies, mortality rates were 10.3%[65] and 11%.[66] In the best series—from Barcelona, Spain—the reporting team had only a 5.5% mortality rate, 5 deaths out of 91 patients managed between 1982 and 1988.[38]

A report on a series of poisonings that occurred in California in 1982 cited 2 deaths in 22 cases—a mortality rate of 9%. The treatment of these patients was conservative and included aggressive intensive care, administration of activated charcoal, and duodenal drainage.[2] This markedly improved outlook in amatoxin-poisoned patients managed with intensive care alone has very important implications for the role of liver transplantation in treating this syndrome (see pages 228–229).

The claim has been made that alcohol taken with the mushroom meal has a protective effect. The idea originates from a series of animal studies that bear little relationship to the way human beings imbibe alcohol or eat mushrooms. Large doses of alcohol were given to mice just before an intraperitoneal injection of powdered, lyophilized mushroom. Most of the mice survived the challenge. Conversely, when purified phallotoxins or amanitins were given by the same route, most of the animals died, so the results were inconclusive.[67] Furthermore, the dose of alcohol given these mice, when adjusted for relative body weight, was considerably more than most individuals would consume at an average meal.

The idea that alcohol may have a hepatoprotective effect is also based on the notions that alcohol competes with the toxins during uptake by liver cells or that alcohol alters the permeability of the liver cell membranes, causing structural modifications of the membrane lipids.

Despite the lack of solid evidence of alcohol's protective effect, this idea persists, perhaps because of supportive anecdotal evidence. For example, it has been claimed that of four patients poisoned in Finland in 1989, the two who drank alcohol during the meal had less liver damage than the two who had not.[68] Although such observations are interesting, dose differences and individual variations in response to toxin could also explain the outcomes. No one should rely on this form of protection from poisonings. At best, it is a convenient excuse to have a glass or two of wine with dinner.

However, alcohol should be taken into account by the treating physician who must evaluate the success of various treatment regimens. Alcohol may modify the effects of some treatments.[69]

CLINICAL FEATURES OF AMATOXIN POISONING

Onset

Onset of symptoms is generally between 8 and 12 hours after ingestion of an amatoxin-containing mushroom. In unusual cases—probably when the amount of toxin ingested is very large—the interval between ingestion and onset can be as short as six hours. Conversely, in mild cases, the first symptoms may not appear until 36 hours after ingestion.

In a few cases, onset is deceptive. When the victim eats a mixture of different species at the same meal, one of which contains a gastrointestinal irritant and another an amatoxin, vomiting begins within two hours. In such a clinical situation, the seriousness of the poisoning may be misjudged. (Preventing this dangerous error is not the only reason to avoid combining wild mushrooms in an epicurean potpourri. Each mushroom has a distinctive flavor and texture. These characteristics cannot be appreciated when multiple species are mixed in the same dish.)

A second situation in which time of onset is misjudged arises when the patient eats the amatoxin-containing mushrooms at two or more meals and the onset of vomiting coincides with the last meal. This circumstance challenges the physician's history-taking skills, because an illness is usually linked to the most proximate event. In this instance, the symptoms might be mistakenly attributed to simple gastrointestinal distress.

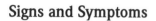

Signs and Symptoms

Classic amatoxin poisoning is a triphasic illness, with the first gastrointestinal phase being followed by a period of relative well-being—sometimes called the "honeymoon" period—before the terminal drama of organ failure is played out.[70–72]

The initial symptoms of amatoxin poisoning constitute a rather non-specific gastrointestinal syndrome characterized by the sudden onset of cramping abdominal pain, vomiting, and watery diarrhea, which is frequently described as "choleralike." The diarrhea can become severe, and dehydration may develop unless fluid replacement therapy is instituted. The diarrheal stool may contain flecks of mucus and blood.

The gastrointestinal symptoms frequently subside spontaneously after about 24 hours. The patient may appear well enough to be discharged from the hospital. There are apocryphal stories of patients dying suddenly at home after discharge.[73] The victim who was never hospitalized tends to believe that he or she has recovered from whatever caused the initial symptoms. Although remission is part of the classic amatoxin-poisoning syndrome, it does not occur in every patient. Indeed, some patients progress rapidly to the next stage without this brief respite.

Abdominal symptoms usually recur by 72 hours postingestion and are accompanied by evidence of hepatic dysfunction and the onset of jaundice. The liver is enlarged and tender to palpation. Hypoglycemia may be seen. Severe diarrhea may recur and lead once again to dehydration. In patients whose liver necrosis progresses rapidly, the ensuing coagulopathy (due to the progressive decrease in the production of coagulation proteins by a dying liver) may be associated with significant gastrointestinal hemorrhage.

Patients with the most severe intoxication progress to acute hepatic failure, with all the expected signs and symptoms. Hepatic encephalopathy develops, with confusion, delirium, convulsions, meningeal signs, and progressive coma. Electroencephalographic data gathered at this time are consistent with liver failure.[74] Death generally occurs 7 to 10 days after onset of symptoms.

Renal failure is common in severe intoxication, but it is seldom prominent early in the course of the illness. Moreover, its symptoms are usually overshadowed by those produced by the failing liver. Kidney failure generally becomes evident five or more days after the onset of the poisoning symptoms, frequently in patients who survive the hepatic necrosis.

In a small number of patients, renal failure may dominate the clinical picture.[75] This is most likely to occur in patients who ingest a sublethal dose of amatoxin and in whom the initial gastroenteritis results in severe dehydration and pre-renal azotemia. If the dehydration is not adequately managed with aggressive fluid replacement, it may exacerbate the toxic effects of the amatoxin on the kidney.[76] These patients develop laboratory evidence of renal compromise 10 to 20 days into the course of the poisoning, when the liver is already showing signs of recovery.

Additional complications are seen on rare occasions. One is paralytic ileus.[77] This complication may be related to an electrolyte imbalance (hypokalemia) or to renal failure. It generally resolves with potassium replacement. A dreaded complication—fortunately, very uncommon—is intestinal perforation. This lesion further complicates management and is often fatal.[78,79] Cardiac manifestations, due both to metabolic disturbances and to direct effects of amatoxin, have been described in patients who survived the organ failures. Abnormalities in rhythm or conduction may last a month or more after the patient has recovered.[80,81] Mesenteric venous thrombosis has also been observed. This disorder may be accompanied by disseminated intravascular coagulation, a condition further impairing the blood's clotting ability.[78,82]

Key Clinical Features of Amatoxin Poisoning— A Triphasic Syndrome

Delayed onset: usually 8–12 hours (range 6–36 hours)

Gastrointestinal phase: 8–48 hours

- sudden-onset, severe, colicky abdominal pain
- vomiting; watery diarrhea
- no fever

"Honeymoon" phase: 48–72 hours

- relative remission of clinical symptoms
- increases in transaminase levels

Terminal phase: 72–96 hours

- recurrence of abdominal symptoms; watery, bloody diarrhea
- onset of jaundice; progressive hepatic failure
- possible renal failure
- encephalopathy, convulsions, and coma
- death in 7–10 days in 10–15% of patients

LABORATORY FINDINGS

In the first 24 hours after mushroom ingestion, all laboratory parameters are normal. The first evidence of impending disaster is an elevation in the serum levels of the transaminase enzymes.[4,83] Levels of alanine aminotransferase (ALT) and aspartate aminotransferase (AST) begin to rise rapidly after 24 hours and generally peak at three to four days after ingestion. They frequently reach values of several thousand. Transaminase levels may then fall gradually over the ensuing days. This decline does not necessarily imply an improving clinical situation; instead, it may indicate that the liver cells have released all the transaminases they contain. In such circumstances, the synthetic function of the liver, as reflected by the prothrombin time, will continue to deteriorate. A few investigators have suggested that the more rapidly the enzyme levels decrease, the worse the prognosis. However, this is not a reliable indicator.

Alkaline phosphatase isoenzymes from bone, liver, and other sites have been examined during the course of amatoxin poisoning by *Lepiota* specimens.[84] Electrolyte abnormalities may be noted, depending on the fluid status of the patient and the success of fluid replacement therapy. In severe cases, a metabolic acidosis may become evident. A recent study of 21 children with moderate poisoning found hypophosphatemia between days 5 and 7. Hypocalcemia occurred in children with severe poisoning.[85]

The presence of hypocalcemia has been corroborated by a more extensive examination of the endocrine system in four patients. Marked abnormalities were found in the levels of hormones regulating glucose, calcium, and the thyroid gland. At the time of admission, insulin and C-peptide levels were elevated, a sign suggesting that the accompanying hypoglycemia was not entirely due to hepatic failure, especially early in the intoxication. The impaired hepatic gluconeogenesis must play a significant role, but the high insulin levels suggest increased production or release, perhaps as a result of pancreatic necrosis. A period of hyperglycemia preceded the hypoglycemia in a number of cases. Calcitonin leve's were elevated in conjunction with hypocalcemia. Parathyroid hormone levels increased with time and then gradually returned to normal. The thyroxine level was decreased in all patients, and triiodothyronine was undetectable in three. Levels of thyroid-stimulating hormone (TSH) never became elevated, a result suggesting an unresponsive pituitary–hypothalamic axis or a euthyroid illness syndrome.[86] In three of the four patients in this study, serum magnesium levels were also mildly elevated. Cortisol levels were markedly elevated. It appears that profound abnormalities in endocrine homeostasis accompany amatoxin poisoning, although the mechanisms underlying the changes are not known.

As liver necrosis progresses, prothrombin time becomes prolonged. This change generally becomes evident at 48 hours and peaks at three to five days. Administration of vitamin K either has no effect or only partially corrects the abnormality. Fibrinogen levels also decrease. The effect of the decreases in all the liver-derived clotting factors is a rapidly evolving coagulopathy.

As noted earlier, hypoglycemia is present after day 3, perhaps as a component of liver failure. Between days 4 and 7, serum ammonia begins to rise.

Evidence of renal failure is usually inconspicuous during the first week, except in patients developing pre-renal azotemia. In fact, blood urea nitrogen may be quite low because hepatic synthesis has been interrupted. It is best to use creatinine level as an index of renal dysfunction.

A number of other abnormalities have been commented on in individual cases. Marrow suppression has been observed, with a modest decrease in platelets, neutrophils, and lymphocytes. The serum creatine phosphokinase level rises in patients that have severe muscle spasms.

PROGNOSIS

Until relatively recently, assessment of the prognosis of patients suffering from amatoxin poisoning was largely an interesting academic exercise, although it was sometimes used to compare various forms of therapy. It has now become a crucial aspect of management—because of the possibility of liver transplantation. Selection of appropriate patients for this heroic surgical treatment now relies on an assessment of the odds that the liver has been irreversibly damaged. Moreover, the sooner this condition can be determined in the course of the poisoning, the greater the likelihood of obtaining a liver and performing successful surgery while the patient is still in operable condition. The early onset of coma is an ominous sign.

Dosage   A definite dose–response relationship has been proved for amatoxin poisoning. The greater the dose, the poorer the prognosis. People who share a meal of toxic mushrooms are the best models for illustrating this effect. Those who have more generous servings generally suffer the more severe consequences. Great care must be exercised in questioning patients about the quantity of a meal eaten, because intake depends on the size of the plate or bowl used, the absolute number of mushrooms eaten, and the number of servings taken. In the case presented at the beginning of this chapter, the patient with the mildest poisoning claimed to have eaten two bowls of the preparation, whereas the more severely affected individuals said they had eaten a single dish. On further questioning, it turned out that the dishes of the most intoxicated patients were large soup plates and the bowl of the least intoxicated patient was a small rice bowl. Furthermore, it is difficult to compare doses ingested in different episodes of poisoning, because the concentration of amatoxins in different collections of mushrooms varies widely. Other signs and symptoms must be considered in assessing dose (see the accompanying table).

Correlation of signs and symptoms with severity of amatoxin poisoning

Parameter	Mild poisoning	Moderate poisoning	Severe poisoning
Onset of symptoms	>3 days	12–24 hours	<8 hours
Transaminases	<1000IU/L (mild rise); peaks at 5 days	1000–5000 IU/L; peaks at 4 days	>5000 IU/L ; peaks at 2 days
Prothrombin	Normal	Mildly abnormal	>100 seconds at 96 hours
Hypoglycemia	None	Present	Present
Fibrinogen	Normal	Normal	Decreased
Metabolic acidosis	None	None	Present
Serum ammonia	Normal	Normal	Elevated by 4–6 days
Bleeding tendency	Not present	Not present	May be present
Renal insufficiency	Not present	May occur by 7–10 days; resolves	May develop late
Outcome	Spontaneous resolution	Recovery after intensive, supportive care	Hepatic failure and death likely, but not inevitable, even without transplantation

Age   The age of the poisoning victim affects the relative dose of amatoxin. When dose is normalized on the basis of body weight, the relative dose of amatoxin ingested by children is often greater than that ingested by adults. Therefore, any amatoxin poisoning in a child should be treated with great concern. Furthermore, children seem to tolerate poisoning much more poorly than adults do. Whereas the mortality rate for adults has decreased to about 10%, it remains at about 50% for children. Whether factors other than dose—for example, intrinsic pharmacokinetic and metabolic differences between children and adults—are responsible for the great difference in mortality is unknown.

Length of Latent Period   The shorter the period between the time of mushroom ingestion and the onset of symptoms, the worse the prognosis. This parameter is one of the most valuable indicators of severity of intoxication. Again, dose is implicated, because generally the time of onset correlates with the dose of amatoxins ingested. In the majority of patients, the first symptoms appear around 12 hours. When onset occurs before eight hours, especially as early as six hours, the prognosis is poor.

Laboratory Results   Several signs that can be assessed by laboratory tests are useful in determining a patient's prognosis (see the table on page 221).

◆ Early rise in transaminase levels and high peak levels indicate a poor outcome. However, there have been many patients with transaminase levels above 5000 IU/L who eventually recovered normal liver function. Therefore, transaminase levels cannot be used as a single or an infallible indicator.
◆ Defects in protein synthesis, especially the synthesis of clotting factors, are associated with a poor prognosis. In the United States, prothrombin time is used as a measure of clotting factor synthesis in the liver. If prothrombin time lengthens significantly within the first 48 hours or if it is greater than 100 seconds at any time, the outlook is poor. Generally, this condition cannot be more than partially corrected with injections of vitamin K. Later in the course of the intoxication, fibrinogen levels also may be decreased. A good prognostic sign is the return of the prothrombin time toward normal. This trend is usually the earliest manifestation of recovery.
   In Europe, the use of the Quick test (thromboplastin time) has been closely correlated with outcome. When it is less than 10% of normal, the prognosis is grim.[87] In one of the largest European

series, 84% of patients with Quick test values of 10% or less of normal died, whereas all those with values above 40% survived.[64]

- Persistent hypoglycemia is an ominous sign. It is indicative of the depletion of hepatic glycogen reserves and the absence of effective gluconeogenesis.
- Metabolic acidosis may be progressive and can be exacerbated by renal failure.
- A rise in serum ammonia usually develops relatively late and is a good indication that the liver has been damaged and is functioning well below normal.

Other laboratory measurements have no role in the assessment of prognosis.

- Bilirubin levels rise quite late in the course of intoxication and can be quite variable. They are not a good indicator of outcome.
- The concentration of amatoxins in either urine or blood has little clinical value. No obvious correlation between serum levels at the time of admission and outcome has been demonstrated. Nor has the persistence of detectable levels of amatoxin been correlated with severity of clinical illness. It is true, however, that a detectable level of amatoxin in the blood is associated with more severe poisoning.[66]

The fact that each case is a paradigm of polypharmacy and polytherapy constitutes one of the great difficulties in assessing both the prognostic factors and the success of any form of therapy. Moreover, because few institutions ever accumulate sufficient experience to develop meaningful statistics, much of the recommended treatment is based on anecdote. Even the factors presaging a poor prognosis are only partially documented.

MANAGEMENT OF AMATOXIN POISONING

The incidence of amatoxin poisoning is much greater in Europe than in the United States, and the experience gained by some of the larger groups of European clinical investigators is an invaluable resource.[38,64,72,88,89] In some countries, special units have been established to treat mushroom poisoning exclusively.[90] Therefore, I have garnered many of the clinical observations and treatment recommendations from European sources. The one problem with this approach is that some drugs, such as silibinin, are not available in the United States.

Because medical personnel in the United States have so little experience in managing cases of amatoxin poisoning, literature on the subject is very limited and sometimes misleading, as the following quotations from articles published in 1989 in an American journal show. After each quotation, my critique appears in brackets.

- "Mycetism or mushroom poisoning is an increasingly common medical emergency in the USA." [This statement is not supported in the article by data and is marginally true, at best.]
- "The mature *Amanita phalloides*, despite its distinctive green cap and white gills, can be mistaken for similarly appearing mushrooms even by the experts. *Agaricus campestris* is indistinguishable from the 'death cap,' as *Amanita phalloides* is known." [The author does not specify which "experts" he is thinking of.]
- "The cholera-type diarrhea seen most commonly is an effect of the phalloidin." [There is no evidence to support this statement, and it is almost certainly untrue, because symptoms are caused by amanitins. Phalloidin plays no role in human poisonings.]
- "50% of patients have pancreatitis" [No adequate data presented.]
- "Once hepatic coma develops in a patient who ingests *Amanita* mushrooms, the chances of survival with medical therapy alone are practically nil." [Although it is true that the prognosis is grim, there are examples of recovery after institution of aggressive intensive care.[91,92] One patient with amatoxin poisoning caused by ingestion of *Amanita virosa* recovered despite having a severe coagulopathy and being in hepatic coma for two days.[93] This prognosis is very important when liver transplantation is being considered.]

Floersheim criticized another article on mushroom poisoning[94] for painting too bleak a prognostic picture and for touting useless treatments.[95] As late as 1988, an article cited a 50–90% mortality for amatoxin ingestion,[96] rather than the 10–15% currently anticipated.

Early, Presymptomatic Treatment

For the patient who arrives in the emergency department within six hours after ingesting amatoxin-containing mushrooms and who is either asymptomatic or barely beginning to develop nausea, vomiting, and diarrhea with abdominal pain, decontamination of the gastrointestinal tract may still be worthwhile. Unfortunately, this clinical situation is a rare occurrence. The recommended procedure includes all or some of the following treatments.

- Gastric lavage. Use a wide-bore tube and physiological saline.
- Charcoal adsorption. An initial dose of 100 g of activated charcoal should be left in the stomach after lavage. If diarrhea has not yet begun, the charcoal can be mixed with 70% sorbitol for adults or 35% sorbitol for children.
- Catharsis. Magnesium citrate can be administered as a cathartic if sorbitol is not used. However, magnesium citrate has been associated with rare fatalities, and some authorities no longer recommend its use.

Unfortunately, the majority of patients present with both vomiting and diarrhea, usually many hours after the mushroom meal, so neither gastric lavage nor use of emetics provides significant benefits.

Removal of Amatoxins from the Duodenum

There is continued uptake of toxin by the liver for the first three days after ingestion. Toxin continues to pass into the bloodstream both from residual mushroom fragments in the gastrointestinal tract and from toxin-contaminated bile. The following measures are designed to eliminate toxin from the gastrointestinal tract.

- Continuous duodenal suction through a single- or double-lumen tube. The tube is passed through the pylorus into the second or third portion of the duodenum. Although there is experimental evidence to support this approach, the clinical utility has yet to be clearly demonstrated.
- Adsorption with activated charcoal. Ten grams of activated charcoal should be added every two to four hours to the duodenum through the suction tube or given orally. The initial dose of charcoal can be combined with sorbitol (1 g of charcoal/kg of body weight in 4.3 mL of 70% sorbitol/kg of adult body weight. For a child, 35% sorbitol should be used.[97]
- Biliary drainage. In some institutions, the common bile duct has been successfully cannulated for constant biliary drainage. If the expertise for this procedure is available, it may be considered as a therapeutic measure in the first 48 hours.[98]

Enhancement of Renal Excretion

Urinary excretion is the major route of toxin elimination. Therefore, many investigators in the past have recommended that urine flow should be maximized.[66,99,100] Alternative procedures for forced diuresis have included both osmotic diuresis or pharmacological diuretics such as furosemide. However, there is no definitive evidence from well-controlled clinical studies to support this treatment. Maintaining an adequate urine output with good hydration is probably sufficient. Aggressive forced diuresis poses additional risks to kidneys that may already be damaged. Although most reference sources still suggest a role for forced diuresis, Faulstich and Zilker have recently backed away from this recommendation.[101]

Vomiting and diarrhea can lead to significant fluid loss, so a patient's fluid balance must be very carefully monitored, especially if forced diuresis has been initiated. When circulatory compromise is evident, forced diuresis is contraindicated until fluid replacement therapy has stabilized the patient's hemodynamic signs. A central venous line is convenient for monitoring.

Supportive Medical Care

Good supportive care has become an important factor in raising survival rates, and it should be vigorously instituted. Not all the following procedures will be required in each patient, but they should all be considered and instituted when appropriate.

- Intake/output monitoring
- Fluid replacement therapy. This treatment should be energetically pursued, because many patients lose large quantities of fluid during the gastroenteritis phase.
- Placement of central venous catheter. This practice facilitates fluid management in the patient who is not hemodynamically stable, and is valuable in a patient with underlying cardiovascular disease.
- Administration of intravenous glucose. Maintenance of a normal blood glucose level prevents the adverse effects of hypoglycemia.
- Monitoring of serum factors. Levels of prothrombin, transaminases, BUN, creatinine, electrolytes, calcium, phosphate, fibrinogen, and ammonia should be monitored on a regular basis (every 12 to 24 hours).
- Administration of fresh frozen plasma. Patients who develop gastrointestinal bleeding may need plasma to maintain blood volume and appropriate levels of clotting factors.
- Blood replacement therapy. This treatment is an alternative to the preceding therapy when gastrointestinal losses have been substantial.

Pharmacological Therapy

Over the years, a number of agents have been tried in an attempt to limit the amount of additional or ongoing damage to the liver. It is difficult to judge the true efficacy of these compounds because of the complexity of care that each patient receives. However, the following agents are still recommended for use.

Intravenous Penicillin G   Dosage: 40 million units/day or 0.5–1 million units/kg adult body weight/day for three days; for a child, 1 million units/kg body weight.[41,102,103] This antibiotic is generally administered as a continuous intravenous infusion. Large doses of penicillin may provoke seizures, so the patient should be carefully monitored during the infusion.

Penicillin G was once thought to interfere with cellular uptake of amatoxin. It is more likely to reduce the reexcretion of the amatoxin into the bile, thus decreasing the amount of toxin entering the entero-hepatic circulation. However, its precise mechanism of action is still not understood. If penicillin G is given as the sodium salt, fluid replacement therapy needs to be even more carefully monitored, because the recommended daily dose may contain the patient's average daily intake of sodium. Some preliminary work in animals has intimated that the cephalosporins may be even more effective than penicillin, but it is still too early to recommend them for the treatment of human poisonings.[104,105]

Silibinin and Silymarin   This therapy is not controversial in Europe, where it is part and parcel of the management of every case of ama-toxin poisoning.[103,106–109] Unfortunately, these compounds are not currently available in the United States, because the purported effica-cious, intravenous form has not been approved by the appropriate gov-ernmental agencies. The compounds silibinin and silymarin are extracts of the milk thistle, *Silybum marianum,* and are widely employed in Europe for their "hepatoprotective" effects. A number of experimen-tal and clinical studies provide reasonable evidence to support its use. One study demonstrated considerable efficacy of this agent in an experimental model (beagles).[108] The pharmacokinetics and pathology of amatoxin intoxication in dogs is very similar to that in humans. Other experimental data have shown a reduction of the amatoxin-induced renal toxicity in silymarin-treated rats.[110]

Silibinin and silymarin appear to interfere with the uptake of ama-toxin into the hepatic cells and interrupt the enterohepatic recircula-tion of the amatoxin.[41] An oral preparation can sometimes be obtained in the United States, but it is said to be much less effective and of little value. The dosage of the intravenous preparation is 20–30 mg/kg body weight/day in three or four divided doses. Each infusion lasts two hours.

Satisfactory and definitive studies on silibinin and silymarin in human amatoxin poisoning have not been undertaken. Although the anecdotal experience is encouraging, the final verdict regarding their efficacy remains to be decided.

Liver Transplantation

Since about 1988, liver transplantation has been employed in the United States as the definitive treatment for a patient with complete hepatic necrosis.[83,111–114] Its use in mushroom poisoning began with a three-year-old girl in California who ate *Amanita ocreata*. She was in coma and on a ventilator at the time of the transplant—undoubtedly, she would have died without the surgery.[115] The liver transplant saved her life, but she developed neurological deficits in the postoperative period.

All subsequent amatoxin-poisoned patients receiving transplants have been adults.[116] All have survived, with only the usual problems and difficulties associated with liver transplantation. The complications have been no worse than those appearing in patients receiving liver transplants for other reasons.

From the standpoint of suitability, many of the amatoxin-poisoned victims are ideal. They are generally young and seldom have other health problems that could complicate the procedure. The two major questions needing to be addressed are, Which patient is the best candidate? When should the operation be performed? Because mortality for all patients with amatoxin poisoning is somewhere between 10 and 30%, very precise criteria must be developed to prevent transplantation from being used inappropriately. Liver transplantation is not a trivial affair. Apart from the huge cost of the procedure, the associated long-term morbidity is still high.

In the presence of overwhelming, rapidly progressing hepatic failure, the accepted criteria for transplantation include very high bilirubin levels (>20 mg/dL), which seldom occur in amatoxin poisoning except very late in the course of the syndrome; a prolonged prothrombin time that is uncorrectable with vitamin K; persistent hypoglycemia requiring continuous administration of intravenous glucose for maintenance; and early (stage I or II) hepatic encephalopathy.[111,117,118] Perhaps more important than these simple criteria are the overall clinical assessment of the patient and the sense of progressive, inexorable deterioration despite the therapy. The hepatologist, gastroenterologist, and transplant surgeon must decide fairly early in the course of the poisoning whether the patient is a transplant candidate so that procurement of a liver can proceed.

It is exceedingly important to manage all amatoxin-poisoned patients with aggressive supportive care and conservative therapy, rather than rely on the possibility of transplantation and the availability of a new liver. Medical centers that manage their poisoning patients with aggressive medical care have reduced the mortality rate to 6–11%.[38,66] Two patients with grade IV hepatic encephalopathy who satisfied all the current U.S. criteria for transplantation were managed in an intensive care unit with supportive care while waiting for suitable donors. Both recovered spontaneously before replacement livers were found. At a one-year follow-up, no clinical evidence of significant residual damage was noted in either patient.[91]

Some members of the medical community in the United States fear that the surgical option of transplantation will supplant traditional forms of therapy. At worst, transplantation should be required in fewer than 20% of all amatoxin-poisoned patients. The pathology of all the resected livers in patients who undergo transplantation needs to be carefully assessed and documented in terms of the remaining percentage of viable cells. These data will help physicians refine their prognostic criteria and prevent unnecessary transplantation.

Key Elements of Treatment of Amatoxin Poisoning

Prevent further toxin absorption

- gastric lavage (only earlier than six hours postingestion)
- activated-charcoal absorption
- *catharsis
- *duodenal intubation and drainage

Aggressive supportive care

- management of fluid, electrolyte, acid–base, and glucose
- management of coagulopathy and hepatic and renal failure

Enhance toxin elimination

- maintain good urine output

Pharmacological therapy

- intravenous penicillin, 0.5–1 million IU/kg body weight/day
- intravenous silibinin (if available), 20–50 mg/kg body weight/day

*Liver transplantation

*Optional treatments. Medical personnel must decide on their use on a case-by-case basis.

OTHER MANAGEMENT OPTIONS

Controversial Treatments

Hemoperfusion   The amatoxins are said to have a high affinity for charcoal surfaces. However, toxicokinetic data gathered in experiments with dogs show that any form of hemoperfusion is most unlikely to provide significant benefits unless used in the first few hours after ingestion and before any symptoms become evident. Amatoxin excretion in urine is many times greater than the maximum removal achieved with hemoperfusion. Despite these theoretical considerations, a number of groups have claimed that this procedure provides some benefit.[119–122] In a number of countries, hemadsorption is a standard approach to most forms of poisoning, at times without rigorous scientific justification.

Plasma Exchange Transfusion   This treatment has been employed in a few desperate cases.[123] From a theoretical standpoint, it is not likely to have a significant impact on the concentration of the toxins, unless performed very early in the course of the intoxication. Amatoxin is present in the bloodstream—but always at a low concentration—for only the first day or two after ingestion. Nevertheless, plasma exchange could be used to normalize metabolic and coagulation imbalances that develop later in the course of the intoxication.

Hyperbaric Oxygen Therapy   Hyperbaric oxygen therapy has been used in a number of European centers.[124] In Floersheim's series, it did appear to provide some benefit, although it is difficult to dissect out its role from those of all the other treatments employed in these patients.[64]

Steroids   The therapies for many life-threatening diseases include administration of steroids. There are numerous claims of efficacy and the occasional dissenting vote. The data available in the literature are inadequate and insufficient to allow us to make an informed judgment concerning their value in the treatment of amatoxin poisoning. In at least one experimental situation, steroids were actually deleterious.[107]

Treatment Methods No Longer Advocated

Thiotic acid (α-lipoic acid) is no longer believed to play a useful role in the management of poisoning. First suggested for use in 1958 by Josef Herink, it became more widely popularized in 1968.[125] For many years, it was one of the mainstays of therapy, and it is still being used in many European medical centers. Numerous reports published after its introduction in the 1960s uniformly claimed beneficial results. However, a critical reanalysis of all patients treated with thiotic acid failed to yield a convincing argument for its use. Most of the improvement in the mortality rate was ascribed to better supportive care. Furthermore, animal studies have not proved its efficacy.[126] It is no longer recommended for use.

Other agents claimed to be beneficial in the treatment of amatoxin-poisoned patients include cytochrome c,[127] cimetidine,[128,129] and N-acetylcystine.[130] These claims are based on a limited number of animal experiments. No well-performed studies have been carried out with human patients. Currently, they are not recommended for use.

LONG-TERM OUTCOME FOR SURVIVORS

For many years, residual liver damage was not thought to occur in individuals recovering from an episode of amanitin poisoning. A recent study casts serious doubt on this assumption. In a series of 64 patients surviving poisoning in the Tuscany area in Italy during 1979, 1980, and 1981 (a bad three-year period for mycophiles), 14 out of 17 patients with moderate to severe intoxication showed biopsy evidence of ongoing hepatic damage or inflammation 6 months after poisoning.[131]

On the basis of laboratory evidence indicating severity of poisoning, Bartoloni and coworkers divided the patients into three groups (see the accompanying table). In the mild group were those whose transaminase enzyme levels never rose above 1000 IU/L and whose prothrombin times were normal. The moderate group included those with transaminase levels between 1000 and 2000 IU/L and some lengthening of prothrombin time. The severely poisoned group had enzyme levels above 2000 IU/L and markedly abnormal prothrombin times. All

Liver pathology in survivors of amatoxin poisoning

Degree of poisoning	Number of patients	Normal liver histology	Lobular hepatitis	Portal and lobular hepatitis	Chronic active hepatitis
Moderate	6	2	2	0	2
Severe	8	0	0	2	6

From Bartoloni et al. (1985).[131]

41 mildly affected patients survived; 1 of the 9 moderately poisoned patients died; and 5 of the 14 severely poisoned patients died. The liver morphology in those survivors on whom liver biopsies were performed is illustrated in the accompanying table.

The long-term outlook for those patient in relationship to further progression and the development of cirrhosis is still unclear. No reports suggest relentless progression of these abnormalities. Indeed, the lack of such cases in the medical literature implies that severe chronic liver disease as a consequence of amanitin toxicity is unusual.

Another Italian group, which managed 160 patients with mushroom poisoning in the period from July to November 1981—of which 40 were amatoxin poisonings—confirmed the development of chronic liver disease in some of the long-term survivors. Fourteen of the 36 survivors had persistently elevated transaminase levels. In addition, some had circulating smooth muscle antibodies.[132] Another study has confirmed that up to half of all survivors who had been moderately or severely poisoned developed chronic active hepatitis. A Spanish group, which has also treated a large number of amatoxin-poisoned patients, reports no clinical problems in long-term survivors, although they have not published biopsy studies.[38]

PATHOLOGY OF AMATOXIN POISONING

Surprisingly few adequate pathological studies of human amatoxin poisoning have been published.[45,133–135] Much of the current knowledge is based on animal studies.

As might be anticipated on the basis of knowledge of the toxicology and mechanism of action of the amanitins, which is to stop all protein synthesis by its inhibition of RNA polymerase, the pathology is not distinctive. In the organs most affected, cells show nonspecific degenerative changes that lead to cell death. There is widespread tissue necrosis. Damage is most striking in the liver, the most severely affected organ. The usual finding are centrilobular necrosis with marked venous and sinusoidal congestion and even intraparenchymal hemorrhage. Fat accumulation is reportedly a major feature in some cases and virtually absent in others. The variation probably reflects timing and severity of injury. An inflammatory response develops, the intensity depending on the timing of biopsy in relation to tissue necrosis. Periportal hepatocytes are those cells most likely to survive and remain available for regeneration. The reticulin architecture is preserved initially but eventually collapses in the presence of massive necrosis.

The early changes in the mucosa of the gastrointestinal tract have not been well described in humans. The epithelium later becomes necrotic and the bowel wall edematous. This pathology has been noted at the time of surgery in the few patients receiving liver transplants.

The kidneys, heart, brain, skeletal muscle, pancreas, and other organs eventually show evidence of cell damage. However, when the amatoxin dose is large enough, most patients die before these pathologies have an opportunity to develop. The evidence for tissue damage may simply be an accumulation of fat within the tissue's cells or other changes indicative of incipient cell death. In the few kidneys adequately examined, both the proximal and distal tubules have shown involvement.

The brain of fatally poisoned individuals is edematous and congested, with punctate hemorrhages. There may be loss of ganglion cells, cellular toxic changes, perivascular inflammation, focal glial proliferation, and fat in the ganglion cells. These observations suggest that the toxins act directly on neurons. However, it is very difficult to separate the direct effects of the toxin from the effects of the metabolic and coagulation derangements associated with liver and renal failure.[136]

No instructive changes in the liver can be observed with the aid of electron microscopy.[133] Neither routine light microscopy nor electron microscopy currently plays a role in the diagnosis or in determining the prognosis. Usually the laboratory parameters of liver dysfunction and the clinical condition of the patient are sufficient for making therapeutic decisions. Of course, this situation may change in the future as alternative forms of therapy become available.

UNANSWERED QUESTIONS ABOUT AMATOXIN POISONING

Although great progress has been made in our understanding of this very serious form of mushroom poisoning, many questions remain unanswered, even unasked. The following list is a selection of issues that need to be resolved.

1. What is the toxin in *Amanita abrupta*? Ingestion of *Amanita abrupta*, the cause of fatalities in Japan, produces a syndrome of acute hepatic necrosis, which from the clinical standpoint is identical to that of amatoxin poisoning. However, at least one investigation has disputed the presence of amanitins in the mushroom. Instead, two novel amino acids have been described, one of which is said to produce

acute hepatic necrosis in animals. The compound is L-2-amino-4-pentynoic acid. This amino acid and another compound (L-2-amino-4,5-hexadienoic acid) are also said to occur in *Amanita smithiana* (=*solitaria*) and *A. pseudoporphyria*. The pentynoic amino acid is thought to interfere with the enzymes involved in cystathione and methionine metabolism.[137] These observations need to be confirmed in another laboratory.

2. How did *Amanita phalloides* reach the United States and is it extending its range?

3. What is the role of silibinin in amatoxin poisoning? How does it work?

4. What are the best prognostic criteria, and how should we select the appropriate patient for transplantation?

5. What is the biological role of the amatoxins and phallotoxins in those fungi containing them?

6. What is the toxin in *Galerina sulcipes*? This mushroom has caused a series of fatal poisonings in Indonesia.[12] The mortality rate has been extraordinarily high and yet the signs and symptoms are not really compatible with amatoxin poisoning. Patients have no vomiting, diarrhea, or apparent liver failure. Instead the symptoms include dizziness, nausea, cardiac dysrhythmias, dyspnea, local anesthesias and parasthesias, coma, and death within two to three days. The lethal compound has not been identified.[13]

REFERENCES

1. K. R. Olson, S. M. Pond, J. Seward, K. Healey, O. F. Woo, and C. E. Becker, "*Amanita phalloides*-type Mushroom Poisoning," *Western Journal of Medicine* (1982) 137(4):282–289.

2. S. M. Pond, K. R. Olson, O. F. Woo, J. D. Osterloh, and R. E. Ward, "Amatoxin Poisoning in Northern California," *Western Journal of Medicine* (1986) 145(2):204–209.

3. M. S. Cappell and T. Hassan, "Gastrointestinal and Hepatic Effects of *Amanita phalloides* Ingestion," *Journal of Clinical Gastroenterology* (1992) 15(3):225–228.

4. W. F. Piering and N. Bratanow, "Role of the Clinical Laboratory in Guiding Treatment of *Amanita virosa* Mushroom Poisoning: Report of Two Cases," *Clinical Chemistry* (1990) 36(3): 571–574.

5. J. F. Ammirati and J. A. Traquair, *Poisonous Mushrooms of Northern United States and Canada* (Minneapolis: University of Minnesota Press, 1985).

6. M. McKenny, D. Stuntz, and J. A. Ammirati, *The New Savory Wild Mushroom*, 3rd ed. (Seattle: University of Washington Press, 1987).

7. M. Isilogulo and R. Watling, "Poisonings by *Lepiota helveola* Bres. in Southern Turkey," *Edinburgh Journal of Botany* (1991) 48(1):91–100.

8. S. Paydas, R. Kocak, F. Erturk, E. Erken, H. A. Zaksu, and A. Gurcay, "Poisoning due to Amatoxin-containing *Lepiota* Species," *British Journal of Clinical Practice* (1990) 44(11):450–453.

9. J. H. Haines, E. Lichstein, and D. Glickerman, "A Fatal Poisoning from an Amatoxin-containing *Lepiota*," *Mycopathologia* (1986) 93(1):15–17.

10. B. C. Johnson, J. F. Preston, and J. W. Kimbrough, "Quantitation of Amanitins in *Galerina autumnalis*," *Mycologia* (1976) 68:1248–1253.

11. H. Okabe, [A heavy poisoning by *Galerina fasciculata* Hongo] (Japanese), *Transactions of the Mycological Society of Japan* (1975) 16:204–206.

12. K. B. Boedijn, "A Poisonous Species of the Genus *Phaeomarasmius* (*Agaricaceae*)," *Bull. Jardi. Bot. Buitenzorg. Serie III* (1938–1940) 16:76–82.

13. R. G. Benedict, "Mushroom Toxins Other Than *Amanita*." In *Microbial Toxins,* Vol. 8, ed. S. Kadis, A. Ciegler, and S. J. Ajl (New York: Academic Press, 1972), pp. 281–320.

14. L. R. Brady, R. G. Benedict, V. E. Tyler, D. E. J. Stuntz, and M. H. Malone, "Identification of *Conocybe filaris* as a Toxic Basidiomycete," *Lloydia* (1975) 38:172–173.

15. T. Wieland, *Peptides of Poisonous Amanita Mushrooms* (New York: Springer-Verlag, 1986).

16. A. Bresinsky and H. Besl, *A Colour Atlas of Poisonous Fungi* (London: Wolfe Publishing, 1990).

17. L. J. Tanghe, "Spread of *Amanita phalloides* in North America," *McIlvainea* (1983) 6(1):4–8.

18. L. J. Tanghe and D. M. Simons, "*Amanita phalloides* in the Eastern United States," *Mycologia* (1973) 65(1):99–108.

19. J. F. Ammirati, H. D. Thiers, and P. A. Horgen, "Amatoxin-containing Mushrooms: *Amanita ocreata* and *A. phalloides* in California," *Mycologia* (1977) 69(6):1095–1108.

20. D. T. Jenkins, *Amanita of North America* (Eureka, CA: Mad River Press, 1986).

21. A. E. Jenkins, "Old Tombstone Inscription Recording the Death of a Family from Mushroom Poisoning," *Mycologia* (1960) 52:521–522.

22. R. R. Yocum and D. M. Simmons, "Amatoxins and Phallotoxins in *Amanita* Species of the Northeastern United States," *Lloydia* (1977) 40:178–190.

23. T. Weiland and H. Faulstich, "Peptide Toxins from *Amanita*." In *Handbook of Natural Toxins,* Vol. 1, ed. R. F. Keller and A. T. Tu (New York: Marcel Dekker, 1983), pp. 585–629.

24. T. Weiland and H. Faulstich, "Amatoxins, Phallotoxins, and Antamanide: The Biologically Active Components of the Poisonous *Amanita* Mushrooms," *Critical Reviews in Biochemistry* (1978) 5:185–260.

25. T. Weiland, "Poisonous Principles of Mushrooms of the Genus *Amanita*," *Science* (1968) 159:946–952.

26. H. Faulstich, Ed., *Modified Amatoxins and Phallotoxins for Biochemical, Biological, and Medical Research* (Oxford: C. A. B. International, 1991).

27. E. Petzinger, J. Burckhardt, M. Schrank, and H. Faulstich, "Lack of Intestinal Transport of [H3]-Demethylphalloidin: Comparative Studies with Phallotoxins and Bile Acids on Isolated Small Intestinal Cells and Ileal Brush Borders," *Naunyn-Schmeidebergs Archives of Pharmacology* (1982) 320:196–200.

28. H. Faulstich, A. Buku, H. Bodenmuller, and T. Wieland, "Virotoxins: Actin-binding Cyclic Peptides of *Amanita virosa* Mushrooms," *Biochemistry* (1980) 19:3334–3343.

29. A. Buku, T. Weiland, H. Bodenmuller, and H. Faulstich, "Amaninamide, a New Toxin of *Amanita virosa* Mushrooms," *Experientia* (1980) 36:33–34.

30. L. Fiume, M. Derenzini, V. Marinozzi, F. Petazzi, and A. Testoni, "Pathogenesis of Gastrointestinal Symptomatology During Poisoning by *Amanita phalloides*," *Experienta* (1973) 29:1520–1521.

31. M. Fischer, H. Faulstich, W. Schmal, and C. Hilber, "Autoradiographic Localization of Amatoxins in Pig Liver." In *Amanita Toxins and Poisoning*, ed. H. Faulstich, B. Kommerell, and T. Weiland (New York: Gerhard Witzstrock, 1980), pp. 98–105.

32. K. D. Kronckë, G. Fricker, P. J. Meier, W. Gerok, T. Weiland, and G. Kurz, "α-Amanitin Uptake into Hepatocytes: Identification of Hepatic Membrane Transport Systems Used by Amatoxins," *Journal of Biological Chemistry* (1986) 261(27):12562–12567.

33. W. H. Adams, R. D. Stoner, D. G. Adams, H. Read, D. N. Slatkin, and H. W. Siegelman, "Prophylaxis of Cyanobacterial and Mushroom Cyclic Peptide Toxins," *Journal of Pharmacology and Experimental Therapeutics* (1989) 249(2):552–556.

34. H. Faulstich, A. Talas, and H. H. Wellhoner, "Toxicokinetics of Labeled Amatoxins in the Dog," *Archives of Toxicology* (1985) 56(3):190–194.

35. J. Homann, P. Rawer, H. Bleyl, K. J. Matthes, and D. Heinrich, "Early Detection of Amatoxins in Human Mushroom Poisoning," *Archives of Toxicology* (1986) 59(3):190–191.

36. A. Jaeger et al., "Amatoxin Kinetics in *Amanita phalloides* Poisoning," *Veterinary and Human Toxicology* (1989) 31:360.

37. A. Jaeger, F. Jehl, F. Flesch, P. Sauder, and J. Kopferschmitt, "Kinetics of Amatoxins in Human Poisoning: Therapeutic Implications," *Journal of Toxicology and Clinical Toxicology* (1993) 31(1):63–80.

38. J. Piqueras, "Hepatotoxic Mushroom Poisoning: Diagnosis and Treatment," *Mycopathologia* (1989) 105:99–110.

39. J. Ses'e, J. Piqueras, G. Morlans, V. Mercade, X. Valls, and A. Herrero, "Intoxicatión por *Amanita phalloides*. Diagnóstico por radioimmunoanálisis y tratamiento con diuresis forzada," *Medicina Clínica (Barcelona)* (1985) 84:660–662.

40. C. Busi, L. Fiume, D. Constantino, M. Langer, and S. Vesconi, "Amanita Toxins in the Gastroduodenal Fluid of Patients Poisoned by the Mushroom *Amanita phalloides*," *New England Journal of Medicine* (1979) 300:800

41. W. Jahn, H. Faulstich, and T. Weiland, "Pharmacokinetics of (3H) Methyldehydroxymethyl-α-amanitin in the Isolated Perfused Rat Liver." In *Amanita Toxins and Poisoning*, ed. H. Faulstich, B. Kommerell, and T. Weiland (New York: Gerhard Witzstrock, 1980), pp. 79–87.

42. H. Faulstich and U. Fauser, "The Course of *Amanita* Intoxication in Beagle Dogs." In *Amanita Toxins and Poisoning*, ed. H. Faulstich, B. Kommerell, and T. Weiland (New York: Gerhard Witzstrock, 1980), pp. 115–123.

43. F. Belliardo, G. Massano, and S. Accomo, "Amatoxins Do Not Cross the Placental Barrier" (letter), *Lancet* (1983) i:1381.

44. V. Dudova, J. Kubicka, and J. Veselsky, "Thiotic Acid in the Treatment of *Amanita phalloides* Intoxication." In *Amanita Toxins and Poisoning*, ed. H. Faulstich, B. Kommerell, and T. Weiland (New York: Gerhard Witzstrock, 1980), pp. 190–191.

45. J. Slodkowska, S. Szendzikowski, J. Stetkiewicz, and J. Muszynski, [The histological picture and mechanism of development of the liver changes in poisoning with *Amanita phalloides*] (Polish), *Patologia Polska* (1980) 31(1):55–66.

46. M. Kaufmann, A. Muller, N. Paweletz, U. Haller, and F. Kubli, [Fetal damage due to mushroom poisoning with *Amanita phalloides* during the first trimester of pregnancy] (German), *Geburtshilfe und Frauenheilkunde* (1978) 38(2):122–124.

47. S. Buttenwieser and W. Bodenheimer, "Ueber den Uebertritt des Knollenblätterschwammgiftes in die Brustmilch," *Deutsche Medizinische Wochenschrift* (1924) 50:607.

48. J. A. Beutler, "Chemotaxonomy of *Amanita*: Qualitative and Quantitative Evaluation of the Isoxazoles, Tryptamines and Cyclopeptides as Chemical Traits," Doctoral thesis, Philadelphia College of Pharmacy and Science, Philadelphia, 1980.

49. W. S. Chilton, "Chemistry and Mode of Action of Mushroom Toxins." In *Mushroom Poisoning: Diagnosis and Treatment*, ed. B. Rumack and E. Salzman (West Palm Beach, FL: CRC Press, 1978), pp. 87–124.

50. P. A. Horgan, J. F. Ammirati, and H. D. Thiers, "Occurrence of Amatoxins in *Amanita ocreata*," *Lloydia* (1976) 39:368–371.

51. V. E. J. Tyler, R. G. Benedict, L. R. Brady, and J. E. Robbers, "Occurrence of *Amanita* Toxins in American Collections of Deadly Amanitas," *Journal of Pharmaceutical Sciences* (1966) 55:590–593.

52. S. H. A. Malak, "Occurrence of Phallotoxins in American Collections of *Amanita virosa*," *Planta Medica* (1976) 29:80–85.

53. H. Faulstich and M. Cochet-Meilhac, "Amatoxins in Edible Mushrooms," *FEBS Letters* (1976) 64:73–75.

54. C. Cessi and L. Fiume, "Increased Toxicity of β-Amanitin When Bound to Protein," *Toxicon* (1969) 6:309–310.

55. H. Faulstich, K. Kirchner, and M. Derenzini, "Strongly Enchanced Toxicity of the Mushroom Toxin α-Amanitin by an Amatoxin-specific Fab or Monoclonal Antibody," *Toxicon* (1988) 26(5):491–499.

56. L. Fiume, S. Sperti, L. Montanaro, C. Busi, and D. Constantino, "Amanitins Do Not Bind to Serum Albumin," *Lancet* (1977) i:1111.

57. J. M. Bauchet, "Treatment of *Amanita phalloides Poisoning*—The Bastien Method," Bulletin of the British Mycological Society (1983) 17(2):110–111.

58. P. Bastien, "Traitement de l'intoxication phalloidienne: Déflexions après ingestion volontaire d'une amanite phalloide crue et directives thérapeutiques," *Annales Médicales de Nancy* (1972) 11.

59. A.-M. Dumont, J.-M. Chennebault, P. Alquier, and H. Jardel, "Management of *Amanita phalloides* Poisoning by Bastien's Regimen," *Lancet* (1981) 1:722.

60. G. L. Floersheim, "Curative Potencies Against α-Amanitin Poisoning by Cytochrome C," *Science* (1972) 117:808.

61. A. Jacqueson, M. Thevenin, J. Warnet, J. Claude, and R. Truhaut, "Sex Influences on the Experimental Fatty Liver Induced by White Phosphorus and *Amanita phalloides* in the Rat," *Acta Pharmacologica et Toxicologica* (1977) 41(suppl 2): 322.

62. C. P. Siegers and O. Strubett, "Failure of Cytochrome C to Cure Experimental *Amanita phalloides* Poisoning," *German Medicine* (1973) 3:103–104.

63. G. L. Floersheim, A. Bieri, R. Koenig, and A. Pletcher, "Protection Against *Amanita phalloides* by the Iridoid Glycoside Mixture of *Picrorhiza kurroa* (kutkin)," *Agents and Actions* (1990) 29:386–387.

64. G. L. Floersheim, O. Weber, P. Tsuchumi, and M. Ulbrich, "Die klinische Knollenblätterpilzvergiftung (*Amanita phalloides*): Prognostische Faktoren und therapeutische Massnahmen," *Schweizerische Medizinische Wochenschrift* (1982) 112:1164–1167.

65. K. Hruby, "Treatment of the Death Cap Poisoning with Silibinin." In *X Congress of the International Society for Human and Animal Mycology—ISHAM*, ed. J. M. Torres-Rodriquez (Barcelona, Spain: Prous Science Publishers, 1988), pp. 361–363.

66. G. Ceravolo, V. Cimellaro, F. Montesano, G. Rossi, C. Busi, and L. Fiume, "Therapy of Cytotoxic Mushroom Intoxication," *Critical Care Medicine* (1985) 13(5):402–406.

67. G. L. Floersheim and L. Bianchi, "Ethanol Diminishes the Toxicity of *Amanita phalloides* Poisoning," *Experienta* (1984) 40:1268–1270.

68. A. Livanainen, M. Valtonen ja Perta, and P.-J. Pentikäinen, [Mushroom poisoning—the disease of autumn] (Finnish), *Duodecim* (1989) 15:1313–1317.

69. G. L. Floersheim, "Influence of Ethanol on Toxicity of Paraquat and *Amanita phalloides*" (letter; comment), *Lancet* (1992) 339(8790):437.

70. H. G. Bivins, R. Knopp, R. Lammers, D. B. McMicken, and O. Wolowodiuk, "Mushroom Ingestion," *Annals of Emergency Medicine* (1985) 14(11):1099–1104.

71. P. Czigan, R. Zimmermann, U. Leuscher, A. Steilh, and B. Kommerell, "Clinical, Biochemical and Morphological Alterations in Patients with *Amanita phalloides* Intoxication." In *Amanita Toxins and Poisoning*, ed. H. Faulstich, B. Kommerell, and T. Weiland (New York: Gerhard Witzstrock, 1980), pp. 131–136.

72. H. Faulstich, "New Aspects of *Amanita* Poisoning," *Klinische Wochenschrift* (1979) 57:1143–1152.

73. B. Passo and D. Harrison, "A New Look at an Old Problem: Mushroom Poisoning," *American Journal of Medicine* (1975) 58(2):505–509.

74. F. Mellerio, M. Gaultier, E. Fournier, P. Gervais, and J. P. Frejaville, "Contribution of Electroencephalography to Resuscitation in Toxicology," *Clinical Toxicology* (1973) 6:271.

75. L. L. Oelsen, "Amatoxin Intoxication," *Scandinavian Journal of Urology and Nephrology* (1990) 24(3):231–234.

76. R. K. Myler, J. C. Lee, and J. Hopper, "Renal Tubular Necrosis Caused by Mushroom Poisoning," *Archives of Internal Medicine* (1964) 114:196–204.

77. L. G. Hanelin and A. A. Moss, "Roentgenographic Features of Mushroom (*Amanita*) Poisoning," *American Journal of Roentgenology, Radiation Therapy, and Nuclear Medicine* (1975) 125:728.

78. P. Sanz, R. Reig, J. Piqueras, G. Marti, and J. Corbella, "Fatal Mushroom Poisoning in Barcelona, 1986–1988," *Mycopathologia* (1989) 108(3):207–209.

79. B. Kendrick, "Mushroom Poisoning—Analysis of Two Cases, and a Possible New Treatment, Plasmapheresis," *Mycologia* (1984) 76:448–453.

80. N. A. Klantsa and I. P. Kuz'min, [Electrocardiographic changes in mushroom poisoning] (Russian), *Vrachebnoe Delo* (1974) 9:62.

81. B. Debiec, W. Bielinska, and T. E. Romer, "Modifications de l'électrocardiogramme au cours d'un empoisonnement par des *Amanites phalloides*," *Pédiatrie* (1964) 19:451.

82. P. Sanz, R. Reig, L. Borrás, J. Martinéz, R. Máñez, and J. Corbella, "Disseminated Intravascular Coagulation and Mesenteric Venous Thrombosis in Fatal *Amanita* Poisoning," *Human Toxicology* (1988) 7:199–201.

83. C. W. Pinson et al., "Liver Transplantation for Severe *Amanita phalloides* Mushroom Poisoning," *American Journal of Surgery* (1990) 159(5):493–499.

84. S. Parra, J. Garcia, P. Martinez, C. de la Pena, and C. Carrascosa, "Profile of Alkaline Phosphatase Isoenzymes in Ten Patients Poisoned by Mushrooms of the Genus *Lepiota*," *Digestive Diseases and Sciences* (1992) 37(10):1495–1498.

85. J. Ryzko, I. Jankowska, and J. Socha, [Evaluation of selected parameters in calcium and phosphate metabolism in acute liver failure following poisoning with *Amanita phalloides*] (Polish), *Polski Tygodnik Lekarski* (1990) 45(49–50):990–992.

86. M. J. Kelner and N. M. Alexander, "Endocrine Abnormalities in *Amanita* Poisoning," *Journal of Toxicology and Clinical Toxicology* (1987) 25(1–2):21–38.

87. T. Zilker, "Diagnosie und Therapie der Pilzvergiftung (Ted II)," *Leber Magen Darm* (1987) 3:173–179.

88. G. L. Floersheim, "Treatment of Human Amatoxin Mushroom Poisoning: Myths and Advances in Therapy," *Medical Toxicology* (1987) 2:1–9.

89. T. Duffy, "Treatment Sheet for Poisoning by *Amanita phalloides* and Other Amatoxin-containing Species," *Mushroom* (1989) October 1988/Winter 1988–89:12.

90. R. Tomasi, [Mushroom poisoning: Establishment of a special treatment center at the Brescia Civil Hospital] (Italian), *Micologia Italiana* (1982) 11(2):23–41.

91. G. Ronzoni, S. Vesconi, D. Radrizzani, C. Corbetta, M. Langer, and G. Iapichino, [Recovery after serious mushroom poisoning (grade IV encephalopathy) with intensive care support and without liver transplantation] (Italian), *Minerva Anestesiologica* (1991) 56(6):383–387.

92. C. Teutsch, "*Amanita* Poisoning with Recovery from Coma: A Case Report," *Annals of Neurology* (1978) 3:177–179.

93. A. S. Genser and S. M. Marcus, "*Amanita* Mushroom Poisoning—An Outbreak of 10 Cases," *Veterinary and Human Toxicology* (1987) 29(6):461–462.

94. J. P. Hanrahan and M. A. Gordon, "Mushroom Poisoning: Case Reports and Review of Therapy," *Journal of the American Medical Association* (1984) 86:1057–1061.

95. G. L. Floersheim, "Treatment of Mushroom Poisoning" (letter), *Journal of the American Medical Association* (1985) 253:3252.

96. T. P. McHugh and N. J. Steward, "Mushroom Poisoning in South Carolina," *Journal of the South Carolina Medical Association* (1988) 84(1):7–12.

97. A. Minocha, E. P. Krenzelok, and D. A. Spyker, "Dosage Recommendations for Activated Charcoal-Sorbitol Treatment," *Journal of Toxicology and Clincial Toxicology* (1985) 23:579–587.

98. I. C. Frank and L. Cummins, "*Amanita* Poisoning Treated with Endoscopic Biliary Diversion," *Journal of Emergency Nursing* (1987) 13(3):132–136.

99. S. Vesconi, M. Langer, G. Iapichino, D. Costantino, C. Busi, and L. Fiume, "Therapy for Cytotoxic Mushroom Poisoning," *Critical Care Medicine* (1985) 13(5):402–406.

100. L. Ghiringhelli, A. Ceriani, G. Lepore, and S. Moda, [Phallin syndrome. Report of 28 cases] (Italian), *Minerva Medica* (1981) 72:2499–2508.

101. H. Faulstich and T. R. Zilker, "Amatoxins." In *Handbook of Mushroom Poisoning*, ed. D. A. Spoerke and B. H. Rumack (Boca Raton, FL: CRC Press, 1994), pp. 233–248.

102. G. L. Floersheim, "Therapie der Knollenblätterpilzvergiftung," *Deutsche Medizinische Wochenschrift* (1983) 108:886–887.

103. G. L. Floersheim, M. Eberhard, P. Tschumi, and F. Duckert, "Effects of Penicillin and Silymarin on Liver Enzymes and Blood Clotting Factors in Dogs Given a Boiled Preparation of *Amanita phalloides*," *Toxicology and Applied Pharmacology* (1978) 46:455–462.

104. K. Neftel et al., [Are cephalosporins more active than penicillin G in poisoning with the deadly *Amanita*?] (German), *Schweizerische Medizinische Wochenschrift* (1988) 118(2):49–51.

105. P. Daoudal et al., [Treatment of phalloid poisoning with silymarin and cef-tazidime] (French), *Presse Médicale* (1989) 18(27):1341–1342.

106. K. Hruby, C. Cosmos, M. Fuhrman, and H. Thaler, "Chemotherapy of *Amanita phalloides* with Intravenous Silibinin," *Human Toxicology* (1983) 2:183–195.

107. G. Halbach and W. Trost, "Zur Chemie und Pharmakologie des Silymarins Untersuchungen am einigen Umsetzungprodukten des Silibins," *Arzneimittel-Forschung* (1974) 24:886–887.

108. G. Vogel, B. Tuchweber, W. Trost, and V. Mengs, "Protection by Silibinin Against *Amanita phalloides* Intoxication in Beagles," *Toxicology and Applied Pharmacology* (1984) 73:355–362.

109. G. Vogel, "The Anti-amanitin Effect of Silymarin." In *Amanita Toxins and Poisoning*, ed. H. Faulstich, B. Kommerell, and T. Weiland (New York: Gerhard Witzstrock, 1980).

110. G. Vogel, R. Braatz, and U. Mengs, "On the Nephrotoxicity of α-Amanitin and the Antagonistic Effects of Silymarin in Rats," *Agents and Actions* (1979) 9(2):221–226.

111. A. S. Klein, J. Hart, J. J. Brems, L. Goldstein, K. Lewin, and R. W. Busuttil, "Amanita Poisoning: Treatment and Role of Liver Transplantation," *American Journal of Medicine* (1989) 86:187–193.

112. T. J. Duffy, "Liver Transplantation for Victims of Amanita Poisoning," *McIlvainea* (1989) 9(1):4–6.

113. G. W. Galler, E. Weisenberg, and T. A. Brasitus, "Mushroom Poisoning: The Role of Liver Transplantation," *Journal of Clinical Gastroenterology* (1992) 15(3):229–232.

114. A. Castiella, L. Lopez-Dominguez, G. Txoperena, A. Cosme, and V. Araburu, [Indication for liver transplantation in *Amanita phalloides* poisoning] (French), *Presse Médicale* (1993) 22(4):117.

115. E. S. Woodle, R. R. Moody, K. L. Cox, R. A. Cannon, and R. E. Ward, "Orthotopic Liver Transplantation in a Patient with Amanita Poisoning," *Journal of the American Medical Association* (1985) 253:69–70.

116. M. Pouyet et al., [Orthotopic liver transplantation for severe *Amanita phalloides* poisoning] (French), *Presse Médicale* (1991) 20:2095–2098.

117. S. Iwatsuki, C. O. Esquivel, and R. D. Gordon, "Liver Transplantation for Fulminant Hepatic Failure," *Seminars in Liver Disease* (1985) 5:325–328.

118. E. Christensen, A. Bremmelgaard, M. Bahnsen, P. B. Andreasen, and N. Tygstrup, "Prediction of Fatality in Fulminant Hepatic Failure," *Scandinavian Journal of Gastroenterology* (1984) 19:90–96.

119. T. Webert, "Mushroom Poisoning" (letter), *Lancet* (1980) ii:640.

120. E. Masini, P. Blandina, and P. F. Mannaioni, "Removal of α-Amanitin from Blood by Hemoperfusion over Uncoated Charcoal: Experimental Results," *Contributions to Nephrology* (1982) 29:76–81.

121. C. Langescheid, H. Schmitz-Salue, H. Faulstich, P. Kramer, and F. Scheler, "In Vitro Elimination of (H3)Methyldehydroxymethyl-α-amanitin by Four Different Extracorporeal Detoxification Methods." In *Amanita Toxins and Poisonings: International Amanita Symposium*, ed. T. Weiland, H. Faulstich, and B. Kommerell (New York: Gerhard Witzstrock, 1980).

122. J. P. Wauters, C. Rossel, and J. Farguet, "*Amanita phalloides* Poisoning Treated with Early Charcoal Hemoperfusion," *British Medical Journal* (1978) 2(6150):1465.

123. R. Ponkivar, J. Drinovic, A. Kandus, J. Varl, A. Gucek, and M. Malvorh, "Plasma Exchange in Management of Severe Acute Poisoning with *Amanita phalloides*," *Progress in Clinical and Biological Research* (1990) 337:327–329.

124. M. Goulon, "Oxygène hyperbare et fonction hépatique," *Annales de l'Anaesthésiologie Français* (1967) 8(special 1):333.

125. J. Kubicka, "Traitement des empoisonnements fongiques phalloidiniens en Tchecoslovaquie," *Acta Mycologica* (1968) 4(2):373–377.

126. W. Trost and W. Lang, "Effect of Thiotic Acid and Silibinin on the Survival Rate of Amanitin- and Phalloidin-Poisoned Mice," *IRCS Journal of Medical Science* (1984) 12:1079–1080.

127. G. L. Floersheim, "Cytochrome c as Antidote in Mice Poisoned with the Mushroom Toxin α-Amanitin," *Current Problems in Clinical Biochemistry* (1977) 7:59–74.

128. S. M. Schneider, D. Borochovitz, and E. P. Krenzelok, "Cimetidine Protection Against α-Amanitin Hepatotoxicity in Mice: A Potential Model for the Treatment of *Amanita phalloides* Poisoning," *Annals of Emergency Medicine* (1987) 16(10):1136–1140.

129. S. M. Schneider, G. Vanscoy, and E. A. Michelson, "Failure of Cimetidine to Affect Phalloidin Toxicity," *Veterinary and Human Toxicology* (1991) 33(1):17–18.

130. S. M. Schneider, E. A. Michelson, and G. Vanscoy, "Failure of N-Acetylcysteine to Reduce α-Amanitin Toxicity," *Journal of Applied Toxicology* (1992) 12(2):141–142.

131. St. Omer F. Bartoloni et al., "*Amanita* Poisoning: A Clinical-Histopathological Study of 64 Cases of Intoxication," *Hepato-gastroenterology* (1985) 32(5):229–231.

132. R. Fantozzi et al., "Clincial Findings and Follow-up Evaluation of an Outbreak of Mushroom Poisoning—Survey of *Amanita phalloides* Poisoning," *Klinische Wochenschrift* (1986) 64(1):38–43.

133. B. J. Panner and R. J. Hanss, "Hepatic Injury in Mushroom Poisoning: An Electron Microscopic Observation in Two Nonfatal Cases," *Archives of Pathology* (1969) 87:35–45.

134. D. C. Harrison, C. H. Coggins, F. H. Welland, and S. Nelson, "Mushroom Poisoning in Five Patients," *American Journal of Medicine* (1965) 37:787.

135. W. Wepler and K. Opitz, "Histologic Changes in the Liver Biopsy in *Amanita phalloides* Intoxication," *Human Pathology* (1972) 3:249.

136. E. Markovitz and B. J. Alpers, "The Central Nervous System in Mushroom Poisoning," *Archives of Neurology and Psychiatry* (1935) 33:53–66.

137. Y. Yamaura, M. Fukuhara, E. Takabatake, N. Ito, and T. Hashimoto, "Hepato-toxic Action of a Poisonous Mushroom, *Amanita abrupta*, in Mice and Its Toxic Component," *Toxicology* (1986) 38(2):161–173.

CHAPTER THIRTEEN

DELAYED-ONSET RENAL FAILURE SYNDROME

Orellanine or Cortinarin Poisoning

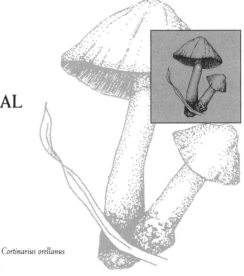

Cortinarius orellanus

ALL THREE PATIENTS were on holiday in Northern Scotland, in August 1979. Mushrooms were gathered and were eaten in a stew by all three. Some of the mushrooms were eaten raw the next morning by patients 2 and 3. . . .

Case 1. A 31 year old, previously healthy man had severe vomiting 36 hours after eating the stew. Over the next 3 days vomiting abated, but nausea and anorexia persisted. During the next week muscle pains, night sweats, rigors, headache, pronounced bilateral loin pain, a severe burning thirst and oliguria developed.

On admission to hospital 10 days after eating the stew he was drowsy, afebrile and anicteric. . . . The urine gave a strongly positive reaction for blood, and red cells were seen by urine microscopy. Plasma urea was 48 mmol/l, sodium 126 mmol/l, potassium 4.9 mmol/l, bicarbonate 16 mmol/l, and creatinine 2945 μmol/l; liver function tests were normal. Hb, WCC and coagulation screen were normal.

Peritoneal dialysis was started on the night of admission and thereafter regular hemodialysis was required. His clinical condition rapidly improved and after seven days of severe oliguria the urine output gradually increased to 2 liters/day. Creatinine clearance, however, never exceeded 4 ml/minute. Retrograde pyelography revealed bilaterally enlarged kidneys. . . . Chronic intermittent hemodialysis was continued and as renal function had not improved and was insufficient to maintain life, renal transplantation was carried out 9 months after the poisoning.[1]

A. I. K. Short et al. (1980)

Patient 2 was a 30-year-old male. His clinical picture was almost identical to that of patient 1, and he also required a new kidney. The third patient, a 25-year-old woman, had transient renal insufficiency for a few days and recovered spontaneously. However, she complained of anorexia, nausea, and night sweats for several weeks.

Although there is almost no resemblance between the collected *Cortinarius* species and the edible chanterelle, beyond the fact that both types of mushrooms grow on the ground in the woods and are yellow to orange, the trio thought that they were picking chanterelles.

DISCOVERY AND INCIDENCE OF ORELLANINE POISONING

The discovery in the early 1950s of the orellanines, an entirely novel type of mushroom poisons producing renal failure, came as quite a surprise to the mycological and medical community. Because wild mushrooms have been collected and eaten for centuries, it was assumed that all the problems they can cause were at least well known, if not well understood. The discovery provided a lesson in humility. While there may be "no new things under the Sun," it is evident we are still unaware of many of them. In the last few years, however, a great deal of information has accumulated about this particular group of mushroom toxins.

The poisons in *Cortinarius* species remained undiscovered for so long because these mushrooms had always been classified as edible[2] and because a long delay precedes the onset of the first symptoms of poisoning; and perhaps most people, especially people in the mycophilic cultures in which many wild mushrooms are eaten annually, do not relate the two events, ingestion and illness. The symptoms produced by the orellanines are entirely different from any others caused by mushroom toxins. Had these mushrooms produced the usual gastrointestinal problems, with an early onset, an association between illness and ingestion might have been made earlier. But we find only what we seek; and we seek only what we know. So it took 102 seriously ill people in Poland—11 of whom died—and the curiosity of an astute epidemiologist to make the connection.

Poisonings in Europe

In 1952, Grzymala[3] recognized that a fungus was probably responsible for the illness that had afflicted a large number of individuals. He was a Polish public health epidemiologist, brought in to investigate this unusual epidemic of renal disease, which was initially thought to be caused by an infectious agent. The only consistent feature in the patients' histories—which Grzymala was perceptive enough to recognize—was the consumption of mushrooms a few days before the onset of symptoms. This observation is actually quite remarkable in a country where the majority of the population consumes wild mushrooms on a regular basis during the fruiting season. The mushroom indicted was *Cortinarius orellanus*.[4]

An unusual feature of the Polish intoxications was the prolonged period—anywhere from three days to three weeks—between consumption of the mushrooms and appearance of symptoms. The kidney was the target organ. All clinical manifestations could be related to kidney damage and progressive renal failure.

Since 1952, reports of similar intoxications have appeared in France[5] and Germany.[6] In 1980, a report from Great Britain described three individuals poisoned by a closely related species, *Cortinarius speciosissimus*.[1,7] This species had already caused illnesses in Norway and Sweden. By 1984, at least 180 poisonings had been attributed to these mushrooms, including reports from Switzerland, Czechoslovakia, Norway, Sweden, and Finland.

The most recent example of mass poisoning by *Cortinarius* species involved 1 officer and 25 enlisted men engaged in a survival exercise in Morbihan, France.[8] The survival manual on which they relied for their foraging information should probably be revised, at least with respect to its description of edible wild mushrooms.

Nonexperimental animal intoxication caused by *Cortinarius* species also has been recorded. A flock of sheep were poisoned in Norway.[9]

Poisonings in North America

Orellanine poisoning has been strangely absent on the North American continent. No convincing, well-documented, proven case has been reported. The first presumptive cases were reported to the toxicology meetings in Toronto in 1991, although a serious element of doubt remains about the nature of the syndrome in these people. One patient, a 55-year-old male, developed symptoms in 24 hours, which is most unusual. He became anuric three days later. The second patient, also a male, developed renal failure five days after a mushroom meal. That meal had included several different species of mushrooms.[10] The specificity of the assay used for the identification of orellanine in these two cases is questionable, because appropriate controls were not reported. At least one of the mushrooms could have been *Amanita smithiana*, now suspected of containing a nephrotoxin.

A third case of kidney failure that supposedly followed a mushroom meal was reported from Alberta, Canada. This case barely fits the pattern of orellanine poisoning; onset of symptoms was eight hours after mushroom ingestion.[11] The authors of this report identified neither the mushroom nor the toxin. Their mycological omission diminished the credibility of the report.

The reasons for the absence of orellanine poisonings in the United States and Canada are not clear, because toxin-containing members of the genus *Cortinarius* populate the continent. It may simply be because few Americans pick and eat wild mushrooms. And for those who do, the genus *Cortinarius* has never been a favored group. It may also be that the medical community is failing to identify orellanine poisoning because it is not a well-publicized cause of renal failure. Few physicians in this country are even aware of the possibility of this diagnosis.

Two other cases of renal failure associated with mushroom poisoning were reported way back in 1964.[12] The first of these is almost certainly an example of amatoxin poisoning in which the patient became so dehydrated from the initial vomiting and diarrhea that he developed hypotension and renal failure. The second case is more intriguing. A 43-year-old native American woman developed acute renal failure with no evidence of liver damage or dehydration. However, the onset of the poisoning episode was most atypical for orellanine. Her symptoms developed two hours after ingestion of unidentified mushrooms and were characterized by urinary urgency and excessive tear formation (both symptoms of muscarine poisoning), followed by vomiting. She was given atropine. Over the next few days, she became oliguric and was in obvious renal failure. She required peritoneal dialysis, and her renal functions returned after 28 days. A biopsy performed 43 days after the mushroom ingestion demonstrated normal glomeruli; tubular atrophy, mainly involving the distal tubules, associated with interstitial fibrosis; and little or no inflammation. The renal biopsy findings were compatible with any form of resolving interstitial nephritis, which orellanine can cause. The mushrooms were never identified. It is possible she ate two different species, the first containing a little muscarine and the second containing orellanine. Once again, *Amanita smithiana* is a prime suspect, especially in view of the short latent period. Perhaps the renal failure was unrelated to mushroom ingestion and the timing was purely fortuitous—or perhaps some other toxin is out there waiting to be discovered.

DISTRIBUTION AND HABITAT OF MUSHROOMS CAUSING ORELLANINE POISONING

Mushrooms of this genus live in the woods and the mountains. They are not common urban mushrooms. They are associated with mixed deciduous and conifer forests; pine trees are notable in some habitats. In Europe, even when the specimens are found in a deciduous forest of oak and beech, a pine tree is usually nearby. *Cortinarius orellanus* and

C. speciosissimus have not yet been definitely identified in North America, although a recent report from Canada suggests that *C. speciosissimus* grows there (J. Ammirati and S. Redhead, personal communication). *Cortinarius rainerensis*, a northwestern representative of the genus, is very closely related to *C. speciosissimus*.

Cortinarius is a very diverse and widespread genus. It is highly probable that other toxin-containing species will be uncovered. Members of this genus usually fruit in the late summer and fall; some fruitings can be very abundant. *Cortinarius* species are not favored as culinary treats in the United States, although *C. violaceus*, a striking purple-violet edible species, is occasionally eaten by the adventurous.

CORTINARIUS TOXINS

The particular mushrooms responsible for causing orellanine poisoning are catalogued in the accompanying box. The taxonomic classification of many of the *Cortinarius* species is difficult and incomplete. The identification of the putative toxins is similarly incomplete.

The initial attempts by Grzymala to identify the toxin causing the Polish epidemic resulted in the isolation of a compound that he dubbed with the trivial name orellanine.[18] This report was followed by a rash of publications that proposed various compounds and structures. Great uncertainty existed, and still exists, about the real nature of the

---

CORTINARIUS SPECIES REPUTED TO CAUSE ORELLANINE POISONING

Species proved to cause orellanine poisoning

C. orellanus	*C. speciosissimus*
C. orellanoides	

Species containing orellanine-related compounds

C. rainerensis (?=orellanoides)[13]

Species that may cause orellaninelike poisoning*

C. splendens[14,15]	*C. atrovirens*
C. venenosus[16]	*C. gentilis*[17]

*There is some doubt about the toxicity of each of these species. Considerably more information is needed before their toxicity is established.

toxins. Two major schools of thought have arisen in recent years, and a lively debate endures. For purposes of discussion, I shall talk of the European school (which proposes a bipyridyl structure for the toxins) and the Scottish school (which proposes a cyclopeptide structure). For those wishing to pursue the chemistry in more depth, the review by K. F. Lampe, completed just prior to his sudden and premature death, is highly recommended.[19]

All investigators agree that the compounds, whatever their chemical nature, show a very strong turquoise or blue fluorescence under ultraviolet light. This fluorescence persists after ingestion by experimental animals. In one child whose kidney biopsy was studied with ultraviolet light, the fluorescence could be demonstrated in the tissues.[20] The suggestion was made that this fluorescence could be the basis for a diagnostic test in biopsy specimens.

Cortinarins

The Scottish school completely disagrees with the idea of a bipyridyl toxin. They have demonstrated a series of compounds—labeled cortinarins A, B, and C—which they believe are the toxic agents.[21–26] Instead of the bipyridyl structure proposed by the European school, their studies suggest that the cortinarins are more closely related to the cyclopeptides (amatoxins) of *Amanita phalloides*. All the cortinarins contain 10 amino acids. Their structures and toxicities have been determined, as well as the relationships among them. Of the three, cortinarin C appears to be nontoxic. It may well be a precursor compound, but it is not responsible for any of the tissue damage. Cortinarin B appears to be the most toxic, especially when it has been metabolized in the liver to its sulfoxide form.

The Scottish investigators found that all the toxicity could be accounted for by the presence of the cortinarins. When they injected extracts of a normally toxic species (*C. callisteus*) from which they had removed both cortinarin A and C, the animals suffered no ill effects. However, this experiment has not been done with extracts of *C. orellanus* or *C. speciosissimus;* nor have toxicity studies in which the experimental animals are challenged with purified preparations of the cortinarins been reported.

Sixty species of *Cortinarius* have been tested for the presence of cortinarin A and B. Cortinarin B was detected in only three species: *C. orellanus, C. speciosissimus,* and *C. orellanoides.* Cortinarin A, however, was present in every species tested except for *Cortinarius violaceus,* which is listed as edible in most field guides. The accompanying table lists the concentrations of the three cortinarins in various *Cortinarius* species;

Concentration of cortinarins in various Cortinarius species

Species	Cortinarin A (% w/v)	Cortinarin B (% w/v)	Cortinarin C (% w/v)
C. speciosissimus	0.47	0.60	0.20
C. orellanus	0.42	0.52	0.24
C. orellanoides	0.45	0.47	0.20
C. pinicola	0.19	—	0.028
C. callisteus	0.20	—	0.19
C. turmalis	0.32	—	0.05
C. mucifluus	0.05	—	0.07
C. betuletorum	0.28	—	0.05

From Tebbet (1985).[21]

concentrations were determined by high-pressure liquid chromatography. (See Tebbet[21] for a complete list of species tested.)

Experimental animals challenged with cortinarin B showed serious renal damage by 24 hours. The Scottish school has suggested that cortinarin B is metabolized to its sulfoxide form in the liver and that this modified compound is responsible for the observed toxicity. This proposal accounts for much of the delay between ingestion and onset of symptoms, namely, time is required for the modification of the ingested poison and for the accumulation of concentrations sufficient to produce renal damage. Even after the active form reaches the kidneys, time is need for the symptoms to develop.

The metabolic pathway proposed by the Scottish school would explain the sex difference that has been noted in clinical responses to poisonings by *Cortinarius* species. A number of observers have pointed

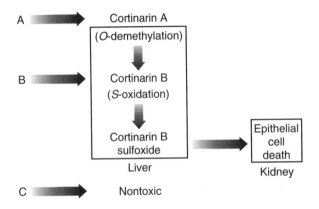

Possible metabolic pathway for the cortinarins

out that women appear to be more resistant to the toxins than men are. If the liver is the major site of conversion of cortinarin A to cortinarin B and of cortinarin B to its sulfoxide, then this process could be more efficient in males than in females. It has been shown that the cytochrome P450 system, which is involved in the conversions, has a different binding capacity in males than in females.[27] This proposal is still unsubstantiated because the details have not been tested experimentally. But the sex difference in susceptibility has also been observed in experimental animals,[28] so model systems are available for studying this phenomenon.

In the group of French soldiers mentioned earlier, all were poisoned with the same dose, but they showed variable clinical responses. Measurements of their livers' abilities to acetylate test compounds and the activities of their cytochrome P450 hydroxylations system were made. No apparent differences between the value for soldiers who developed renal failure and the values for soldiers who were barely symptomatic were found. Perhaps other hepatic mechanisms that alter toxicity play an important role in the toxic response.

Pretreating an experimental animal with phenobarbitone, an inducer of a number of hepatic enzymes, enhances the effect of the cortinarins.[29] This finding supports the idea that the liver may be involved in the conversion of a precursor compound into a more toxic product. Alternatively, such treatment could also affect the liver's detoxification abilities—and produce the same outcome.

The Scottish school has also proposed that the toxicity of the cortinarins may be similar to the effects of vasopressin, to which the cortinarins are structurally and pharmacologically similar. However, their arguments for this activity are rather vague and not very convincing.

Orellanine

The European school has postulated a bipyridyl structure for the compound they called orellanine, which they believed to be the major toxin.[30–34] In support of this hypothesis, they cited a structural similarity between orellanine and bipyridyl herbicides. Some investigators showed that orellanine had herbicidal activity similar to that of paraquat and diquat and a toxic effect on cultured kidney cells.[35–37] They suggested that orellanine reduces the amount of NADPH in the cells and makes them more vulnerable to an attack by free radicals, which cause peroxidation and destruction of lipid membranes. Other investigators suggested that orellanine causes defects in pinocytosis,[38] although in most experiments this function was unimpaired. The proposed role of orellanine in disrupting oxidation–reduction reactions also has critics and skeptics.[39,40]

The most recent proposal by investigators of the European school for orellanine's mechanism of action is based on the ability of a metabolite of orellanine to directly inhibit protein synthesis in cultured kidney cells. To produce this effect, native orellanine had to be preincubated with a hepatic microsomal system, which presumably metabolized it to the active toxin.[41] The lag time between the ingestion of an orellanine-containing mushroom and the onset of clinical symptoms might then represent the time needed for the following processes to occur: (1) uptake of orellanine by the target cells; (2) metabolic conversion to the active form; (3) accumulation of the toxic metabolite to inhibitory levels; (4) inhibition of protein synthesis; (5) progressive cellular damage as protein turnover diminishes.

In other experiments, orellanine inhibited the growth of the bacterium *Escherichia coli* and the slime mold *Dictyostelium discoideum*.[42] These experimental models may be useful in the search for the mechanism of action involving the inhibition of protein synthesis.

The evidence garnered from experiments with orellanine in animal models is tantalizing but not sufficient to produce a clear picture of the toxin's action. In some experiments, purified orellanine showed about 40% of the toxicity of an equivalent dose of dried mushrooms. This difference was attributed to an inefficient extraction procedure. An alternative explanation was that orellanine may not be the major or the only toxin. Other experiments showed that low doses of orellanine administered orally to mice were ineffective; however, identical doses exposed to daylight before use were toxic.[43] Still other work rather convincingly demonstrated the nephrotoxicity of orellanine in mice.[40] In rats, both proximal and distal tubules were involved. Morphological changes could be demonstrated in the tubule cells, primarily along the brush border, 12 hours after dosing. Orellanine was found in the urine only during the first 24 hours.[44]

The European community is currently convinced that the active components in *Cortinarius* species are orellanine and two related compounds, orelline and orellinine. These latter two chemicals are photodecomposition products of the primary orellanine.

The debate between the European and Scottish schools continues unabated. A reinvestigation of the issue in 1988 revealed two toxins. One was deemed irrelevant because it was active only when given intraperitoneally; it had no effect when given orally. The other was compatible with orellanine. None of the fluorescent cortinarin compounds could be identified.[44,45] Unlike some of the previous experiments in which total toxicity could not be accounted for by orellanine alone, these investigators stated that in this particular study the entire extent of the toxicity could reasonably be ascribed to orellanine. They discounted the presence of the cortinarins.

In even more recent investigations, a group in Göttingen were unable to reproduce any of the previous findings on the cortinarins.[46–48] For the cortinarin story to be taken seriously, the Scottish observations need independent confirmation by other groups, in other laboratories. Certainly at this time the weight of scientific evidence favors the bipyridyl compound, orellanine, as the most likely candidate.

TOXICITY OF CORTINARIUS POISONS

The toxicity of *Cortinarius* specimens is not altered by cooking, canning, or drying. Indeed, the toxins have been detected in 60-year-old herbarium specimens.[49] Despite the number of individuals poisoned in the past three decades, little is known about the quantity of mushrooms a human must consume to induce symptoms. One psychiatric patient who ate two fruiting bodies in an attempt to commit suicide developed symptoms on the eighth day and was in renal failure by the tenth. She recovered normal kidney function by the end of a month.[50] It is estimated that 100–200 g (3–10 caps) of fresh mushrooms may be sufficient to produce irreversible kidney damage in an adult.

As noted in the preceding section, not all individuals exposed to the same dose develop the same degree of renal impairment. This clinical evidence mirrors the results of many of the animal experiments in which only a fraction of the animals manifested pathology, despite the fact that all animals were given identical doses. The reasons for this variability in resistance or susceptibility are unknown. Again, as noted earlier, women seem to be relatively resistant to the toxins. This observation needs to be confirmed in a large series of patients in which the dose consumed by females can be compared with that consumed by males.

The dose of orellanine that kills 50% of the experimental animals to which it is administered (the LD_{50}) has been determined to be 4.9 mg/kg of body weight in the cat, 8.3 mg/kg in the mouse, and 8 mg/kg in the guinea pig. The histological appearance of the kidneys is said to be identical to that seen in orellanine-poisoned human kidneys. The problem with such a statement is that no one knows precisely what orellanine-induced lesions in the human kidney look like in the earliest phases of poisoning and at different sites along the nephron.

Toxicity studies for the cortinarins have been done in mice. The lethal dose of cortinarin A was 5 mg (given intraperitoneally). In 1-mg doses, renal damage developed but was reversible.

CLINICAL FEATURES OF ORELLANINE POISONING

Onset

One of the characteristic features of orellanine poisoning is the long delay between ingestion and the first evidence of intoxication. In the usual situation, symptoms do not appear before the third day following ingestion and may not appear for three weeks.[51] In a few exceptional cases, onset was much sooner, no more than 12 hours after ingestion. The average presymptomatic interval is approximately eight days.

Signs and Symptoms

In most patients, the presenting symptoms are nausea, vomiting, and anorexia (see the accompanying table). These are soon followed, in hours to a few days, by symptoms indicating kidney involvement: frequent urination (polyuria) and an intense burning thirst (polydypsia).[19,39,52–54] Unless the patient is able to keep up with the fluid losses, which may require several liters of water a day, the mouth feels dry and the lips and tongue burn. These symptoms are often associated with a sense of exhaustion, lethargy, and a lack of appetite. Gastrointestinal disturbance may follow, with either a watery diarrhea or constipation. Many patients have commented on a sensation of coldness and shivering, but fever is generally absent. A persistent headache is common. Pains may develop in the lumbar triangles over the kidneys, and patients may experience generalized musculoskeletal and joint discomfort. Mild hypertension has been noted in a few patients. With further

Frequency of signs and symptoms in orellanine poisoning

Symptom	Frequency*	Symptom	Frequency*
Thirst	98	Headache	66
Nausea	81	Chill	58
Dry mouth	73	Flank pain	57
Vomiting	70	Tinnitus	37
Abdominal pain	57	Extremity pain	32
Constipation	26	Somnolence	31
Diarrhea	21	Visual defects	22

From Grzymala (1959).[54]

*Frequency is expressed as the percentage of patients displaying the symptom. Total number of patients surveyed was 132.

progression of the renal failure, the urine output decreases (oliguria) and may finally cease altogether (anuria).

In rare instances, hepatic and neurological symptoms have been reported.[5,55] The latter included convulsions and coma. These additional organ involvements have not been confirmed in any of the subsequent clinical descriptions over the last decade.

Key Clinical Features of Orellanine Poisoning

- Delayed onset: 2–21 days
- Vomiting and diarrhea
- Marked polyuria and polydypsia
- Headache
- Sensation of coldness with shivering
- Lethargy, anorexia, and generalized muscle and joint pain
- Evidence of progressive renal failure; oliguria and anuria

LABORATORY FINDINGS

Changes in the usual clinical laboratory tests are all consistent with an interstitial nephritis and renal failure. Urinalysis demonstrates various degrees of proteinuria, hematuria, leukocytes, and renal casts. Hyposthenuria is commonly noted as renal failure progresses. Blood urea nitrogen and creatinine levels become progressively elevated as the renal failure worsens and the glomerular filtration rate falls.

DETECTION OF ORELLANINE IN MUSHROOMS AND BODY FLUIDS

Outside the world of the experimental toxicologist's laboratory, two clinical situations exist in which it would be nice to know whether a particular mushroom or blood sample contained the *Cortinarius* toxins: arrival in the emergency department of an asymptomatic child—and a sample mushroom or piece of mushroom; and admission of a adult patient exhibiting signs of renal failure and volunteering a history of recent mushroom ingestion.

Tests for Orellanine in a Mushroom Specimen   A rapid qualitative test for the presence of orellanine in a mushroom sample has been proposed.[56] A small piece of fresh or dried mushroom is crushed in water and allowed to stand for 10 minutes before being filtered. The filtrate is mixed with an equal volume of 3% ferric chloride hexahydrate dissolved in 0.5 N hydrochloric acid. If the solution turns a dark gray-blue (looking like dilute ink), the mushroom is presumed to contain orellanine.

A more rigorous detection method requires extraction with methanol and separation of the components of the extract by thin layer chromatography.[43,49]

Tests for Orellanine in Animal or Human Specimens   There is one report of blue fluorescence noted in the proximal tubules obtained by renal biopsy from a child who subsequently died from orellanine poisoning.[20] This finding has been replicated in experimental animals but needs to be confirmed in more human specimens. Other investigators have noted an orange, not a blue, fluorescence.[15]

Animal studies have shown that orellanine cannot be detected in urine samples after 24 hours postingestion. This finding implies that urine is not a useful specimen for diagnostic tests designed to detect toxins. Blood and tissue samples represent the best specimens for these tests, and the toxin can be detected by thin layer chromatography,[43] spectrofluorometry,[45,57] and high-pressure liquid chromatography.[57,58]

In one study, orellanine was detected in both blood and renal tissue by thin layer chromatography and direct spectrofluorometry. Orellanine was present in serum at 10 days and was still detectable in renal tissue three months after ingestion.[50] This finding suggests a very slow release of the compound that is bound to the cells.

PROGNOSIS

In mild to moderate cases of orellanine poisoning, damage to the kidneys may be reversible, with a transient period of mild renal failure followed by complete recovery. In more severe cases, the damage may be irreversible, and the victim may require either lifelong hemodialysis or kidney transplantation. Of the 132 orellanine poisonings in Poland in the 1950s, 15 patients (11%) died.[39,54] With renal dialysis now generally available, deaths are extremely rare. Renal transplantation or chronic dialysis is required in 10–15% of cases.

Outcome appears to be directly related to the amount of toxin ingested, although a characteristic of this syndrome is the great individual variability in the amount of renal damage. This variability was best demonstrated in the mass poisoning of the soldiers in France, all of whom were said to have ingested similar quantities of mushrooms. When the presentations of these victims were analyzed, it was found

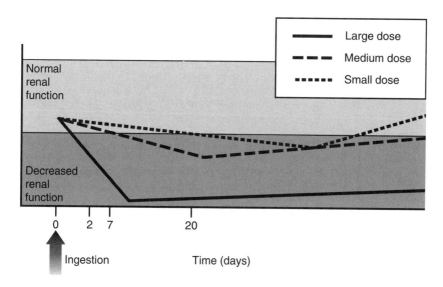

Renal function in orellanine poisoning

that they divided into two groups. One group (N = 12) presented with a large number of symptoms at two to four days and developed renal failure. The rest (N = 14) had far fewer symptoms and did not report ill for three to nine days. This second group did not progress to renal failure.[8]

The two best indicators of prognosis are the length of the latent period and the development of oliguria. The shorter the latent period, the worse the prognosis. In the 1950s, before the advent of dialysis, patients with the most severe involvement died in renal failure in approximately three to four weeks. Patients in whom the latent period is longer than 10 days seldom have any long-term sequelae. Many do not even require hospitalization; and most have only polyuria and polydypsia, which resolve spontaneously within a few days. Symptoms in patients in the intermediate group (onset between 8 and 10 days) usually disappear in three to four weeks, although the renal concentrating defect may last for a number of months. In the French series, 8 of the 12 soldiers who went into renal failure eventually recovered normal renal function; the remaining 4 required chronic dialysis or transplantation.[8] Interestingly, continual and gradual improvement in renal function was noted in some patients for up to a year—the length of the reported follow-up. Further examination of these patients in the future will continue to be instructive.

MANAGEMENT OF ORELLANINE INTOXICATION

Specific treatment of this form of poisoning, beyond the standard management of renal failure, is not yet available. Most patients with severe but not irreversible damage begin to recover renal function between two and four weeks after the onset of symptoms. The medical team must wait for the patient to recover as much renal function as possible before making a decision about long-term dialysis or renal transplantation.

Therapies with No Proven Benefits   Hemadsorption with extracorporeal hemoperfusion through charcoal or resin has no appreciable role in the management of these cases. Nor is there any reason to consider plasma exchange transfusion. Each of these approaches has been advocated at one time or another in the literature. Indeed, the European studies contain a huge spectrum of therapies, and a single patient frequently received several.[50,59] But pharmacokinetic data do not support the use of any of these therapies.

Recently, steroids were administered to a group of severely poisoned French patients in an effort to limit the damage caused by inflammation. No discernible benefits were obtained, but the physicians speculated that this treatment failure was due to the delayed initiation of steroid therapy.[8] The inflammatory response per se probably plays little, if any, role in this syndrome. The poisonous compounds appear to act directly on epithelial cells, an action that is not affected by steroids.

Contraindicated Therapy   The diuretic furosemide should not be administered to victims of orellanine poisoning, because it has been shown to enhance orellanine toxicity in animals[60] and may have a similar effect in humans.

PATHOLOGY OF ORELLANINE POISONING

Human Biopsy Material   Few biopsy reports from human patients with orellanine poisoning are available. The published record is so scanty and cryptic that only the presence of a tubulointerstitial nephritis can be positively associated with this syndrome. In patients with severe intoxication, there is healing with interstitial fibrosis. The few published photographs show necrosis of the tubular epithelial cells, of both the proximal and distal tubules, accompanied by a modest inflammatory cell infiltrate of neutrophils, monocytes, and lymphocytes. Eosinophil infiltration is not a feature of this syndrome and is usually absent. Moderate interstitial edema separates the tubules. Whether or not Henle's loops or the collecting ducts are involved is impossible to determine.[8] A second biopsy taken from one patient at three months postonset showed a persistently evolving picture, with ongoing inflammation and tubular damage. It is unclear whether this disturbing finding is due to continuing toxicity or is a result of very gradual resolution of the initial damage. Glomeruli and vessels are uniformly normal. Immune reactants are regularly absent, an observation that excludes the likelihood that the damage is mediated by an immune mechanism.

Animal Studies Confusion about the precise site of action of the purported toxins persists. Experimental studies have produced inconsistent findings, no doubt because of species differences. In some animals, the proximal convoluted tubules are involved, whereas the distal tubules are the target structure in others. The data from many of these studies are presented in two reviews.[19,61]

The primary damage in rats challenged with cortinarins appears to be in the epithelium of the distal convoluted tubules and the collecting ducts. The epithelial cells of the glomeruli and the proximal convoluted tubules are largely spared. Early in the course of the intoxication, the epithelial cells show degenerative changes that rapidly progress to necrosis. The change observed in rats appears to be similar to that thought to occur in humans. Because of the portion of the nephron involved, greatest damage is noted at the inner cortical and the outer medullary region.[62] Within three to four days, this damage is accompanied by an inflammatory cell infiltrate. In animals with severe intoxication, frank necrosis is followed by reparative scar formation with fibrous tissue and typical interstitial fibrosis. Interestingly, considerable variation in response to the toxin is evident in the experimental animals; only a fraction of the animals are affected.

The compound orellanine is said to affect the proximal tubules. In rats, there is vacuolation of the proximal tubular epithelium about 12 hours after challenge with mushroom or purified orellanine. Protrusion of the cell's cytoplasm into the lumen and destruction of the brush border are some of the first changes to develop. A number of nonspecific changes observed with electron microscopy are merely indicative of cellular degeneration and subsequent cell death. Occasional distal tubules may be involved. The glomeruli are always spared. In the mouse, however, primarily the distal tubules are affected. Orellanine has been shown to inhibit oxidative phosphorylation in mitochondria, which become swollen. Vacuolation of the endoplasmic reticulum is another common, nonspecific feature. The alkaline phosphatase along the brush border of the epithelial cells is inhibited. Many of these experimental studies have been done in isolated or cultured cells; their relevance to the intact animal, especially human victims, is unclear.

The two purported toxins, cortinarin and orellanine, seem to attack very different portions of the kidney. It should be possible to re-solve the issue of which compound is the toxin responsible for poi-soning human kidneys by examining renal biopsy materials from both humans and animals at regular intervals after the ingestion of whole mushrooms.

POISONINGS BY CORTINARIUS SPLENDENS

There are only two reports of poisoning by *Cortinarius splendens*, although each involved a number of victims. In France, 17 individuals of various ages were poisoned. They had evidently mistaken the col-lected mushrooms for the edible *Tricholoma flavovirens* (man-on-horse-back). All the victims suffered symptoms resembling those caused by *Cortinarius orellanus*. The one difference was that a single meal was insuf-ficient to cause noticeable renal damage—several meals eaten within a few days were required to produce poisoning. This dose effect, or cumulative toxic effect, needs to be validated by other reports. One of the victims, a 47-year-old woman who was already in poor health, died as a result of the poisoning. Four other patients lost renal function, and one eventually received a renal transplant.[63,64]

The illness appeared to require repeated meals, and orellanine could not be demonstrated in the suspected mushrooms by thin layer chro-matography.[65] Therefore, Bresinsky and Besl have distinguished between *C. splendens* poisoning and the orellanine syndrome;[66] Azéma has suggested that *C. splendens* is simply not responsible for human poi-sonings.[67] If a few careless or adventurous mycophagists eat this mush-room, then the issue may be resolved. A few well-controlled animal experiments could also assist in clarifying the situation.

UNANSWERED QUESTIONS ABOUT THE ORELLANINE SYNDROME

Despite a great deal of work since the 1950s, many of the fundamen-tal aspects of this form of mushroom poisoning are still unclear. The following questions still need answers:

1. What are the *Cortinarius* toxins that are responsible for human poisoning?
2. What is their site of action in the human kidney?
3. What is the biochemical mechanism of their action?
4. What is the role of the liver in enhancing or reducing toxicity of the native toxin?
5. What is the distribution of toxins in the *Cortinarius* species in the United States?
6. Do cases of orellanine poisoning occur in the United States? Physicians treating patients with idiopathic kidney failure should consider this syndrome in the differential diagnosis.
7. Do repeated ingestions of members of the genus *Cortinarius* result in any long-term renal problems? What are the consequences of chronic low-level ingestion?
8. What is the mechanism responsible for the differential sensitivity (or resistance) of various individuals to identical doses of mushrooms? Are women really more resistant to this form of poisoning than men are?
9. What is the toxin in *Cortinarius splendens?*

At least 200 publications relating to poisoning with selected members in the genus *Cortinarius* have appeared since the 1950s, but we are still woefully ignorant of some of the most basic aspects of the toxicity. Many of the recent articles attempt to define the molecular mechanisms of the toxins, even though the precise nature of the poison is still in doubt. Some investigators make comments such as, "The changes are similar to those in humans," when, in fact, we have little idea of the evolution and development of the related pathology in the human kidney. It is time to combine the efforts of the mycologist, toxicologist, pathologist, analytical chemist, and clinician to resolve the uncertainties. Each one of these groups working in isolation is blindly feeling only one part of the elephant.

Only a few things about orellanine poisoning can currently be asserted with some degree of assurance:

♦ *Cortinarius orellanus, C. speciosissimus,* and *C. orellanoides* cause an interstitial nephritis in humans.
♦ The response of many individuals to the same dose is variable.
♦ Symptoms develop slowly (usually 2–21 days after ingestion).
♦ A general dose–response relationship is evident. The higher the dose, the shorter the latent period and the more severe the renal damage.

All the rest is still open to debate and reassessment.

REFERENCES

1. A. I. K. Short, R. Watling, M. K. MacDonald, and J. S. Robson, "Poisoning by *Cortinarius speciosissimus*," *Lancet* (1980) ii(Nov):942–944.

2. E. Martin-Sans, *L'empoisonnement par les champignons* (Paris: Lechevalier, 1929).

3. S. Grzymala, "Massenvergiftung durch Orangefuchsigen Hauptkopf," *Zeitschrift für Pilzkunde* (1957) 23:139–142.

4. A. Skirgiello and A. Nespiak, "Erfahrungen mit *Dermocybe orellana* (Fr.) in Polen: *A. Cortinarius (Dermocybe) orellanus* Fr. non Quel.—Cause d'intoxications fongiques en Pologne en 1952–55," *Zeitschrift für Pilzkunde* (1957) 23:138–139.

5. J.-F. Marichal, F. Triby, J.-L. Weiderkehr, and R. Carbiener, "Insuffisance rénale chronique après intoxication par champignons de type *Cortinarius orellanus* Fries: (deux cas d'intoxication familiale)," *Nouvelle Presse Médicale* (1977) 6:2973–2975.

6. D. Färber and S. Feldmeier, "Orellanus-Pilzvergiftung im Kindesalter," *Anasthesiol Intensivmed Prax* (1977) 13:87–92.

7. R. Watling, "*Cortinarius speciosissimus*: Cause of Renal Failure in Two Young Men," *Mycopathologia* (1982) 79:71–78.

8. J. Bourget et al., "Acute Renal Failure Following Collective Intoxication by *Cortinarius orellanus*," *Intensive Care Medicine* (1990) 16:506–510.

9. J. Överas, M. J. Ulvund, S. Bakkevig, and R. Eiken, "Poisoning in Sheep Induced by the Mushroom *Cortinarius speciosissimus*," *Acta Veterinaria Scandinavica* (1979) 20:148–150.

10. B. Moore, B. J. Burton, J. Lindgren, F. Rieders, E. Kuehnel, and P. Fisher, "*Cortinarius* Mushroom Poisoning Resulting in Anuric Renal Failure," *Veterinary and Human Toxicology* (1991) 33(4):369.

11. E. Raff, P. F. Halloran, and C. M. Kjellstrand, "Renal Failure after Eating 'Magic' Mushrooms," *Canadian Medical Association Journal* (1992) 147(9):1339–1341.

12. R. K. Myler, J. C. Lee, and J. Hopper, "Renal Tubular Necrosis Caused by Mushroom Poisoning," *Archives of Internal Medicine* (1964) 114:196–204.

13. H. Keller-Dilitz, M. Moser, and J. Ammirati, "Orellanine and Other Fluorescent Compounds in the Genus *Cortinarius*, Section Orellani," *Mycologia* (1985) 77:667–673.

14. B. Schliessbach, S. Hasler, H. P. Friedli, and U. Müller, "Akute Niereninsuffizienz nach Pilzvergiftung mit *Cortinarius splendens* (Fries) oder 'schlöngelbem Klumpfuss' (sog. Orellanus Syndrome)," *Schweizerische Medizinische Wochenschrift* (1983) 113(4):151–153.

15. A. Gérault, "Intoxication collective de type orellanine provoquée par *Cortinarius splendens*," *Bulletin de la Société Mycologique de France* (1981) 97:67–72.

16. S. Kawamura, *Nihon kinrui zukan (Icones of Japanese Fungi)*, Vol. 5 (Tokyo: Kazamashobo, 1970).

17. M. Möttönen, L. Nieminen, and H. Heikkilä, "Damage Caused by Two Finnish Mushrooms, *Cortinarius speciosissimus* and *Cortinarius gentilis*, on the Rat Kidney," *Zeitschrift für Naturforschung* (1975) 30(5):668–671.

18. S. Grzymala, "L'isolement de l'orellanine, poison du *Cortinarius orellanus* Fries, et l'étude de ses effets anatomo-pathologiques," *Bulletin de la Société Mycologique de France* (1962) 78:394–404.

19. K. F. Lampe, "Human Poisoning by Mushrooms of the Genus *Cortinarius*." In *Handbook of Natural Toxins: Toxicology of Plant and Fungal Compounds*, Vol. 6, ed. R. S. Keeler and A. T. Tu (New York: Marcel Dekker, 1991), pp. 497–521.

20. I. Bouska, "Detection of UV Fluorescence in Renal Tissue Following Poisoning with *Cortinarius orellanus*," *Česká Mykologie* (1980) 34(4):188–189.

21. I. R. Tebbet, "Mushroom Toxins of the Genus *Cortinarius*," Doctoral thesis, University of Strathclyde, Glasgow, Scotland, 1985.

22. B. Cady, C. B. M. Kidd, I. R. Robertson, W. J. Tilstone, and R. Watling, "*Cortinarius speciosissimus* Toxins—A Preliminary Report," *Experientia* (1982) 38:1439–1440.

23. I. R. Tebbet and B. Cady, "Toxins of the Genus *Cortinarius*," *Experientia* (1984) 40:441–446.

24. I. R. Tebbet and B. Cady, "Analysis of *Cortinarius* Toxins by Reversed-phase High-Pressure Liquid Chromatography," *Journal of Chromatography* (1984) 283:417–420.

25. I. R. Tebbet, C. B. M. Kidd, B. Cady, J. Robertson, and W. J. Tilstone, "Toxicity of the *Cortinarius* Species," *Transactions of the British Mycological Society* (1983) 81:636–638.

26. D. Michelot and I. R. Tebbet, "Poisoning by *Cortinarius* Mushrooms," *Cryptogamie et Mycologie* (1988) 9(4):345–362.

27. R. Kato, "Sex-related Differences in Drug Metabolism." In *Drug Metabolism Reviews*, Vol. 3, ed. F. J. DiCarlo (New York: Marcel Dekker, 1975), pp. 2–29.

28. L. Nieminen and K. Pyy, "Sex Differences in Renal Damage Induced in the Rat by the Finnish Mushroom *Cortinarius speciosissimus*," *Acta Pathologica et Microbiologica Scandinavica* (1976) A84:222–224.

29. L. Nieminen, "Effects of Drugs on Mushroom Poisoning Induced in the Rat by *Cortinarius speciosissimus*," *Archiv für Toxicologie* (1976) 35:235–238.

30. M. Tiecco, M. Tingoli, D. Testaferri, D. Chiamelli, and E. Wenkert, "Total Synthesis of Orellanine, the Lethal Toxin of *Cortinarius orellanus* Mushroom," *Tetrahedron* (1986) 42:1475–1485.

31. E. V. Dehmlow and H. J. Schulz, "Synthesis of Orellanine, the Lethal Poison of a Toadstool," *Tetrahedron Letters* (1985) 26:4903–4906.

32. W. Z. Antkowiak and W. P. Gessner, "The Structures of Orellanine and Orelline," *Tetrahedron Letters* (1979) 20:1931–1934.

33. W. Z. Antkowiak and W. P. Gessner, "Isolation and Characteristics of Toxic Components of *Cortinarius orellanus* Fries," *Bulletin de l'Académie Polonaise des Sciences, Série des Sciences Chimiques* (1975) 23:729–733.

34. J.-M. Richard and J. Ulrich, "Mass Spectrometry of Orellanine, a Mushroom Toxin, and of Related Bipyridine-*N*-oxides," *Biomedical and Environmental Mass Spectrometry* (1989) 18(1):1–4.

35. K. Høiland, "Extracts of *Cortinarius speciosissimus* Affecting the Photosynthetic Apparatus of *Lemna minor*," *Transactions of the British Mycological Society* (1983) 81:633–635.

36. K. Høiland, "Spiss giftslørsopp-soppen som inneholder ugrasmiddel," *Naturen* (1983) 3:105–110.

37. G. Gstraunthaler and H. Prast, "Studies on the Nephrotoxicity of *Cortinarius orellanus* (Fr.) Fr.: The Effect of Dipyridyls on Renal Epithelial Cell Cultures," *Sydowia Annales Mycologici* (1983) 36:53–58.

38. J. Ahlmen, J. Holmdahl, J.-O. Josefesson, and L. Nassberger, "Inhibition of Pinocytosis by *Cortinarius speciosissimus* Toxins," *Acta Pharmacologica et Toxicologica* (1983) 52(3):238–240.

39. D. Michelot and R. Tebbet, "Poisoning by Members of the Genus *Cortinarius*: A Review," *Mycological Research* (1990) 94:289–298.

40. J.-M. Richard, J. Louis, and D. Cantin, "Nephrotoxicity of Orellanine, a Toxin from the Mushroom *Cortinarius orellanus*," *Archives of Toxicology* (1988) 62:242–245.

41. J.-M. Richard, E. E. Creppy, J. L. Benoit-Guyod, and G. Dirheimer, "Orellanine Inhibits Protein Synthesis in Madin-Darby Canine Kidney Cells, in Rat Liver

Mitochondria, and in vitro: Indication for Its Activation Prior to in vitro Inhibition," *Toxicology* (1991) 67(1):53–62.

42. G. Klein, J.-M. Richard, and M. Satre, "Effect of a Mushroom Toxin, Orellanine, on the Cellular Slime Mold *Dictyostelium discoideum* and the Bacterium *Escherichia coli*," *FEMS Microbiology Letters* (1986) 33:19–22.

43. C. Andary, S. Rapior, A Fruchier, and G. Privat, "Cortinaires de la section *Orellani*: Photodécomposition et hypothèse de la phototoxicité de l'orellanine," *Cryptogamie et Mycologie* (1986) 7:189–200.

44. H. Prast and W. Pfaller, "Toxic Properties of the Mushroom *Cortinarius orellanus* (Fries): II. Impairment of Renal Function in Rats," *Archives of Toxicology* (1988) 62:89–96.

45. H. Prast, E. R. Werner, W. Pfaller, and M. Moser, "Toxic Properties of the Mushroom *Cortinarius orellanus*: I. Chemical Characterization of the Main Toxin of *Cortinarius orellanus* (Fries) and *Cortinarius speciosissimus* (Kühn & Romagn.) and Acute Toxicity in Mice," *Archives of Toxicology* (1988) 62:81–88.

46. L. Matthies and H. Laatsch, "Cortinarins in *Cortinarius speciosissimus*? A Critical Revision," *Experientia* (1991) 47(6):634–640.

47. L. Matthies, H. Laatsch, and W. Paetzold, "Fluorescent Constituents of *Cortinarius rubellus* Cke.: Steroids, Not Nephrotoxic Cyclopeptides," *Zeitschrift für Mykologie* (1991) 57(2):273–280.

48. H. Laatsch and L. Matthies, "Fluorescent Compounds in *Cortinarius speciosissimus*: Investigation for the Presence of Cortinarins," *Mycologia* (1991) 83(4):492–500.

49. S. Rapior, C. Andary, and G. Privat, "Chemotaxonomic Study of Orellanine in Species of *Cortinarius* and *Dermocybe*," *Mycologia* (1988) 80:741–747.

50. S. Rapior, N. Delpech, C. Andary, and G. Huchard, "Intoxication by *Cortinarius orellanus*: Detection and Assay of Orellanine in Biological Fluids and Renal Biopsies," *Mycopathologia* (1989) 108(3):155–161.

51. S. Grzymala, "Étude clinique des intoxications par les champignons du genre *Cortinarius orellanus* Fr.," *Bulletin of Medico-legal Toxicology* (1965) 8:60–70.

52. R. Flammer, "Das Orellanus-Syndrom: Pilzvergiftung mit Nierensuffizienze," *Schweizerische Medizinische Wochenschrift* (1982) 112:1181–1184.

53. K. F. Lampe and J. F. Ammirati, "Human Poisoning by Mushrooms of the Genus *Cortinarius*," *McIlvainea* (1990) 9(2):12–25.

54. S. Grzymala, "Zur toxischen Wirkung des orangefuchsigen Hautkopfes (*Dermocybe orellana* Fr.)," *Deutsche Zeitschrift für die Gesamte Gerichtliche Medizin* (1959) 49:91–99.

55. J.-F. Marichal, R. Carbiener, J.-L. Weiderkehr, and F. Triby, "Insuffisance rénale chronique après intoxication par champignons de type *Cortinarius orellanus* Fries," *Collect. Med. Leg. Toxicol. Med.* (1978) 106:161–165.

56. T. Schumacher and K. Høiland, "Mushroom Poisoning Caused by Species of the Genus *Cortinarius* Fries," *Archives of Toxicology* (1983) 53:87–106.

57. S. Rapior, C. Andary, and D. Mousain, "*Cortinarius* section *Orellani*: Isolation and Culture of *Cortinarius orellanus*," *Transactions of the British Mycological Society* (1987) 89:41–44.

58. J. Holmdahl, J. Ahlmén, S. Bergek, S. Lundberg, and S. Å. Persson, "Isolation and Nephrotoxic Studies of Orellanine from the Mushroom *Cortinarius speciosissimus*," *Toxicon* (1987) 25:195–199.

59. G. Busnach et al., "Plasma Exchange in Acute Renal Failure of *Cortinarius speciosissimus*," *International Journal of Artificial Organs* (1983) 6:73–74.

60. L. Nieminen, K. Pyy, and Y. Hirsimäki, "The Effects of Furosimide on the Renal Damage Induced by Toxic Mushroom *Cortinarius speciosissimus* in the Rat," *British Journal of Experimental Pathology* (1976) 57:400–403.

61. M. Moser, "What Do We Know About the Action of Orellanine?" *Documents Mycologiques* (1989) 77(Oct):71–77.

62. L. Nieminen, M. Möttönen, R. Tirri, and S. Ikonen, "Nephrotoxicity of *Cortinarius speciosissimus*: Histological and Enzyme Histochemical Study," *Experimentelle Pathologie* (1975) 11:239–246.

63. S. Colon et al., "*Cortinarius splendens* Intoxication and Acute Renal Failure: A Clinicopathologic Study," *Kidney International* (1982) 21:1221–1222.

64. S. Colon et al., "Acute Renal Insufficiency During Collective Intoxication by *Cortinarius splendens*: Anatomic-clinical Study," *Néphrologie* (1981) 2:199.

65. A. Gérault, "Point de vue et compléments," *Bulletin de la Société Mycologique de France* (1981) 97:76–77.

66. A. Bresinsky and H. Besl, *A Colour Atlas of Poisonous Fungi* (London: Wolfe Publishing, 1990).

67. R. C. Azéma, "Mémoire sur la toxicité de *Cortinarius splendens*," *VIIèmes Journées Européennes du Cortinaire à Dôle* (1989) 24ème Septembre au 1er Octobre, pp. 1–12.

CHAPTER FOURTEEN

GYROMITRIN POISONING
Effect of Hydrazines

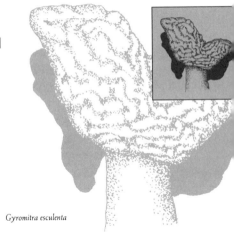

Gyromitra esculenta

THE FATHER, an intelligent man, assured me that in former seasons he had eaten the fungus, locally know as "elephant ears," without experiencing any noticeable ill effects. In preparing them, he had rejected all but fresh, clean specimens and as usual had put them to soak overnight in salt water to be drained off before cooking. Two members of the family had eaten rather sparingly and had not been very ill. He attributed his own recovery and that of his wife and the fatal consequences to his son, aged sixteen, to the fact that in their cases vomition had occurred sooner than in the son's. Death ensued in 48 hours after the meal.

The situation where the collection [of mushrooms] was made was at the top and half-way down the side of a somewhat sandy bank, thinly wooded with pine, hemlock, oak, witch hazel, etc. The day after the funeral, there was still a good crop of lorchels on the slope. The majority of them were of the usual size and bay red color and looked like clean wholesome food.[1]

John Dearness (1924)

It is ironic that the mushroom *Gyromitra esculenta*, whose very name means edible, is so poisonous under certain circumstances. A number of factors conspired against the discovery of this mushroom's poison, called gyromitrin. It was not characterized until 1969,[2,3] for reasons summarized by Lincoff and Mitchel.[4] First, only a few consumers of this mushroom, all of whom might eat the same quantity of the same mushroom, became ill. Because of this inconsistency, these poisonings were ascribed to "allergy" or "individual idiosyncrasy." Similarly, occasionally a dedicated mycophagist who had eaten *G. esculenta* specimens for years without ill effects would suddenly and unaccountably have a toxic reaction. Such an event also was attributed either to a newly developed allergy, to an identification error (a poisonous specimen consumed by mistake), or to the bacteria in a spoiled specimen. Furthermore, although *G. esculenta* caused many poisonings in Europe, a seemingly identical species collected in the western United States appeared to be harmless. All sorts of explanations were proposed to explain this discrepancy, including seemingly fanciful ones such as the suggestion that Americans cook their vegetables longer. For some years, the clues contained in these anecdotes stories were missed, despite the fact that it was well known that gyromitrin was destroyed by thorough cooking.

   Gyromitra esculenta **is an edible mushroom that sometimes kills.** Perhaps the best advice—next to "Never eat this mushroom," which is clearly the very best advice—was given by John Trestrail:[5] "Persons who decide to continue with this gastronomic gamble should have the numbers of their regional poison center permanently engraved on their eating utensils."

INCIDENCE

Poisonings by *Gyromitra* Species

   In eastern Europe, poisoning by *Gyromitra* species is still a significant problem. *Gyromitra esculenta* fruits in abundance in that area and is highly sought after. Both fresh and dried specimens are found in many markets. In western Europe, however, gyromitrin poisonings are less frequent.

   Apart from a major episode of intoxications recorded in Minnesota in the 1930s, very few gyromitrin poisonings have been reported in the medical literature in the United States.[1,6,7] A few cases have been reported by regional poison control centers and by the North American Mycological Association. Usually the reported symptoms were relatively mild and deaths extremely rare.[5]

   Certain investigators have estimated that in the United States about 100,000 persons eat *G. esculenta*, but the original source of this figure is unknown. It is highly likely that the true number of *G. esculenta* consumers is considerably less. In a survey of the dietary habits of the Puget Sound Mycological Society, which is one of the largest amateur groups in the country (membership of 500–600) and includes some very aggressive and adventurous mycophagists, only one member admitted to eating this mushroom, and then only infrequently. However, on the basis of the number of minor incidents reported to regional poison centers, it is clear that *G. esculenta* is eaten in many of the western states and in the Great Lakes region.[8] Those persons not associated with mycology in any way may be the ones who persist in flirting with danger. The reality is that we have no reliable estimates of the popularity of this mushroom or of the incidence of *Gyromitra*-caused poisonings.

The reasons for the relative paucity of reported poisonings in the United States are not entirely clear. Obviously, the number of persons eating wild mushrooms in North America is considerably smaller than that in Europe, thereby reducing the exposure risk here. In addition, these mushrooms are rarely sold in fresh produce markets in the United States. Occasionally, *Gyromitra esculenta* specimens are available in the dried form in the United States, primarily in small specialty food stores and gourmet food catalogues, but most North American mycophagists obtain their *Gyromitra* specimens by foraging in the mountains and forests. It is widely believed that the concentrations of the toxins in some *Gyromitra* species are lower in specimens picked west of the Rocky mountains than in specimens from the east side of the mountains or from Europe. This idea has not been formally documented.

The mushrooms known to be responsible for gyromitrin poisoning are listed in the accompanying box. Their taxonomy is a topic of continuing, hot debate. These species have at various times and places been listed as dangerous or suspect even though reliable data on their edibility and toxicity are sorely lacking. Some, such as those in the *Gyromitra* "gigas" complex, have been eaten for many years by many people with no ill effects, but a few collections of some of these species have been shown to contain traces of hydrazines, the important toxins in this genus.

GYROMITRA SPECIES REPUTED TO CAUSE GYROMITRIN POISONING

Species proved to cause poisoning

G. esculenta   *Common names: False morel, brain mush-room, beefsteak mushroom, elephant ears, turban fungus, lorchel*

G. ambigua[9,10]   *(Toxin not formally determined)*

G. infula[9,10]   *(Toxin not formally determined)*

Species suspected of causing poisoning

G. "gigas" complex; includes G. montana and G. korfii

G. fastigiata (=brunnea)[11]

G. californica

G. sphaerospora

Gyromitra montana, part of the *G. "gigas"* complex, is one of the species that has been eaten for many years in western North America without any obvious toxicity. Rare poisonings attributed to this mushroom have been reported from eastern Europe. However, because of taxonomic and identification uncertainties, these reports cannot be regarded with great confidence. An eastern member of the *G. "gigas"* complex, *Gyromitra korfii* is also listed as edible. The few specimens tested to date do not contain toxins.

Gyromitra fastigiata (*=brunnea*), another eastern species, does not appear to pose any appreciable risk. Both *G. californica* and *G. sphaerospora* have been listed as toxic, although there is no documentable proof of their culpability.

Poisonings by Related Species

Some species of closely related genera—namely, *Helvella, Sarcosphaera, Peziza, Disciotis,* and *Verpa*—may also be poisonous. The edibility of a certain *Helvella* species, the elfin saddle, has always been in doubt, and recent studies have demonstrated the presence of toxins in this genus.[12] Many guides include suggestions for blanching and discarding the water prior to the final cooking step. Other guides recommend that only small amounts of *Helvella* and *Verpa* specimens be consumed, whether in one meal or in several meals. Still other guides recommend that only young, prime specimens be eaten.

In the Pacific Northwest, *Verpa bohemica*—called the early morel—has been eaten by many people for years. Although very few reports of significant toxic effects after eating either *V. bohemica* or *V. conica* have been published, mild gastrointestinal symptoms occur in a small number of individuals that eat them. Fresh specimens of these two species are sometimes sold as morels to unwitting market customers.

Whenever the subject of *V. bohemica* toxicity is raised at a lecture, I am both derided and besieged by many staunch mycophagists who accuse me of being an alarmist. Most eschew the boiling step; they prepare and eat their *V. bohemica* directly. But not all collections of these mushrooms are safe. A former chairman of the New York Mycological Society, D. Simons, reported that he developed incoordination and could not write. This effect, which he rightfully found to be most distressing, lasted for almost five hours.[8] Alexander Smith, another highly respected mycologist, has experienced similar effects. Others, including persons who have eaten *V. bohemica* for many years without noticeable distress, have reported gastrointestinal upsets. The symptoms were mild, consisting largely of a sense of abdominal bloating and diarrhea.[13] Surveys of members of both the San Francisco and Puget Sound

Mycological Societies indicate that *V. bohemica* specimens are responsible for more cases of gastrointestinal discomfort than can be accounted for by chance.

Despite the lack of evidence for major toxicity, most mycologists—being conservative by nature, at least when it comes to labeling mushrooms as edible—recommend against the excessive consumption of *V. bohemica* and *V. conica*. Certainly, prudence suggests that this approach is the most reasonable one.

DISTRIBUTION AND HABITATS OF MUSHROOMS CAUSING GYROMITRIN POISONING

Gyromitra esculenta is a terrestrial fungus, fruiting in the early spring soon after the snows have thawed. It is commonly associated with the temperate conifer forests throughout the Northern Hemisphere. It is especially associated with pine, although other conifers may also be present; it has even been reported in association with aspen. Morel hunters will frequently recognize *G. esculenta*, because it precedes the main morel fruiting by a few days to as long as a month, and it is generally around for the majority of the morel season. At higher elevations, it continues to fruit even later in the year.

Gyromitra montana is one of the *Gyromitra "gigas"* complex and is called snow bank mushroom or snow morel. It is found in the western states; it is especially common in the Rocky Mountains and is frequently found in the Sierra Nevada and Cascade ranges.

Gyromitra californica grows along the West Coast.

Gyromitra sphaerospora is a rare eastern species that extends as far west as Montana. Other eastern species include *Gyromitra korfii*, which is a member of the *G. "gigas"* complex, and *G. fastigiata* (=*brunnea*).

TOXICOLOGY OF GYROMITRIN AND HYDRAZINES

Attempts to identify and purify the toxic principles from *Gyromitra* species were unsuccessful until 1969.[2,3] The investigators found between 1.2 and 1.6 g of a compound with the trivial name of gyromitrin (*N*-methyl-*N*- formylhydrazone acetaldehyde) in each kilogram of fresh mushroom. Since the original report, at least three other hydrazones [a hydrazone is a hydrazine condensed with an aldehyde or a ketone] have been detected in specimens of this genus, sometimes in concentrations greater than that of gyromitrin.[14–16]

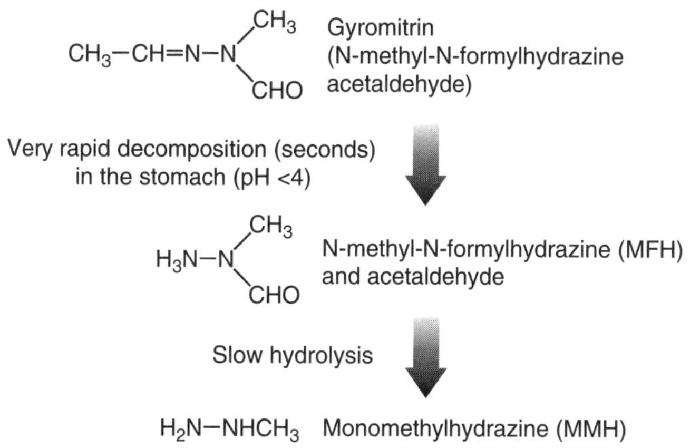

$$CH_3-CH=N-N{\overset{\displaystyle CH_3}{\underset{\displaystyle CHO}{<}}}$$

Gyromitrin
(N-methyl-N-formylhydrazine
acetaldehyde)

Very rapid decomposition (seconds)
in the stomach (pH <4)

$$H_3N-N{\overset{\displaystyle CH_3}{\underset{\displaystyle CHO}{<}}}$$

N-methyl-N-formylhydrazine (MFH)
and acetaldehyde

Slow hydrolysis

$H_2N-NHCH_3$ Monomethylhydrazine (MMH)

Metabolism of gyromitrin to MMH

Gyromitrin is very unstable and undergoes a two-step hydrolysis to monomethylhydrazine (MMH),[17-19] which appears to be the major toxic component of these mushrooms (see the accompanying figure). Of course, the other derivatives and metabolites of gyromitrin may also contribute to the symptoms of the gyromitrin syndrome. This conjecture remains to be proved.

Readers may recognize the chemical monomethylhydrazine, because it has been employed as a rocket fuel. Reports of symptoms very similar to those of gyromitrin poisoning were recorded in the early days of the aerospace industry. Some of the best studies of MMH were done by the military–industrial complex, which was concerned about the hazards MMH posed to workers.[20]

Monomethylhydrazine has a boiling point of 87.5°C, so it is more volatile than water and rapidly dissipates into the atmosphere when the mushrooms containing it are boiled. However, in an enclosed space—such as a poorly ventilated kitchen—MMH vapors can accumulate. At least one cook has developed headache and vomiting after inhaling the vapors from a pot of cooking mushrooms (quoted by K. Lampe[21]). Individuals who dry these mushrooms for commercial purposes have also taken ill. The recommended exposure limits for humans are 0.3 ppm for five hours per day, three to four days per month.[22] It must be stressed, however, that despite the volatility of MMH, enough of the compound remains in the cooking liquid and the mushroom itself to pose a definite threat to the consumer. Furthermore, MMH is readily absorbed through the unbroken skin.

Hydrazines such as MMH and gyromitrin interfere with the action of pyridoxine (vitamin B_6), an important cofactor in numerous enzymatic processes and amino acid metabolism. This activity may account for a number of symptoms, including those associated with neurotoxicity. A relative deficiency of γ-aminobutyric (GABA) can develop as a result of the antipyridoxine effect.[23,24] Pyridoxine administered to challenged animals inhibits gyromitrin-induced convulsive fits in rats but has no effect on toxin-induced liver damage. Also in rats, the renal response—an initial diuresis lasting 12 hours, followed by a relative oliguria—was modified by pyridoxine administration.[25,26]

Other metabolites and mechanisms may be important in the dysfunction of organs, especially the liver. Increased lipid peroxidation has been suggested as the causal mechanism of the liver damage.

In some animals, including the rat, hydrazines have been shown to have a significant inhibiting effect on GABA transaminase in the brain. This action could affect the GABA neurotransmitter system and also could be the cause of neurotoxicity and convulsions.[27] In monkeys, evidence suggests that lower doses of MMH, while not producing the usual signs of toxicity, impair the performance of animals trained to perform complex tasks. This effect is far more subtle than the convulsions and coma produced by high doses and is most likely due to the action of MMH on the central nervous system. It may last a number of hours. No one is sure whether mental impairment occurs in humans following the occasional meal of a few mushrooms.

Gyromitrin and its metabolites also affect the mixed-function oxidase systems in the liver. Because of this action, the metabolism and pharmacokinetics of drugs used in the management of the poisoning, such as barbiturates, are altered.

Inhibition of a diamine oxidase in the intestinal mucosa may be responsible for the gastrointestinal phase of the illness.[28]

Some people are "rapid acetylators" and may convert gyromitrin to an acetylating agent that damages the liver rather than excreting it largely unchanged or as a hydrazone. Mitchell postulates that these people are the ones most at risk.[29] However, the opposite view has also been expressed, because the acetyl derivatives of gyromitrin are demonstrably nontoxic.[30,31] Experimental evidence in rabbits convincingly demonstrates the greater susceptibility of "slow acetylators" to the central nervous system toxicity associated with hydrazines such as isoniazid. Only primary hydrazines combine with pyridoxal phosphate (vitamin B_6), thereby creating a functional vitamin B_6 deficiency.[32] It has also been suggested that "slow acetylators" in whom larger amounts of N-methylhydrazine are formed could be at a greater risk for carcinogenetic events, because in such subjects methylation of the DNA in cell nuclei has been demonstrated.[33]

ENIGMATIC TOXICITY OF GYROMITRIN

One of the most fascinating aspects of these mushrooms is the wide variation in the sensitivity of its individual human consumers. Animal experiments have also demonstrated a wide variation in sensitivity from one species to the next.[34] Benedict has described a family in which the 69-year-old mother died within five days of eating the meal, the 36-year-old daughter recovered rapidly and spontaneously after mild symptoms of vomiting, and the father, two children, and a son-in-law were entirely asymptomatic.[35]

Of the very first cases of gyromitrin poisoning reported in the European literature during the last century, almost a third of the victims died.[36] Today, the mortality rate is very much lower, but people still do die of this syndrome. In a 1967 report of a series of 513 cases, about 14% of the poisonings were fatal.[37]

In a single subject, the margin between an asymptomatic dose and a toxic dose is very narrow. For example, 5 mg of monomethylhydrazine given to monkeys causes only vomiting, whereas 7 mg is lethal.[38] This characteristic of gyromitrin has been used to explain the fact that several people who apparently eat the same quantity of mushroom may range from entirely asymptomatic to very severely affected. It may also account for the observation that certain individuals have been able to eat this mushroom for many years without any ill effects and then very suddenly have a profoundly adverse reaction. It has also been postulated that toxin accumulates in an individual who consumes these mushrooms over a period of days.[39]

Schmidlin-Mészáros has calculated that the lethal dose is 20–50 mg/kg body weight in adults and 10–30 mg/kg body weight in children.[40] This dose corresponds to 0.4–1.0 kg (2–5 cups) of fresh mushrooms in adults and 0.2–0.6 kg for children. The lethal dose of monomethylhydrazine itself is much lower than that of gyromitrin (1.6–4.8 mg/kg body weight in children and 4.8–8.0 mg/kg body weight in adults).[12]

Studies on the amount of gyromitrin in individual mushroom specimens have produced very different estimates. At the high end, specimens contained 1.2–1.6 g/kg fresh weight.[2,15] At the low end, specimens contained 60–320 mg of gyromitrin per kilogram fresh weight and smaller quantities (50–60 mg/kg) of N-methyl-N-formylhydrazone.[11] Drying removes most of the gyromitrin,[40] and cooking in boiling water for 10 minutes eliminates over 99% of the gyromitrin.[15]

The concentration of monomethylhydrazine in the average crop of mushrooms varies from 50 to 300 mg/kg fresh mushrooms.[41] If one assumes that 1 kg of fresh mushroom contains 150 mg of monomethyl-hydrazine, then a 70-kg human male would need to consume about 2.5 kg of mushrooms to induce a fatal illness. This rather absurd quantity suggests that the toxicity estimates are not very precise and have a broad range.

A study done in France has demonstrated that the concentration of toxin is lower in mushroom specimens growing at higher elevations than at lower elevations. Whether elevation alone is the reason for the observed concentration difference is unknown, but it does indicate an environmental influence on toxin concentration. Other studies have shown a variation in the quantity of toxins in different cultured strains.

Many reports from the countries in which this poisoning is more common suggest that children have an enhanced sensitivity to the tox-ins. Children usually ingest a relatively higher dose because of their smaller body mass. In addition, children's metabolic activity and enzyme-system maturity are different from those of adults, factors that may contribute to their increased vulnerability. As mentioned before, evidence indicates that toxin accumulates, because many of the indi-viduals who manifest symptoms only do so after the second or third meal of mushrooms within a few days.

The pharmacodynamics of the toxins have not been well studied in humans, either young or old. The rates of metabolism, excretion, and distribution of the toxins among body compartments and tissues are not precisely known.

Eating *Gyromitra* has been likened to playing Russian roulette.[42] Even the doyen of mycophagy, Charles McIlvaine, had second thoughts about this mushroom:

> It is not probable that in our great food-giving country anyone will be nar-rowed to G. [Gyromitra] esculenta for a meal. Until such an emergency arrives, the species would be better let alone.
>
> Charles McIlvaine and Robert K. MacAdam,
> *One Thousand American Fungi* (1900)

Other health risks associated with the consumption of *Gyromitra* species have already been discussed in Chapter 7. Gyromitrin and its derivatives are mutagenic,[43,44] teratogenic,[45,46] and carcinogenic in ani-mal experimental systems[47,48]—not a good combination, even for a mushroom that is only eaten on an occasional basis. Although the amount of toxin in well-dried specimens is very much reduced from that contained in fresh specimens—certainly below the level that could cause acute toxicity—enough remains to pose a risk for long-term carcinogenic effects. However, no human or epidemiological studies linking these mushrooms either to congenital defects or to cancer have yet been designed, commissioned, funded, or done.

CLINICAL FEATURES OF THE GYROMITRIN SYNDROME

Poisoning with this species tends to follow a biphasic course. It looks similar to amatoxin poisoning, especially in the early phases of the illness. However, gyromitrin poisoning has a number of additional features that allow a differential diagnosis.[7,12,26,37,41,49–51] Perhaps the most distinguishing feature is its springtime or early summer occurrence. Furthermore, a simple description of the mushroom ingested is usually sufficient to establish the diagnosis. Although some cases of amatoxin poisoning do occur early in the year, the basic anatomy of the respective mushrooms are so distinctive that differentiating them is seldom difficult. The majority of amatoxin poisonings, however, occur later in the year.

Onset

The latent period between ingestion of the mushroom and the first symptoms is not long—between 6 and 12 hours. The most severe intoxications, however, may present with symptoms as early as two hours after ingestion, whereas very mild intoxication may not produce noticeable symptoms until 24 hours or more after ingestion.

Signs and Symptoms

As noted earlier, gyromitrin poisoning may follow a biphasic course, a gastrointestinal phase followed by hepatorenal and neurological involvement.

Gastrointestinal Phase The initial phase in gyromitrin poisoning is decidedly gastrointestinal in character. The onset is frequently quite sudden, with a rapid progression of symptoms. Patients express a sense of abdominal fullness or bloating accompanied by nausea, vomiting, and diarrhea. The latter, although not always present, is watery and may contain a little blood. The patient feels dizzy, lethargic, and exhausted. Diffuse abdominal pain and severe headaches are common. If the diarrhea is severe, the usual manifestations of dehydration may develop.

This presentation is very nonspecific, without any unique features to help the physician differentiate it from many other forms of mushroom poisoning or gastroenteritis. However, fever is more frequently present in gyromitrin poisoning than in amatoxin poisoning.

In the majority of the patients with gyromitrin intoxication, the illness never progresses beyond this stage, and spontaneous recovery occurs within a few days.[52] Patients may continue to have gastrointestinal symptoms for up to a week before showing improvement.

Hepatorenal and Neurological Phase Patients who are more severely poisoned develop signs of liver toxicity by about 36–48 hours; early evidence of jaundice becomes apparent and progressively deepens. The liver enlarges and becomes sensitive to palpation. The spleen may be mildly to moderately enlarged. In most patients, the two phases of the illness merge imperceptibly; in a few individuals, however, a period of relative well-being may precede the onset of hepatic toxicity.

In a small number of patients, intravascular hemolysis occurs. Free plasma hemoglobin increases, the extent of which depends on the rate of red cell destruction. The increase in free plasma hemoglobin is accompanied by hemoglobinuria and a decrease in the serum concentration of haptoglobin.

Renal failure develops is some patients. It is not clear whether this sequela is a direct result of the toxicity of the ingested mushroom or an indirect result of the severe intravascular hemolysis or the dehydration, hypotension, and pre-renal azotemia that sometimes complicate the initial phase of the illness. Renal failure occurs so rarely in this illness that sufficient clinical data are not available for analysis.

Methemoglobinemia is another well-described phenomenon associated with gyromitrin poisoning. It should be suspected whenever a cyanosis unresponsive to oxygen therapy is encountered. Symptoms, such as shortness of breath, occur once the blood concentration of methemoglobin rises over 30%. Laboratory analysis readily confirms the presence of methemoglobin in the bloodstream. Blood drawn during phlebotomy may have a dark brown color.

The development of the hemolytic component of gyromitrin poisoning is still poorly understood. It occurs in a minority of patients. Azéma suggests that it is more likely to occur in individuals deficient in the enzyme glucose-6-phosphate dehydrogenase.[42] If this suggestion is correct, gyromitrin-induced hemolysis is somewhat similar to the hemolytic process suffered by some inhabitants of the Mediterranean region after they ingest fava beans (favism). However, an earlier study reported that persons with apparently normal levels of glucose-6-phosphate dehydrogenase developed hemolysis.[53] And experiments with animal models have shown that dogs are particularly susceptible to the development of hemolysis and subsequent methemoglobinemia and methemoglobinuria. The reason for this is not known.

Patients who ingest large doses of gyromitrin go through a terminal phase showing neurological manifestations. These patients often progress rapidly to coma, circulatory collapse, and respiratory arrest. The neurological signs include delirium, tremulousness, muscle fasciculations or spasms, tonic/clonic seizures, mydriasis, and stupor. They may be the consequence of metabolic derangements typical of hepatic encephalopathy and renal failure. However, they often develop before liver and kidney functions have deteriorated to levels normally provoking such symptoms. Therefore, most of the central nervous system symptoms have been ascribed to the direct effect of the toxins on pyridoxine metabolism, especially in neural tissue.

Autopsies have demonstrated cerebral edema, necrosis and fatty degeneration of the liver, renal damage, and scattered petechiae and other small hemorrhages.[54]

Key Clinical Features of Gyromitrin Poisoning

Onset: 6–12 hours

Gastrointestinal phase: 6–48 hours

* Feeling of fullness and bloating

* Abdominal pain

* Vomiting with/without diarrhea

* Severe headache

* Fever may/may not be present

The majority of patients recover in two to seven days, with no further progression of damage.

Hepatorenal and neurological phase: 36–48 hours

* Jaundice

* Intravascular hemolysis in some patients; methemoglobinemia may occur

* Fever

* Delirium, convulsions

* Coma

MANAGEMENT OF GYROMITRIN POISONING

Only one review of gyromitrin poisoning has appeared in English in the last 20 years.[41] However, both French and German scientists have written excellent reviews of this syndrome.[42,49-51] Because North Americans have such limited experience with gyromitrin poisoning, most of the preceding observations and following therapeutic recommendations are based on the European practice and wisdom.

Decontaminating the gastrointestinal tract has little value in the management of gyromitrin-poisoned patients because they generally present many hours after the meal and frequently vomit at the onset of symptoms. If the patient does appear at a medical facility within six hours of ingestion, then decontamination of the gastrointestinal tract, including the use of activated charcoal, is appropriate.

Laboratory tests should include measurements of serum levels of transaminases, creatinine, methemoglobin, hemoglobin, haptoglobin, prothrombin time, BUN, and urinalysis.

Treatment depends on the severity of the case and the organs most involved. Fluid replacement may be required if there has been severe diarrhea. Should evidence of liver dysfunction become apparent, the appropriate management approach is that taken for patients with acute hepatitis. Intravascular hemolysis, if massive, may require exchange transfusions. Any renal failure should be treated with dialysis. Symptomatic methemoglobinemia can be treated with methylene blue: 0.1–0.2 mL/kg body weight of a 1% solution in an intravenous infusion given over five minutes. Careful monitoring should accompany this treatment because, in the presence of glucose-6-phosphate dehydrogenase deficiency, intravascular hemolysis may be induced. Exchange transfusion may be required in severe cases of iatrogenic hemolysis.

Because of the effect gyromitrin and other hydrazines have on enzymes employing pyridoxal phosphate as a cofactor, the use of pyridoxine (100 mg/day) has been strongly advocated.[55] The central nervous system manifestations react beneficially to this additional vitamin B_6. CAUTION: Overdoses of pyridoxine have been associated with acute sensory neuropathy and should be avoided.[56]

Folic acid (20–200 mg/day) has also been recommended as part of the therapy for gyromitrin poisoning. Hydrazines may inhibit the transformation of folic to folinic acid,[50] thereby potentially interfering with the function of this vitamin.

If treatment with pyridoxine is insufficient, diazepam may be required for the management of gyromitrin-induced seizures.

Key Elements of Treatment of Gyromitrin Poisoning

- ◆ Activated charcoal
- ◆ Fluid replacement
- ◆ Pyridoxine (vitamin B6): 100 mg/day
- ◆ Folic acid: 20–200 mg/day
- ◆ Management of liver failure as for acute hepatitis
- ◆ Methylene blue when excess methemoglobin is detected
- ◆ Blood transfusion/exchange transfusion for intravascular hemolysis
- ◆ Diazepam for seizures
- ◆ Renal dialysis for kidney failure

PREPARATION OF FALSE MORELS

Almost all the instances of acute poisoning by gyromitrin have been associated with the consumption of incorrectly prepared fresh specimens. To be rendered potentially edible, this mushroom must be parboiled—at least twice according to most recommendations—and the water in which they are boiled must be discarded. There is sufficient toxin in the cooking water to cause intoxication, so it must never be included in the final dish.[57] Because gyromitrin is very volatile, little remains in the mushrooms if they have been correctly dried. These specimens are said to be safe; certainly, they will not produce acute poisoning.

Although no one should condone or support the practice of eating members of this group of fungi, they are still widely consumed by many people, most of whom either have not heard of the inherent dangers or have failed to be impressed by the evidence. To ensure some element of safety for those who insist and persist in playing the game of mushroom roulette, the following guidelines must be observed.

- ◆ NEVER eat a raw false morel.
- ◆ Parboil fresh specimens twice—10 minutes for each boiling. Discard the cooking water.

◆ Ventilate the kitchen or room where the cooking is done.
◆ Cook dried specimens well.
◆ Avoid eating false morels on consecutive days, because the toxic effect is cumulative.
◆ Eat only young, firm, fresh specimens. Do not eat old, soft, or decomposing mushrooms.
◆ Never overindulge. Small quantities are always safest.

Regardless of the method of preparation, these mushrooms are potentially dangerous. Even with meticulous preparation, this mushroom may still cause symptoms.

ADVERSE EFFECTS OF MORELS

How about the true morels? These mushrooms are consumed in vast numbers each year in many countries across the globe. They are unquestionably safe mushrooms when cooked, even though an occasional person is sensitive to them. These unfortunate souls will never know the pleasures of a morel cream sauce or a *blanquette de veau aux morilles*.

Uncooked morels, however, are toxic to many people. They should never be eaten raw. In fact, morels require prolonged, gentle cooking to fully release the volatile compounds responsible for their exquisite flavor. It is not a mushroom whose texture is of great importance.

Raw or partially cooked morels provoke a typical gastrointestinal syndrome of nausea, vomiting, and abdominal pain. This reaction generally resolves spontaneously in a day.

That raw morels are toxic is known to most amateur mycologists in the United States and to those who live in countries in which wild mushrooms are a frequent dietary component. However, this fact is not well known to young chefs in America, for whom wild mushrooms are a welcome novelty. More than a few unsuspecting restaurant patrons have been poisoned by raw or undercooked morels. A chef at a prestigious Seattle hotel admitted privately to me that he had been responsible for an illness suffered by at least eight guests before he became aware of the toxicity of raw morels.

The most sensational case of accidental poisoning occurred in Vancouver, British Columbia. A pasta salad, including raw morels, shiitake, and *Agaricus brunnescens* was served at an important civic function. The guests included the chief of police, for whom this was a retirement dinner, and a number of members of the city health department. Many of the guests spent the latter part of the evening "worshipping at the ceramic shrine" or at a nearby emergency room. Of the 483 persons present, 77 experienced some symptoms.

UNANSWERED QUESTIONS ABOUT GYROMITRIN POISONING

Gyromitra esculenta is still a highly esteemed edible mushroom in some regions of the world, so it would be a significant boon if someone were to produce a toxin-free, commercially grown *Gyromitra*. Currently, investigators are attempting to achieve this goal. The concentration of the toxin is very similar in the mycelium and in the fruiting body, and very low concentration strains are known to exist. In addition, the fungus fruits in artificial culture.[58] Of course, the problem of potential carcinogenicity will also have to be faced if all the hydrazines cannot be bred out of this mushroom species.

While we await the results of these breeding and genetic engineering experiments, numerous questions about this group of mushroom and their toxins should be answered.

1. Are the differences in toxicity in United States due to separate chemotaxonomic or genetic strains or to environmental and substrate conditions?

2. What is the mechanism causing the individual variation in susceptibility to the effects of the toxin?

3. What provokes the hemolytic reaction in some people and not others?

4. What is the mechanism of the hemolytic reaction?

5. What is the mechanism of the hepatotoxicity in humans and why is this response absent in most experimental animals?

6. Are these mushrooms responsible for increased rates of cancer or birth defects in humans?

7. Do low doses of the mushroom cause any impairment of mental functions in humans?

REFERENCES

1. J. Dearness, "*Gyromitra* Poisoning," *Mycologia* (1924) 16:199.

2. P. H. List and P. Luft, "Nachweis und Gehaltbestimmung von Gyromitrin in frischen Lorchel," *Archiv der Pharmazie* (1969) 302:143–146.

3. P. H. List and P. Luft, "Gyromitrin, das Gift der Frühjahrslorchel," *Archiv der Pharmazie* (1968) 301:294–305.

4. G. Lincoff and D. H. Mitchel, *Toxic and Hallucinogenic Mushroom Poisoning* (New York: Van Nostrand Reinhold, 1977).

5. J. H. Trestrail, "Gyromitrin-Containing Mushrooms—A Form of Gastronomic Roulette," *McIlvainea* (1993) 11(1):45–50.

6. H. V. Hendricks, "Poisoning by the False Morel (*Gyromitra esculenta*)," *Journal of the American Medical Association* (1940) 114:1625.

7. J. O. Cottingham, "Notes on *Gyromitra esculenta* (Fr.)," *Proceedings of the Indiana Academy of Sciences* (1955) 62:210–211.

8. D. M. Simons, "The Mushroom Toxins," *Delaware Medical Journal* (1971) 43:177–187.

9. V. L. Wells and P. E. Kempton, "Studies on the Fleshy Fungi of Alaska II," *Mycologia* (1968) 60:888–901.

10. H. Harmaja, "Another Poisonous Species Discovered in the Genus *Gyromitra*: *G. ambigua*," *Karstenia* (1976) 15:36–37.

11. H. Viernstein, J. Jurenitsch, and W. Kubelka, "Verleich des Giftgehaltes der Lorchelarten *Gyromitra gigas*, *Gyromitra fastigiata* and *Gyromitra esculenta*," *Ernährung* (1980) 4:392–395.

12. C. Andary and G. Privat, "Variations of Monomethylhydrazine Content in *Gyromitra esculenta*," *Mycologia* (1985) 77(2):259–264.

13. V. L. Wells and P. E. Kempton, "Studies on the Fleshy Fungi of Alaska," *Lloydia* (1967) 30:258–268.

14. H. Pyysalo, "Some New Toxic Compounds in the False Morels, *Gyromitra esculenta*," *Naturwissenschaften* (1975) 62:395.

15. H. Pyysalo, "Identification of Volatile Compounds in Seven Fresh Mushrooms," *Acta Chemica Scandinavica* (1976) 30B:235–244.

16. H. Pyysalo and A. Niskanen, "On the Occurrence of N-Methyl-N-formylhydra-zones in Fresh and Processed False Morel, *Gyromitra esculenta*," *Journal of Agricultural and Food Chemistry* (1977) 25:644–647.

17. D. Nagel, L. Wallcave, B. Toth, and R. Kupper, "Formation of Methylhydrazine from Acetaldehyde N-Methyl-N-formylhydrazone, a Compound of *Gyromitra esculenta*," *Cancer Research* (1977) 37(9):3458–3460.

18. R. Braun, U. Greeff, and K. J. Netter, "Indications for Nitrosamide Formation from the Mushroom Poison Gyromitrin by Rat Liver Microsomes," *Xenobiotica* (1980) 10:557–564.

19. A. von Wright, H. Pyysalo, and A. Niskanen, "Quantitative Evaluation of the Metabolic Formation of Monomethylhydrazine from Acetaldehyde-N-methyl-N-formylhydrazone," *Toxicology Letters* (1978) 2:261–265.

20. P. N. Magée, "Toxicité et métabolisme de la méthylhydrazine." In *Proceedings of the Fifth International Congress on Pharmacology*, ed. R. Loomis (San Francisco: Karger, 1972), pp. 140–149.

21. K. Lampe, "Pharmacology and Therapy of Mushroom Intoxications." In *Mushroom Poisoning: Diagnosis and Treatment*, ed. B. H. Rumack and E. Salzman (West Palm Beach, FL: CRC Press, 1978), pp. 125–169.

22. C. B. Shaffer and R. C. Wands, "Guides for Short Term Exposure Limits to Hydrazines," *Proceedings of the Annual Conference on Environment and Toxicology* (1973), AD-781031, pp. 235–242.

23. H. J. Klosterman, "Vitamin B_6 Antagonists of Natural Origin," *Journal of Agricultural and Food Chemistry* (1974) 22:13–16.

24. R. Garnier, F. Conso, M. L. Efthymiou, G. Riboulet, and M. Gaultier, "L'intoxication par *Gyromitra esculenta*," *Toxicological European Research* (1978) 1:359–364.

25. R. Braun, J. Kremer, and H. Rau, "Renal Functional Response to the Mushroom Poison Gyromitrin," *Toxicology* (1979) 13:187–196.

26. R. Braun, U. Greeff, and K. J. Netter, "Liver Injury by the False Morel Poison Gyromitrin," *Toxicology* (1979) 12:155–163.

27. W. S. Chilton, "Chemistry and Mode of Action of Mushroom Toxins." In *Mushroom Poisoning: Diagnosis and Treatment*, ed. B. H. Rumack and E. Salzman (West Palm Beach, FL: CRC Press, 1978).

28. T. Bieganski, R. Braun, and J. Kusche, "N-Methyl- N-formylhydrazine: A Toxic and Mutagenic Inhibition of the Intestinal Diamine Oxidase," *Agents and Actions* (1984) 14:351–355.

29. J. R. Mitchell et al., "Increased Incidence of Isoniazid Hepatitis in Rapid Acetylators: Possible Relation to Hydrazine Metabolites," *Clinical Pharmacology and Therapeutics* (1975) 18:70–79.

30. R. Braun, G. Weyl, and K. J. Netter, "The Toxicology of 1-Acetyl-2-methyl-2-formylhydrazine (Ac-MFH0)," *Toxicology Letters* (1981) 9:271–277.

31. A. Bresinsky and H. Besl, *A Colour Atlas of Poisonous Fungi* (London: Wolfe Publishing, 1990).

32. D. W. Hein and W. W. Weber, "Relationship Between N-Acetylator Phenotype and Susceptibility Toward Hydrazine- induced Lethal Central Nervous System Toxicity in the Rabbit," *Journal of Pharmacology and Experimental Therapeutics* (1984) 228:588–592.

33. K. Bergman and K. E. Hellenas, "Methylation of Rat and Mouse DNA by the Mushroom Poison Gyromitrin and Its Metabolite Monomethylhydrazine," *Cancer Letters* (1992) 61(2):165–170.

34. A. Niskanen and A. V. Wright, "New Aspects of the Toxicity of the False Morel," *Proceedings of the 10th International Conference on the Science and Cultivation of Edible Fungi* (France, 1978), pp. 777–784.

35. R. G. Benedict, "Mushroom Toxins Other Than Amanita." In *Microbial Toxins*, Vol. 8, ed. S. Kadis, A. Ciegler, and S. J. Ajl (New York: Academic Press, 1972), pp. 281–320.

36. E. Bostroem, "Ueber die Intoxicationen durch die essbare Lorchel (Stockmorchel *Helvella esculenta*). Eine experimentelle Untersuchung," *Deutsche Archiv für Klinische Medizin* (1883) 32:209.

37. S. Franke, U. Freimuth, and P. H. List, "Über die Giftigkeit der Frühjahrslorchel, *Gyromitra (Helvella) esculenta*," *Archiv für Toxikologie* (1967) 22:293–332.

38. K. C. Back and M. L. Pinkerton, "Toxicology and Pathology of Repeated Doses of Monomethylhydrazine in Monkeys." In Report AMRL-TR 66–199 (1967), Aerospace Medical Research Laboratory, Wright Patterson Air Force Base, Dayton, Ohio.

39. M. Coulet and J. Guillot, "Poisoning by *Gyromitra*: A Possible Mechanism," *Medical Hypotheses* (1982) 8(4):325–334.

40. J. Schmidlin-Mészáros, "Gyromitrin in Trockenlorcheln (*Gyromitra esculenta* sicc.)," *Mitteilungen aus dem Gebiete der Lebensmitteluntersuchung und Hygiene* (1974) 65:453–465.

41. D. Michelot and B. Toth, "Poisoning by *Gyromitra esculenta*—A Review," *Journal of Applied Toxicology* (1991) 11(4):235–243.

42. R. C. Azéma, "Mémoire sur la toxicité des gyromitres," *Documents Mycologiques* (1979) 10:1–28.

43. A. von Wright, A. Niskanen, and H. Pyysalo, "Mutagenic Properties of Ethylidene Gyromitrin and Its Metabolites in Microsomal Activation Tests and in Host-Mediated Assay," *Mutation Research* (1978) 54(2):167–173.

44. A. von Wright and L. Tikkanen, "The Comparative Mutagenicities of Hydrazine and Its Mono- and Dimethyl Derivatives in Bacterial Test Systems," *Mutation Research* (1980) 78(1):17–23.

45. T. von Kreybig, R. Preussmann, and I. von Kreybig, "Chemische Konstitution und teratogene Wirkung bei der Ratte. III. N-Alkylcarbohydrazide, weitere Hydrazin-derivate," *Arzneimittel-Forschung* (1970) 20:363–367.

46. B. Toth, "Teratogenic Hydrazines: A Review," *In Vivo* (1993) 7(1):101–110.

47. B. Toth, K. Patil, H. Pyysalo, C. Stessman, and P. Gannett, "Cancer Induction in Mice by Feeding the Raw False Morel Mushroom *Gyromitra esculenta,*" *Cancer Research* (1992) 52(8):2279–2284.

48. B. Toth, "Carcinogenic Fungal Hydrazines," *In Vivo* (1991) 5(2):95–100.

49. M. Coulet and J. Guillot, "Des champignons toxiques au sein des discomycètes: Le problème des gyromitres," *Revue Scientifique et Nature d'Auvergne* (1978) 44:33–45.

50. J.-C. Chénieux, "Les gyromitres toxiques," *Collect. Med. Leg. Toxicol. Med.* (1978) 106:151–159.

51. R. Flammer, "Hämolyse bei Pilzvergiftungen: Fakten und Hypothesen," *Schweizerische Medizinische Wochenschrift* (1983) 113:1555–1561.

52. K. Lampe, "Toxic Fungi," *Annual Reviews of Pharmacology and Toxicology* (1979) 19:125–169.

53. E. Breuer and O. Stahler, "Hämolytischer Ikterus bei Lorchelvergiftung," *Medizinische Welt* (1966) 18:1013–1016.

54. G. V. Giusti and A. Canevale, "A Case of Fatal Poisoning by *Gyromitra esculenta,*" *Archives of Toxicology* (1974) 33(1):49–54.

55. A. von Wright, A. Niskanen, H. Pyysalo, and H. Korpela, "Amelioration of the Toxic Effects of Ethylidene Gyromitrin (False Morel Poison) with Pyridoxine Chloride," *Journal of Food Safety* (1981) 3(3):199–203.

56. R. L. Albin et al., "Acute Sensory Neuropathy–Neuronopathy from Pyridoxine Overdose," *Neurology* (1987) 37(11):1729–1732.

57. T. J. Duffy and P. P. Vergeer, "California Toxic Mushrooms." In Report of Toxicology Committee, Mycologic Society of San Francisco (1977).

58. P. H. List and G. Sundermann, "Achtung! Frühjahrslorcheln," *Deutsche Apotheker Zeitung* (1974) 114:331–332.

CHAPTER FIFTEEN

"ANTABUSE" SYNDROME
Coprine Poisoning

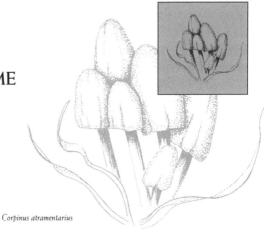

Coprinus atramentarius

A STRONG PLUMP peasant woman, 35 years old, from the vicinity of Geneva, ate with her husband and son at noon a dish of *Coprinus atramentarius*. She was used to drinking a fair amount of wine. After eating, no special symptoms. In the evening the woman and her husband again ate the mushroom dish; the son did not eat the mushrooms again nor did he drink any alcohol, and he remained well. The woman did not feel well that night, but not so unwell as to be unable in the morning to drive to the market with the vegetables. After the market she drank a cup of black coffee and mounted the cart. On the way back she became unwell. She felt a congestion of blood in the face, pressure in the head, and freezing. But she drove home, and at noon ate and drank wine. Again heat and freezing. The husband was affected in the same way. As both thought that they had the stomach overloaded with the mushroom dish from which they had eaten in great quantity each of them drank 3–4 small glasses of liqueur. As the alcohol did not help, the woman drank one litre of milk straight off and at once vomited the half digested mushrooms. No diarrhoea, but palpitations of the heart and unconsciousness followed so that she did not even remember the arrival of the doctor and being taken to the hospital. She was given a purgative and a subcutaneous camphor injection. The second day her physical state was good, and she was discharged from the hospital and allowed to go home. In the husband the illness showed similar, but much less violent symptoms, and after two days the symptoms disappeared completely.

Reported by Thellung in Geneva, Switzerland, in 1940[1]

At the time of this report, the interaction between the common inky cap mushroom and alcohol had been known for only a few years. The compound responsible for this effect is called coprine. In later years, this type of poisoning has been termed the Antabuse syndrome. Antabuse, the commercial name for the compound disulfiram, has been widely used in the management of alcoholism. The combination of disulfiram and alcohol produces the same unpleasant clinical effects as does the combination of coprine and alcohol. Disulfiram therapy (for the alcoholic) is designed to discourage further drinking.

Only one common species of mushroom has been proved to interact with alcohol in this way. *Coprinus atramentarius*, whose common names includes tippler's bane and the common inky cap, is the known culprit. *Clitocybe clavipes*[2] (reviewed by Imazeki[3]) has also been reported to produce similar symptoms. However, this mushroom has been assayed and does not appear to contain coprine.[4] Its activity may be due to a metabolite of coprine or some other as yet unidentified moiety.

Unconfirmed clinical reports have associated a reaction to alcohol with a number of mushrooms: *Coprinus micaceus*,[9] *C. insignis*, *C. quadrifidus*, *C. variegatus*, *C. erethistes*, and *C. fuscescens*.[5-7] Of these, the only ones known to contain coprine are *C. insignis*, *C. quadrifidus*, and *C. variegatus*. These are not often used as food, even in the eastern United States,

---

SPECIES ASSOCIATED WITH THE ANTABUSE SYNDROME

Coprinus species known to contain coprine

C. atramentarius   Common name: Common inky cap, tippler's bane

C. insignis[5-7]
C. quadrifidus[5-7]
C. variegatus[5-7]

Other species implicated in an alcohol-related poisoning syndrome

Coprinus erethistes[5-7]   *Coprinus fuscescens*[5-7]
Clitocyte clavipes[2,3]   *Boletus luridus*[8]

where they fruit more commonly than in the west. Coprine has never been found in either *C. comatus* (shaggy mane or lawyer's wig mushroom) or *C. micaceus*.[4]

In 1935 Suss bravely experimented on himself. He ate a small dish of inky caps (*Coprinus atramentarius*) on two successive days without drinking any alcohol at the same time and suffered no evidence of poisoning. On the third day, he drank a glass of beer. Within 10 minutes, he felt hot, developed marked reddening of the face, and a slight feeling of anxiety; his heart rate increased up to 100 beats per minute. After an hour, the symptoms disappeared entirely. The next day he drank a glass of cider and a small glass of fruit brandy. Once again the symptoms appeared but were much less severe in character.

The most curious case of a mushroom–alcohol reaction occurred in cows. This incident was reported by Bresinsky and Besl:[10] "A farmer noticed that his cows, which had previously been given fodder containing fungi, had developed flatulence. He therefore gave them his old trusty household remedy—home made cherry brandy— whereupon the animals became so ill that they had to be slaughtered immediately."[11]

DISTRIBUTION AND HABITATS OF SPECIES RESPONSIBLE FOR COPRINE POISONING

The group of mushrooms called inky caps are common urban mushrooms, often growing where the ground has been disturbed. They are frequently found along roadways, in parking lots, in lawns and parks, and even pushing their way up between paving stones or rocks. They fruit after a rainfall in both spring and fall.

Their spore dispersal system involves the dissolution of the gill tissue into an inky liquid that drops to the ground. This process is also termed deliquescing, and the liquid, with the black spores, can be and has been used as ink. Generally, these mushrooms develop very rapidly—a beautiful, firm, white specimen can change into a mass of dissolving gray-black tissue within a matter of hours to days.

One of the edible favorites in the group is the shaggy mane (*Coprinus comatus*), also called the lawyer's wig in England. Because members of this species are so readily available and often fruit in great abundance, they are eaten by many. And because of their very rapid growth habit, they are treated like corn—the butter should be melted in the pan before one picks them.

The fruiting body of *Coprinus atramentarius* is not as evanescent as some of the other species of inky caps. It lasts a number of days before curling its edges and deliquescing. *Coprinus atramentarius* is a ubiquitous, cosmopolitan mushroom. It is found around the globe in temperate climes. It often occurs in large clusters wherever organic material has accumulated. It is common in urban areas, in gardens, near compost heaps or dumps of organic material, near decaying trees, and along roadsides. It will often push its way up through hard-packed soil. Although favoring the cool moist season of fall, it does fruit at other times, including spring.

Clitocybe clavipes tends to grow in conifer forests or in mixed woods and fruits in the late autumn or winter. Single specimens are found widely scattered across the forest floor. Its name derives from its distinctive club-shaped stem.

TOXICOLOGY OF COPRINE

The compound in *Coprinus atramentarius* that causes the Antabuse syndrome has been given the trivial name coprine. It is not disulfiram, as was first thought. Coprine has been found only in *Coprinus atramentarius* and a few other rarely picked or eaten *Coprinus* species. Although the symptoms associated with ingestion of *Clitocybe clavipes* and alcohol are most suggestive of the same agent, coprine has not been identified in this species.

The structure of coprine, which is an unusual cyclopropylglutamine, has been determined.[12,13] Its mechanism of action is identical to that of disulfiram. Both of these agents interfere with the further conversion of acetaldehyde to acetate.[14] Acetaldehyde is the first compound to which alcohol is metabolized. When its breakdown is thus inhibited, it accumulates in the bloodstream. Acetaldehyde tends to react with the β-adrenergic receptors of the autonomic nervous system, thereby producing the violent and unpleasant vasomotor responses associated with coprine intoxication.

The mechanism by which coprine interferes with alcohol metabolism was initially unclear, because it does not inhibit aldehyde dehydrogenase in vitro, as disulfiram does. Disulfiram competes for the NAD-binding active centers on the enzyme, thereby decreasing the rate of destruction of acetaldehyde by oxidation. It is now known that the active agent in coprine intoxication is a metabolite of coprine. This metabolite is 1-aminocyclopropanol,[4] and it does inhibit the enzyme aldehyde dehydrogenase both in vivo and in vitro.[15] Cyclopropanone hydrate, which is derived from coprine by hydrolysis, also inhibits aldehyde dehydrogenase, both in the test tube

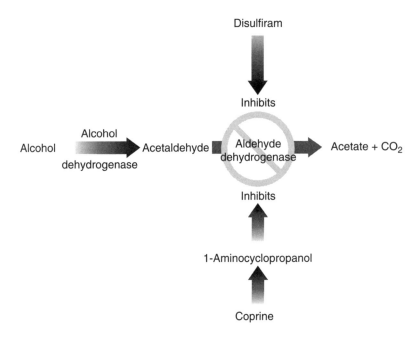

Mechanism of interaction of coprine and alcohol

and in the intact organism.[16] In rats, coprine itself is a potent inhibitor of aldehyde dehydrogenase, with a rapid onset and a long-lasting effect.[17]

A variety of other effects have been noted in animals, including hypotension in rabbits,[18,19] prolongation of the sleep cycle in mice,[20] and swelling of the face with increased tear production in rats.[21] In rats and dogs given repeated oral doses of coprine, damage to spermatozoa and testes was reported.[22] The pathology appeared to be a direct effect of the coprine on the germ cells, similar to that produced by some alkylating agents. It only occurred when the dose was large and administered for a prolonged time.

No parallel situation in which humans have consumed huge amounts of mushrooms for many consecutive days are known. Most of us are quite happy if we come across a nice fruiting of this species a few times during the season. Specimens of *Coprinus atramentarius* contain approximately 160 mg coprine/kg wet weight of mushroom. The dogs showing germ cell damage were fed a minimum of 25 mg coprine/kg body weight/day for 28 days. Consequently, a 70-kg human male would have to eat about 6 kg (20 cups) of mushrooms a day for a month to be similarly affected, assuming that humans show the same type of toxic reaction.

CLINICAL FEATURES OF COPRINE INTOXICATION

Onset

How long it takes to become sensitive to alcohol following a mushroom meal is unknown, but the onset of the symptoms is generally within 5 to 10 minutes after the ingestion of alcohol. When small quantities of alcohol and coprine-containing mushrooms are ingested simultaneously and no further alcohol is drunk, symptoms do not develop.[23] Presumably, in this situation the alcohol has been metabolized before the coprine has had time to exert its effect. It has been reported that a glass of beer drunk with the meal is enough to provoke the effect,[10] but I am unable to confirm this. The earliest confirmed time of onset is 30 minutes after the mushroom meal and alcohol were consumed.

Animal species differ in their sensitivity to alcohol and coprine. For example, mice become maximally sensitized to alcohol between three and six hours,[20] whereas rabbits become sensitive in 30 minutes.[18] No systematic studies have been undertaken in humans, despite much self-experimentation with this mushroom. It is probable that several hours are required to develop maximum sensitivity in humans, but the precise meaning of "several" is not known. Obviously, if enough alcohol is drunk with a meal, a measurable blood alcohol level will still be present two to three hours later, and symptoms can appear then.

Signs and Symptoms

The signs and symptoms[4,24,25] are virtually identical to those produced by disulfiram (Antabuse). Symptoms include flushing and swelling of the face, accompanied by a sensation of warmth; a tingling sensation in the arms and legs; nausea and vomiting; a metallic taste in the mouth; tachycardia (rapid heart rate) associated with palpitations; severe headache; sweating, anxiety, vertigo; confusion; and hypotension and collapse. It is quite uncommon for any of these symptoms to be severe, although the combination can be very unpleasant.

Symptoms begin when the blood alcohol rises above 5 mg/dL, become marked at concentrations of 50–100 mg/dL, and can produce unconsciousness when levels of 125 mg/dL are exceeded. Alcohol alone may alter consciousness at this high level.

In the uncomplicated case, in which no further alcohol is imbibed, the symptoms usually resolve by themselves in two to three hours. However, the severity and the time course can be quite variable, depending somewhat on the quantity of alcohol consumed. When large amounts have been drunk, the effects may last up to eight hours. The symptoms recur if the patient drinks alcohol again within 72 hours of the mushroom meal. The severity of the symptoms in each succeeding episode usually diminishes.[26]

An additional complication, rarely observed, is that of cardiac arrhythmia. This sequela is well described in patients exhibiting disulfiram–alcohol interaction. A 37-year-old male in good health developed supraventricular tachycardia with atrial fibrillation lasting nearly three days before disappearing. His illness followed a meal of mushrooms and a few after-dinner drinks.[27]

In the report linking ingestion of *Clitocybe clavipes* and alcohol,[2,3] whiskey was associated with more severe effects than was either gin or vodka. Whiskey produced a severe, throbbing, prostrating headache that lasted a number of hours. No systematic study on the effects of different forms of spirits, beer, and wine has been undertaken. Nor is it known whether blended whiskey and single-malt liquors produce different effects.

Key Clinical Features of Coprine Intoxication

- Onset occurs 5–10 minutes after alcohol ingestion by a person who has eaten common inky caps 30 minutes to 3 days previously
- Sensation of warmth, flushing, and possible swelling of the face
- Sensation of tingling in arms and legs
- Nausea and vomiting
- Metallic taste in mouth
- Tachycardia and palpitations
- Severe headache
- Sweating, anxiety, vertigo
- Confusion
- Hypotension and collapse

PROGNOSIS

The outcome is uniformly good. In the majority of cases of coprine intoxication, the effects are relatively mild and self-limited. Even though the symptoms have been quite distressing in a few patients, no deaths have been attributed to this syndrome. Fatalities have been incorrectly ascribed to the ingestion of mushrooms and alcohol, but these reports were all based on a series of cases relating to the use of disulfiram in chronic alcoholics who had a host of other confounding factors.[28]

The possibility of experiencing coprine poisoning is certainly not grounds for temperance. Wine is such an important accompaniment to a fine meal that we should be especially critical and skeptical of any report that brings this marriage into question.

MANAGEMENT OF THE ANTABUSE SYNDROME

In the vast majority of cases, reassurance is the most important aspect of therapy, because the symptoms can be quite disconcerting and the patient may well believe that he or she is acutely and severely poisoned. An emetic is not indicated, because the mushrooms are usually well digested by the time the patient presents for treatment. If the reaction has been severe, the patient will have vomited already. In those instances in which the patient has vomited excessively and has become hypotensive, rehydration with intravenous solutions is indicated. Any drugs with β-adrenergic effects should be avoided because they can increase the tachycardia. Should any serious cardiac dysrhythmias or arrhythmias develop, they should be treated appropriately.

Key Elements of Treatment of Coprine Intoxication

- Reassurance
- Avoid emetics
- Avoid β-adrenergic drugs
- Rehydration (either oral or intravenous) if the patient is significantly dehydrated
- Monitor for cardiac arrhythmias

UNANSWERED QUESTIONS ABOUT ALCOHOL–MUSHROOM INTERACTION

1. What mushrooms other than *C. atramentarius* cause the Antabuse syndrome? *Coprinus* species implicated in the Antabuse syndrome

have already been mentioned on page 284. But other unrelated mushrooms have been suspected of causing a toxic interaction with alcohol. The following episodes have been reported.

Two persons developed severe nausea, diarrhea, and vomiting about four hours after sharing a lunch of morels.[29] The symptoms developed 15–20 minutes after drinking rye whiskey. Because morels are so widely eaten, and alcohol such a ubiquitous drink among the gourmands who enjoy this mushroom, any association at all seems most unlikely. Other mechanisms could well account for this isolated case. The collection appears to have been contaminated with at least some *Verpa bohemica*, the false morel, which may have caused the illness.

The report of another episode was prompted by the preceding account about morels. The indicted mushrooms were *Pholiota squarrosa*, and the symptoms were vomiting, diarrhea, and shock in one patient.[30] These symptoms developed about four hours after a meal of mushrooms and a couple of "high-balls." Since this 1965 report, *Pholiota squarrosa* has been implicated in a number of other intoxications in which alcohol was not present.

Boletus luridus has been incriminated in three instances of a disulfiramlike interaction with alcohol.[8] Many mycophagists do not consider *B. luridus* to be edible, and it appears that their judgment is correct. Whether or not *B. luridus* contains coprine or a toxin with a similar action remains to be proved?

2. Do raw mushrooms have the same effect as cooked ones? It may seem superfluous to even ask this question, having repeatedly insisted that mushrooms should always be cooked. But continuing controversy surrounds this particular mushroom (*C. atramentarius*), which for some reason is consumed raw by certain people.

Another episode of self-experimentation is worth reporting. This experiment was an attempt to determine whether or not the mushrooms need to be cooked to produce the adverse reaction with alcohol. The following extract is Lampe's translation of the original German report.[31,32]

> Each of us gathered 250 grams of medium sized mushrooms to eat. Two of the investigators ate the finely chopped mushrooms prepared as a salad, the third consumed them as a cooked mushroom dish. Directly after the meal each drank 100 ml of red wine and after that 40 ml of dry gin. None of the investigators perceived effects associated with the mushrooms.
>
> The investigator who had eaten the cooked mushrooms drank a glass of beer 16 hours after the meal, after which he felt slightly unwell, but to no significant degree. The next day, about 24 hours after the meal he drank a glass of wine. Shortly thereafter followed severe nausea which

progressed to vomiting; in addition there was a headache passing to a severe cold feeling in the arms and legs. The cold feeling followed a variable course through the night with fever and chills. All symptoms were accompanied with a feeling of great uneasiness. In the morning all complaints were gone.

The two other investigators each drank wine 24 hours after the meal, yet felt no reaction. One of us repeated the investigation with raw mushrooms and had the same negative result.

In experimental animals, there is no difference between the effects of cooked and raw mushrooms.[20]

3. Do the mushrooms need to be heated to "activate" the coprine so that it can be absorbed by humans? This question should be easy to resolve, with the appropriate volunteers. Coprine is water soluble, and any extract of mushrooms, cooked or raw, is likely to contain the compound. In some of the experiments attempting to resolve the issue of the effects of raw versus cooked mushrooms, boiling water was used as part of the extraction procedure, so the issue remains unresolved. It is also possible that parboiling and discarding the water may be another way to rid the mushrooms of the toxic agent, if anyone is willing to go to all the trouble and effort to produce a tasteless and rather textureless meal.

4. In humans, how long is the interval between a mushroom meal and the beginning of the sensitivity to alcohol? Anecdotal data indicate that this interval is about 30 minutes, but more controlled experiments need to be done. Any volunteers?

REFERENCES

1. A. Pilát and O. Ušák, *Mushrooms* (London: Spring Books, 1954).

2. K. W. Cochran and M. W. Cochran, *"Clitocybe clavipes*: Antabuse-like Reaction to Alcohol," *Mycologia* (1978) 70:1124–1126.

3. H. Romagnesi, "Champignons toxiques au Japon," *Bulletin de la Société Mycologique de France* (1964) 80:iv–v.

4. G. M. Hatfield and J. P. Schaumberg, "The Disulfiram-like Effects of *Coprinus atramentarius* and Related Mushrooms." In *Mushroom Poisoning: Diagnosis and Treatment*, ed. B. H. Rumack and E. Salzman (West Palm Beach, FL: CRC Press, 1978).

5. O. Miller, Jr., *Mushrooms of North America* (New York: E. P. Dutton, 1972).

6. A. Rinaldi and V. Tynaldo, *The Complete Book of Mushrooms* (New York: Crown, 1974).

7. R. Heim, *Les champignons toxiques et hallucinogènes* (Paris: N. Boubee, 1963).

8. H. Budmiger and F. Kocher, [*Boletus luridus* and alcohol. Case report], *Schweizerische Medizinische Wochenschrift* (1982) 112(34):1179–1181.

9. C. M. Christensen, *Molds, Mushrooms and Mycotoxins* (Minneapolis: University of Minnesota Press, 1975), p. 23.

10. A. Bresinsky and H. Besl, *A Colour Atlas of Poisonous Fungi* (London: Wolfe Publishing, 1990).

11. H. Clémençon, "Antabus-Wirkung bei Kühnen?" *Schweizerische Zeitschrift für Pilzkunde* (1962) 40:170–172.

12. P. Lindberg, R. Bergman, and B. Wickberg, "Isolation and Structure of Coprine, a Novel Physiologically Active Cyclopropanone Derivative from *Coprinus atramentarius* and Its Synthesis via 1-Aminocyclopropanol," *Journal of the Chemical Society, Chemistry Communications* (1975), pp. 946–947.

13. G. M. Hatfield and J. P. Schaumberg, "Isolation and Structural Properties of Coprine, the Disulfiram-like Constituent of *Coprinus atramentarius*," *Lloydia* (1975) 38:489–496.

14. B. B. Coldwell, K. Genest, and D. W. Hughes, "Effect of *Coprinus atramentarius* on the Metabolism of Ethanol in Mice," *Journal of Pharmacy and Pharmacology* (1969) 21:176–179.

15. H. Marchner and O. Tottmar, "A Comparative Study on the Effects of Disulfiram, Cyanamide, and 1-Aminocyclopropanol on the Acetaldehyde Metabolism in Rats," *Acta Pharmacologica et Toxicologica* (1978) 43:219–232.

16. J. S. Wiseman and R. H. Abeles, "Mechanism of Inhibition of Alcohol Dehydrogenase by Cyclopropanone Hydrate of the Mushroom Toxin Coprine," *Biochemistry* (1979) 18(3):427–435.

17. O. Tottmar and P. Lindberg, "Effects on Rat Liver Acetaldehyde Dehydrogenases in vitro and in vivo by Coprine, the Disulfiram-like Constituent of *Coprinus atramentarius*," *Acta Pharmacologica et Toxicologica* (1977) 40:476–481.

18. R. Barkman and P. Perman, "Supersensitivity to Ethanol in Rabbits Treated with *Coprinus atramentarius*," *Acta Pharmacologica et Toxicologica* (1963) 20:43–46.

19. A. Carlsson et al., "On the Disulfiram-like Effect of Coprine, the Pharmacologically Active Principle of *Coprinus atramentarius*," *Acta Pharmacologica et Toxicologica* (1978) 42:292–297.

20. K. Genest, B. B. Coldwell, and D. W. Hughes, "Potentiation of Ethanol by *Coprinus atramentarius* in Mice," *Journal of Pharmacy and Pharmacology* (1968) 20:102–106.

21. P. Lindberg, R. Bergman, and B. Wickberg, "Isolation and Structure of Coprine, the in vivo Aldehyde Dehydrogenase Inhibitor in *Coprinus atramentarius*: Synthesis of Coprine and Related Cyclopropanone Derivatives," *Journal of the Chemical Society, Perkins Transactions* (1977) 1:684–691.

22. M. Jönsonn, L. Lindquist, S. Plöen, S. Ekvärn, and T. Kronevi, "Testicular Lesions of Coprine and Benzcoprine," *Toxicology* (1979) 12:89–100.

23. G. P. Child, "The Inability of *Coprini* to Sensitize Man to Ethyl Alcohol," *Mycologia* (1952) 44:200–201.

24. W. A. Reynolds and F. H. Lowe, "Mushrooms and a Toxic Reaction to Alcohol: Report of Four Cases," *New England Journal of Medicine* (1965) 272:630–631.

25. J. H. Meyer, J. E. Herlocher, and J. Parisian, "Esophageal Rupture after Mushroom-Alcohol Ingestion," *New England Journal of Medicine* (1971) 285:1323.

26. T. J. Duffy and P. P. Vergeer, *California Toxic Fungi*, Toxicology Monograph No. 1, Mycological Society of San Francisco (1977).

27. M. J. Caley and R. A. Clark, "Cardiac Arrhythmia after Mushroom Ingestion," *British Medical Journal* (1977) 2:1633.

28. A. R. Alha, E. Hjelt, and V. Tamminen, "Disulfiram-Alcohol Intoxication. Investigation of Five Fatal Cases and the Chemical Determination of Disulfiram and Acetaldehyde," *Acta Pharmacologica et Toxicologica* (1975) 13:277–288.

29. J. W. Groves, "Poisoning by Morels When Taken with Alcohol," *Mycologia* (1964) 56:779–780.

30. R. L. Scaffer, "Poisoning by *Pholiota squarossa*," *Mycologia* (1965) 57:318–319.

31. P. H. List and H. Reith, "Der Faltentintling, *Coprinus atramentarius* Bull., und seine dem Tetraathylthiuramdisulf ähnliche Wirkung, "*Arzneimittel-Forschung* (1960) 10:34–40.

32. K. Lampe, "Pharmacology and Therapy of Mushroom Intoxications." In *Mushroom Poisoning: Diagnosis and Treatment*, ed. B. H. Rumack and E. Salzman (West Palm Beach, FL: CRC Press, 1978), pp. 125–169.

1 *Amanita phalloides*

2 *Amanita phalloides*

3 *Amanita virosa*

4 *Galerina autumnalis*

5 *Lepiota* species (toxic)

6 *Gyromitra esculenta*

7 *Coprinus
atramentarius*

8 *Amanita muscaria*

17 *Agaricus praeclarisquamosus*

18 *Agaricus xanthodermus*

19 *Boletus satanas*

20 *Armillaria mellea* complex

21 *Entoloma lividum*

22 *Hebeloma crustiliniforme*

23 *Laetiporus sulphureus*

24 *Lactarius rufus*

25 *Omphalotus olivascens*

26 *Russula emetica*

27 *Pholiota squarrosa*

28 *Auricularia auricula*

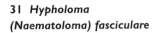

30 *Paxillus involutus*

29 *Amanita smithiana*

31 *Hypholoma
(Naematoloma) fasciculare*

32 *Ramaria gelatinosa* var. *oregonensis*

CHAPTER SIXTEEN

INEBRIATION OR PANTHERINE SYNDROME
Effects of Isoxazole Derivatives on the Central Nervous System

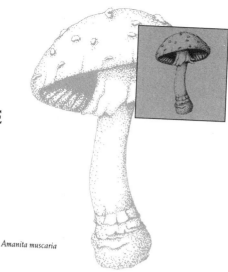

Amanita muscaria

"One side will make you grow taller, and the other side will make you grow shorter."

"One side of what? The other side of what?" thought Alice to herself.

"Of the mushroom," said the Caterpillar, just as if she had asked it aloud; and in a moment it was out of sight.

Lewis Carroll, *Alice in Wonderland* (1865)

Mushrooms capable of producing a state of inebriation similar to but distinguishable from that of alcohol are part of the mythology, rituals, and religion of many cultures. The term *pantherine* syndrome also has been used to describe the symptoms of this type of mushroom poisonings. The name is derived from *Amanita pantherina* (the panther mushroom), one of several common species associated with the inebriation syndrome. The toxins responsible are isoxazole derivatives, and they work their "magic" by interfering with neurotransmission in the brain.

Another common mushroom responsible for the inebriation syndrome is *Amanita muscaria*, a large colorful mushroom probably more associated with folklore than any other. "Color is one of its most distinguishing characteristics. Its cap ranges from a golden orange to a dazzling crimson. It looks like a candy apple with a bad case of dandruff."[1] It is the first mushroom most children learn about from their illustrated fairy stories. It is the place where the elves live. Not surprisingly, it has spawned a copious literature, both lay and scientific, reflecting interest as well as careful study. Entire books have been written about this one mushroom, and professional reputations have been staked on its presumed role in history and culture.[2–6]

The common English name for *A. muscaria* is the fly agaric, because it was reputedly used both for attracting and for poisoning common houseflies. This role is mentioned in some of the earliest herbals, but these references have not prevented a prolonged, vigorous debate among the cognoscenti about the true origin of the common name. In fact, the common name is appropriate; the mushroom contains 1,3-diolein, a compound that flies find irresistible. Then, after they have dined on the juice of the mushroom, the main isoxazole toxin, ibotenic acid, stuns or kills them.

HISTORICAL REPORTS OF THE EFFECTS OF AMANITA MUSCARIA

Amanita muscaria and Religion

No one has gone further in dissecting the name, role, and place of this mushroom in Western civilization than has John Allegro.[7] In *The Sacred Mushroom and the Cross*, he posits that the foundations of Christianity rest on an ancient cult that used and worshipped this mushroom. His argument is a difficult, complex, and convoluted etymological and historical one, and, quite naturally, its publication was followed by angry attacks by the Church and others. In truth, few people have the requisite knowledge to criticize his contentions, but he has not been very successful in persuading others to his view.

More accessible is the work of Gordon Wasson and his Russian wife, Valentina. After their first major work, *Mushrooms, Russia and History*, they undertook a study of the ancient Rig Vedas, the Sanskrit hymns and songs that became the foundation of Hinduism. These religious elements originated about 800 B.C. with the Aryans who had invaded the northern parts of India centuries before. These archaic poems contained a description of an amazing power, the Soma, which literally means "the pressed one." The juice of the Soma, the plant-god or god-plant, was both drunk by the priests and offered in sacrifice. This narcotic was clearly of vegetable origin; and the holy inebriant was worshipped and used in magicoreligious rites. The source of this strange and wonderful power has been a great puzzle, and a variety of substances have been proposed to account for all the facts. The Wassons's conclusion, based on a careful evaluation of the botanical, toxicological, linguistic, and cultural comments in these poems, was that the Soma of the Rig Vedas was none other than *Amanita muscaria*. This determination was reached after considering a number of other possibilities.[8] The characteristics of *A. muscaria* seemed to account for every reference to the Soma they could find. Admittedly, the evidence is highly circumstantial, and accepting some of it requires imagination and an element of faith and trust; but the study is, on the whole, quite persuasive and interesting. Moreover, *Soma: The Divine Mushroom of Immortality*[4] is a valuable addition to the ethnobotanical literature, whether or not one accepts its premises. Like any religious treatise that arouses emotion, this work has been variously praised, condemned, and dismissed. It cannot be ignored.

Use of *A. muscaria* as an inebriant or for religious purposes in other cultures is virtually unknown. It has been claimed that the Ojibway Indians in Michigan used it for ceremonial purposes.[3] Reports of its use in Central America have been largely discounted.

Use of *Amanita muscaria* as an Inebriant

Although the effects of *A. muscaria* on flies had been known for centuries, not only in the West, but also in Japan and elsewhere, the effects on humans were not well publicized until the 1700s, following the return of a Swedish colonel who had spent 12 years as a prisoner of war in Siberia.[9] Von Strahlenburg's account of the use of this mushroom by the local population has been repeated down the years in almost every book every written on mushrooms. I can think of no reason to break the long tradition:

> The Russians who trade with them [the Koryak], carry thither a Kind of Mushrooms, called in the Russian tongue, Muchumur, which they exchange for Squirrel, Fox, Hermin, Sable and other furs: Those who are rich among them, lay up large Provisions of these Mushrooms, for the winter. When they make a Feast they pour Water upon some of these Mushrooms and boil them. Then they drink the Liquor, which intoxicates them. The poorer Sort, on these Occasions post themselves around the Huts of the Rich, and watch for the Opportunity of the Guests coming down to make Water; and they hold a Wooden Bowl to receive the Urine, which they drink off greedily, as having still some Virtue of the Mushroom in it, and by this Way they also get Drunk.[9]

This report was translated into English and became widely known through the writing of Oliver Goldsmith. In his book *Citizens of the World*, published in 1762, the intoxicating effect of these mushrooms is well described. More detailed accounts of the use of the fly agaric by the peoples of Siberia and especially the Kamchatka peninsula became available as travelers returned with rather astounding tales. An excellent description was written by a Russian visitor, S. P. Krasheninnikov, in 1755:[10]

> Sometimes for the enjoyment they also use the mukhumor, the well known mushroom that we ordinarily use for poisoning flies. It is first soaked in the must of kiprei (fire-weed) which they drink, or else the

mushrooms are rolled and swallowed whole, which method is very popular. The first and usual sign by which one can recognize of a man under the influence of the mukhumor is the shaking of the extremities which will follow after an hour or less after which the person thus intoxicated have hallucinations, as if in a fever: they are subject to various visions, terrifying or felicitous, depending on difference in temperament: owing to which some jump, some dance, others cry and suffer great terrors, while some might deem a small crack to be as wide as a door, and a tub of water as deep as the sea. But this applies only to those who overindulge, while those who use a small quantity experience a feeling of extraordinary lightness, joy, courage, and a sense of energetic well-being.

Humans are not the only species known to use A. muscaria; certain other animals seem partial to it as well. Early reports described reindeer that apparently were addicted to these mushrooms and showed signs of intoxication. Eating the flesh of one of these animals was supposedly as effective as recycling the urine from a tipsy companion.[11] Other examples of animals becoming inebriated on natural products are well known. Perhaps the best observed are African elephants drunk on fermented marulo fruit. It is not only humanity that "put an enemy in their mouth to steal away their brain" (Shakespeare, 1604).

An interesting feature commented on by many observers is the difference between the effects of the mushroom and that of alcohol:

> The mushroom produces only a feeling of great comfort, together with the outward signs of happiness, satisfaction, and well-being. Thus far the use of the fly-agaric has not been found to lead to any harmful results, such as impaired health or reduced mental powers.[12]
>
> B. G. von Maydell (1893)

> Mukhomor eaters describe the narcosis as most beautiful and splendid. The most wonderful images, such as they never see in their lives otherwise, pass before their eyes and lull them into a state of the most intense enjoyment. Among the numerous persons whom I myself have seen intoxicated in this way, I cannot remember a single one who was raving or wild. Outwardly the effect was always thoroughly calming—I might say, comforting. For the most part the people sit smiling and friendly, mumbling quietly.[13]
>
> C. von Dittmar (1900)

> Mushroom intoxication had a quite different effect from alcoholic drunkenness, since the former put the Kamchatka into a peaceful and gentle (skromno in Russian) mood, and they had seen how differently the Russians were affected by spirits.[14]
>
> A. Erman (1933)

Counterculture's Experimentation with
Amanita muscaria

The Wassons coined the term *bemushroomed* for the state of inebriation produced by *A. muscaria* to differentiate the mushroom's effect from the drunkenness of alcohol. Lest anyone be tempted by these wonderful descriptions to go out and pick a mess of mushrooms for experimentation, the effects described are not fully reproducible with the majority of the strains available in the United States. Although some of the features of intoxication are similar, they seldom reach the heights attributed to the Kamchatka mushrooms and are more often associated with the most annoying side effects. This difference came as a rather unpleasant shock to those who in the 1960s and 1970s were looking for the hoped-for effects. So different was the experience in the United States with the fly agaric that A. McDonald was prompted to undertake a more formal investigation of the strains available in California.[15] He administered the mushroom to six different volunteers, carefully recording their reactions. The overwhelming sensation was that of nausea; the most underwhelming was the occurrence of pleasant visions of any variety. The side effects were so pronounced that plain human decency and consideration brought an end to the investigation. Even the Wassons were disappointed with the effects, and they were clearly primed for a pleasurable experience:

> In 1965 and again in 1966 we tried out the Fly-agarics repeatedly on ourselves. The results were disappointing. We ate them raw, on empty stomachs. We drank the juice on empty stomachs. We mixed the juice with milk, and drank the mixture, always on empty stomachs. We felt nauseated and some of us threw up. We felt disposed to sleep, and fell into a deep slumber from which shouts could not arouse us, lying like logs, not snoring, dead to the outside world. When in this state I once had vivid dreams, but nothing like what happened when I took the *Psilocybe* mushrooms in Mexico, where I did not sleep at all. In our experiments at Sugadaira there was one occasion that differed from all others, one that could be called successful. Rokuya Imazeki took his mushrooms with mizo shiru, the delectable soup that the Japanese usually serve with breakfast, and he toasted the mushroom caps on a fork before an open fire. When he arose from the sleep that came from the mushrooms, he was in full elation. For three hours he could not help but speak; he was a compulsive speaker. The purport of his remarks was that this was nothing like the alcoholic state; it was infinitely better, beyond all comparison.[4]

> R. G. Wasson (1968)

The variability of the effects produced by *A. muscaria*, even in the same person, is quite striking. It is probably very dangerous to draw any conclusions on the basis of a few comments of observers many years ago. If one listens to the experience of a single individual, the problem of drawing any general conclusions becomes clear:

I have eaten the fly agaric three times. On the second of those occasions I experienced nothing but a slight nausea. The other times I got gloriously, colossally drunk.

I say "drunk" rather than "high" because I was illuminated by none of the sweet oceanic electricity that it has been my privilege to conduct after swallowing mescaline or LSD-25. On acid, I felt that I was an integral component of the universe. On muscaria I felt that I was the universe. There was no sense of ego loss. Quite to the contrary: I was a superhero who could lick any archangel in town and the rusty boxcar it hoboed in on.

I wasn't hostile, understand, but felt invincibly strong and fully capable of dealing with the furniture that was breaking apart and melting into creeks of color at my feet. Although my biceps are more like lemons than grapefruit, I would have readily accepted a challenge from Muhammad Ali, and even in the sober light of two years after, I believe that I could have given him a good tumble.

Euphoric energy was mine aplenty, but at both the onset and the termination of the intoxication I fell fitfully asleep. . . . if not actually the godhead, is holistic awareness of the godhead. But it does not do this gently. Instead of slipping one into the cosmic fabric like a silver needle, it drives one in like a wooden stake. And of course, a stake is blunted in the driving. It was not mere psychedelic fickleness that prompted both the olden Greeks and the Mexicans to drop *Amanita muscaria* cold when they discovered that the innocuous looking little *Psilocybe* made up in grace what it lacked in flamboyance.[1]

Tom Robbins (1976)

The sentiments of Tom Robbins were shared by many others who experimented with the psychedelic compounds that became available in the 1970s. *Amanita muscaria* fell into relative disfavor, to be used only if one were desperate. Even the Siberians, who had relied on *A. muscaria* for generations for their escapes from reality, welcomed the introduction of Russian vodka. How much mushroom eating still persists is not known. A few lamentable souls continue to seek the appropriate strain of the Siberian *Amanita muscaria*, hoping to reproduce the effects described in the historical records.

Nausea, a frequent symptom with the strains in the United States, was largely unreported in the classical descriptions. A 1903 report observed that one of the Siberian peoples, the Koryak, give the dried mushrooms to their women for a preliminary chew. Whether this was to soften up the mushroom to make it more palatable or to ensure that the nausea was experienced by the wife, as is suggested by the writer, is not known.[11]

It is also a mystery why *Amanita pantherina*, which is a rather pale brown, drab cousin of the fly agaric and has the same constituents, sometimes in even higher concentrations, has never found similar favor for those trying to get in touch with the universe. I am not suggesting that it has not been used for such purposes; certainly it has been tried in western Washington.[16] The key to the relative unpopularity of *A. pantherina* may lie in its color or in the fact that it is generally more toxic. Additional compounds not present in *A. muscaria* have been identified in *A. pantherina*.

Role of *Amanita muscaria* in Aggression and Warfare

One unproved historical use of *A. muscaria* involves its role in warfare. One of its common names, one not widely known, is woodpecker of Mars.[1] Although most of the outside observers who visited the Kamchatka peninsula highlighted the beneficent aspects of mushroom eating, the uncontrollable frenzy it sometimes produced was often commented on. A 1908 report mentions that the Koryak used the mushrooms to increase their strength and stamina.[17] This property has become a part of the Koryak myths and legends, thereby supporting the truth of this notion. Crunwdell recorded the Koryak legend of Big Raven:[18]

> Big Raven . . . had caught a whole whale and could not send it home because he was unable to lift the bag containing its traveling provisions. He appealed to the Existence to help him. The deity said *"Find the white soft stalks with the spotted hats—these are the spirits wa'pag."* Big Raven found and ate the fungus, lifted the bag and sent the whale home.

The controversy surrounding this purported property of *A. muscaria* was precipitated in 1784 when Odman attempted to explain the behavior of the group of Norse warriors called the Berserkers.[19] These fearsome soldiers were referred to in the Norse sagas as "raging, half mad, insensates who went into battle without armour—mowing down everything in their path, immune from fire or iron." Odman suggested that they might have been using *A. muscaria*, although no illustrations or any literary references support the concept. Gordon Wasson absolutely rejected the idea, as have others, but the intriguing references remain. Although the idea has been thoroughly discounted in the last few decades, the controversy lingers on. By repeating the story here, I hope to keep the argument simmering. It is certainly true that when children are poisoned, they often appear to be hypermanic, at least for a short time.

Amanita muscaria in English Literature

Amanita muscaria was not used solely in the illustrations of children's books. Lewis Carroll seems to have had its effects in mind when he wrote *Alice in Wonderland*, and he may have learned about it from M. C. Cooke, who published articles about British fungi in the *Gardener's Chronicle* and *Agricultural Gazette* in October 1862.

Both Charles Kingsley and H. G. Wells wrote about it. In Wells's story *The Purple Pileus*, one of the characters, a small, mousy, insignificant little man, a Mr. Coombes, employs the mushroom to change his fortune. He is henpecked, irritated by his wife who loves to spend more than he earns as a draper, and dislikes the friends she entertains. One Sunday he leaves the house to escape the domestic harassment and goes for a walk in the woods:

> It was late October and the ditches and heaps of fir needles were gorgeous with clumps of fungi. . . . They were wonderful little fellows, these fungi, thought Mr. Coombes, and all of them the deadliest poison, as his father had often told him. [Intending to kill himself and, like most English-speaking persons, thinking that all mushrooms were deadly, he ate some.] . . . He was no longer dull—he felt bright and cheerful. And his throat was afire. He laughed in the sudden gaiety of his heart. Had he been dull? He did not know; but at any rate he would be dull no longer. He got up and stood unsteadily, regarding the universe with an agreeable smile. He began to remember. He could not remember very well, because of a steam roundabout that was beginning in his head. And he knew he had been disagreeable at home, just because they wanted to be happy. They were quite right; life should be as gay as possible. He would go home, and make it up, and reassure them. And why not take some of this delightful toadstool with him, for them to eat. A hatful, no less. Some of the red ones with the white spots as well, and a few yellow. He has been a dull dog, an enemy to merriment; he would make up for it. It would be gay to turn his coat sleeves inside out, and stick some yellow gorse into his waistcoat pockets. Then home—singing—for a jolly evening.
>
> H. G. Wells (1895)

Only a few authors have used mushrooms in their fiction for the purpose of murder. Considering how lethal most English-speaking writers believe the toadstool to be, this is a little surprising. It probably reflects their discomfort with the subject. When they have employed a mushroom to dispatch a victim, their limited knowledge has resulted in some rather bizarre or fantastic results, which bear little reality to mushroom poisoning. The classic example of this, recorded in many mushroom texts, is the book by the doyenne of mystery writers, Dorothy Sayers. Her novel *The Documents in the Case* was a joint effort with a physician, Eustace Rawlins.[20] He was evidently asked to assist her to ensure the scientific accuracy of the story. The scientific literature available at the time was full of inconsistencies and errors that took a number of years to unravel. It was fully 27 years after the novel was written before the true structure of muscarine was finally known. Perhaps, like many people, including physicians today, the initial description of the discovery of muscarine in this mushroom was where his knowledge ended, because it was subsequently determined that the important toxins are isoxazole derivatives and not the trivial amounts of muscarine present.

Nevertheless, the plot is a typically elegant Sayers construction. The victim is a renowned plant expert. His wife, the foremost villain, convinces her lover to dispose of her husband. Taking advantage of her husband's passion for spending his vacations collecting, cooking, and eating wild mushrooms, the murderer adds some synthetic muscarine, stolen from a nearby college, to a dish of wild mushrooms that is subsequently avidly consumed by the victim. To other characters in the book, it appears that he either committed suicide with a dish of *Amanita muscaria* or misidentified the red blusher, the edible *Amanita rubescens*. All goes well for the villain and her lover until a nephew returns from abroad. The young man doubts that his uncle, a known botanical expert, could have made such a fundamental mistake in identification. The murder plot is finally uncovered when the optical properties of the toxin in the dish show it to be a synthetic muscarine. For the mycologically naive reader, the story is a good one. But the story is spoiled for the mycologically sophisticated by the fact that Sayers used an unreliable poison, the wrong mushroom (the amount of muscarine in *A. muscaria* is vanishingly small), the wrong symptoms, and questionable chemistry.

INCIDENCE OF ISOXAZOLE POISONING

The two major population groups poisoned by *A. muscaria* and *A. pantherina* are children, usually toddlers, and those seeking a chemical escape from reality. Fortunately, the use of these mushrooms for self-induced intoxication never became popular, despite promotion in the underground drug literature of the 1960s and 1970s—perhaps because the central nervous system effects never lived up to the promotion and because other, more potent drugs were available and did not cause nausea and vomiting that seemed to be the necessary precursor to the "high." Remarkable descriptions from the Russian North and Siberia suggest that a distinct chemical "race" of mushrooms existed there; or perhaps compounds other than the isoxazoles played a role in producing the more beneficent effects. Currently, in the 1990s, these mushrooms are rarely eaten intentionally, except by the most daring experimentalists.

In the Pacific Northwest and other regions of the United States, however, A. pantherina is the most common cause of serious poisoning in toddlers. Frequent spring poisonings caused by A. pantherina were noted in the Seattle area in the 1930s,[21] and the situation remains the same 60 years later.[22] At that time of year, after the short, cold, dark days of winter, children are playing in yards and parks and sample the large, statuesque mushroom. For example, in 1994 I learned of 12 mushroom poisoning-related hospital admissions of children in the Puget Sound area alone.

Some adult victims were poisoned because they mistook the button stage of A. muscaria and A. pantherina for puffballs; others have been immigrants who were unaware of the toxicity of these fungi.

Although a number of other mushrooms contain isoxazole derivatives, there are only a handful of well-documented reports of human poisoning by these other species. Poisonings caused by Amanita crenulata are well established. The clinical features of poisoning by A. crenulata are very similar to those produced by A. pantherina, although one patient had profuse sweating, an atypical sign of the inebriation syndrome.[23]

MUSHROOMS RESPONSIBLE FOR THE INEBRIATION SYNDROME

Amanita species containing ibotenic acid
A. muscaria   Common name: fly agaric
A. pantherina   Common name: panther mushroom

Amanita species reported to contain similar toxins[24,25]
A. cothurnata[26]   A. crenulata[23]
A. frostiana   A. strobiliformis
A. gemmata (may occur in some collections)[27,28]

Other species containing ibotenic acid derivatives
Tricholoma muscarium (a common Japanese mushroom; contains tricolomic acid)

DISTRIBUTION AND HABITATS OF THE ISOXAZOLE-CONTAINING SPECIES

Like most (but not all) amanitas, both A. *muscaria* and A. *pantherina* are mycorrhizal. They associate with a number of different trees, depending on the geography.

Amanita muscaria is widespread throughout the United States and Canada. It is a common mushroom and one that most people recognize. The color of the cap varies from a brilliant scarlet through orange and yellow to a pale buff, with different distributions of colors across the country. The color itself is not responsible for any significant difference in toxicity, because even within a single region in which the color of the cap is the same, considerable variation in toxicity has been noted. In the East, A. *muscaria* fruits in summer and then again in the fall. In the western mountains, it fruits in late summer and early fall; and along the Pacific Coast, it has a fall–early winter fruiting pattern. Occasionally, it is found in the late spring and early summer, accompanying the fruiting of *Boletus edulis*. Sometimes it fruits in obvious fairy rings, but it can also be solitary or scattered. *Amanita muscaria* is usually associated with conifers, frequently pine, but it occurs with beech and aspen as well. It can be found in many habitats, on roadsides, in parks, in open forests, and on the edges of pastures.

Amanita pantherina is more common in the western United States than in the East; it is especially common in the western coastal plain. It often has two fruitings a year, one in the spring and early summer and a second in the fall. It, too, is generally associated with conifers, although it occasionally associates with hardwoods.

Amanita crenulata is primarily an East Coast mushroom, ranging from western Pennsylvania up through Maine and Vermont to Quebec. It fruits in late summer and fall in mixed woods.

TOXICOLOGY OF ISOXAZOLE DERIVATIVES

Myth of Muscarine

The history of the identification of the chemical compounds in A. *muscaria* has linked this mushroom irrevocably to muscarine, which was the very first mushroom toxin described.[29] The name *Amanita muscaria* is unfortunate, because muscarine is present in this mushroom in minute amounts—from the clinical standpoint, its muscarine content is insignificant. This fact, repeatedly stated in many articles and every mycology book since the 1960s, has not reached all those who should know. The fifteenth edition of *Encyclopedia Britannica* in 1974 still claimed that *A. muscaria* poisoning was due to muscarine. This claim became the excuse for the totally inappropriate but routine administration of atropine in many cases of mushroom poisoning.

So pervasive was the belief that muscarine was the cause for all the symptoms that "response to muscarine" was used to explain many unlikely situations. The most unusual incident involved a group of 86 physicians' wives who lunched at a Polynesian restaurant. Following lunch, 55 of the women developed headache, a sensation of bitemporal pressure, malaise, muscle aches, vertigo and lightheadedness, and mild visual disturbances. This conglomeration of symptoms we would now interpret as a bad case of the Asian restaurant syndrome, a condition due to the overenthusiastic use of the flavor-enhancing agent monosodium glutamate. However, at the time, the illness was diagnosed as muscarine poisoning caused by canned Asian mushrooms in the soup—despite the fact that muscarinic signs or symptoms such as excessive perspiration, tearing, and salivation were entirely absent. The diagnosis was a classic example of forcing the clinical findings to fit the prejudice of the day.

The most recent misinformation about *A. muscaria* appeared in 1989 in the journal *Postgraduate Medicine*.[30] The authors described six cases of intoxication, of which five were probably caused by a muscarine-containing *Clitocybe*. The sixth was, as they suggested, probably caused by *A. pantherina*. However, they claimed that muscarine poisoning was the most common cause of mushroom poisoning and recommended the use of atropine to treat victims of *A. pantherina* intoxication. Anecdotes about muscarinic symptoms associated with *A. muscaria* still circulate within the mycological networks, but these stories have not been scientifically documented.

Muscarine plays no significant clinical role in poisonings caused by *A. muscaria* or *A. pantherina*.

Ibotenic Acid and Muscimol— The Real Toxins

Three different groups, working independently in England, Japan, and Switzerland, identified the toxins causing the inebriation syndrome at about the same time. The Japanese workers isolated the parent toxin—ibotenic acid—while investigating the flavor-enhancing properties of mushrooms. A second compound—muscimol—was reported the following year,[31–34] and was soon shown to be no more than the decarboxylation product of ibotenic acid. Since then, further details of the structure and synthesis of ibotenic acid have been published.[35]

The word *ibotenic* in ibotenic acid is derived from *ibo-tengu-take* (*ibo*, "with warts"; *tengu*, "long-nosed goblin"; *take*, "mushroom"), which itself is derived from the Japanese words for *Amanita strobiliformis*, the mushroom in which the toxin was first identified.

A variety of other compounds have been isolated from *A. muscaria* and *A. pantherina* over the years. These have some biochemical interest but do not appear to affect the basic toxicology of these species. They include stizolobic acid, stizolobinic acid, and aminohexadienoic acid.[36,37] In Europe, the compound muscazone has been detected in *Amanita* species. This is derived from ibotenic acid by ultraviolet irradiation.[38]

The occurrence of ibotenic acid in *A. strobiliformis* has not been confirmed by other laboratories, and it has been suggested that the taxonomic identification of the original *Amanita* specimen was erroneous. So, both the chemistry and nomenclature in this group of mushrooms are based on errors. On the one hand, we have *Amanita muscaria* donating its name to muscarine, despite the fact that muscarine is toxicologically insignificant in this species; and on the other hand, the trivial chemical label *ibotenic acid* is derived from the name of a mushroom that might not even contain that compound.

The compound responsible for the most of the symptoms in humans is muscimol, which is formed by the decarboxylation of ibotenic acid. Because of the structural similarities between muscimol and γ-aminobutyric acid (GABA), the mechanism of action of muscimol appears to be its competitive binding to the GABA receptors in the brain, an interaction that leads to disordered neurotransmission.[39–42] The GABA system plays an important inhibitory role in the nervous system, suppressing the activity of certain cholinergic, dopaminergic, serotonergic, and adrenergic synapses and pathways.

The precise details of the mechanism of action of muscimol is not fully understood, because GABA does not show identical effects. Receptors other than the GABA receptors are probably also involved, because muscimol binds to more sites than can be accounted for by the GABA receptors alone.[43,44] Electroencephalographic changes resemble those caused by anticholinergic drugs like atropine,[45] but the effect cannot be blocked by physostigmine; nor does muscimol have any of the typical peripheral effects that would suggest anticholinergiclike activity. However, as mentioned earlier, some patients do manifest occasional atropinelike effects. Once the muscimol binds to the receptors, it is not degraded or removed as normal neurotransmitters are. Thus, the pharmacological action is prolonged and persistent. This feature may play a significant role in the toxicity of muscimol.

Ibotenic acid and muscimol are both able to cross the blood–brain barrier more readily than either dietary glutamic acid or GABA. The mechanism by which they do this is unknown, but it may involve active transport. In an experimental model, ibotenic acid excited or stimulated neurons, whereas muscimol had a depressant or inhibitory effect.

The action of ibotenic acid is more closely related to that of glutamic acid and its receptors than to the GABA system. The relative amounts of ibotenic acid and muscimol that actually reach the central nervous system are unknown. However, because ibotenic acid has an excitatory effect, it could account for the fluctuating symptoms— stimulation by ibotenic acid and depression by muscimol.

Ibotenic acid is also thought to be a neurotoxin, causing neuronal cell death through a mechanism known as excitoneurotoxicity.[46] However, no studies have been undertaken with humans to determine whether chronic consumption of A. muscaria is associated with long-term neurological or psychological sequelae.

As would be expected from the stories of urine-drinking emanating from the depths of Siberia and the northern Urals, muscimol and its metabolites are excreted by the kidneys. Only 30–35% of an ingested dose of muscimol is excreted unchanged; the rest is excreted as conjugates and oxidation products. Conversely, most of an ingested dose of ibotenic acid is excreted unchanged. This excretion is very rapid, much of the ibotenic acid being cleared from the body within 90 minutes of ingestion.[47] It is interesting to note that the major symptoms occur after most of the ibotenic acid has been excreted.

The similarities of the properties of ibotenic acid, glutamic acid, and a third compound—tricholomic acid—isolated from the Japanese mushroom *Tricholoma muscarium* are interesting. Tricholomic acid is structurally similar to ibotenic acid; its scientific name is L-erythrodihydroibotenic acid. All three compounds have the ability to stimulate or inhibit selected neurons, stun flies, and enhance flavor. In fact, ibotenic acid has been patented as a flavor enhancer in the United States (U.S. Patent 3,466,175, Sept. 9, 1969), although it has never reached the supermarket shelves. The metabolism, relationships, and pharmacological effects of these compounds are illustrated in the accompanying figure.

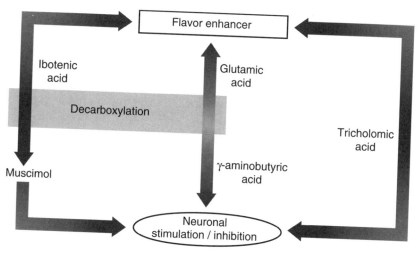

Metabolism of ibotenic acid, glutamic acid, and tricholomic acid

TOXICITY OF THE ISOXAZOLE DERIVATIVES

In an experimental situation, 6 mg of muscimol cause the central nervous system effects, whereas 5–10 times more ibotenic acid is required to produce the same level of effects.[48] This quantity is contained in a single cap of A. pantherina and in many specimens of A. muscaria. In human volunteers given such a modest dose, symptoms develop within one hour and last for approximately four hours. Residual effects, including a severe headache for up to 10 hours, may occur.[49] However, 10 isoxazole-poisoned children did not have hangover headaches; in fact, they all appeared to be completely recovered by 12 hours after the ingestion.[22]

The fatal dose is approximately 15 caps, but even this dose has been tolerated by some gluttons. Toxin concentration varies considerably from one specimen to the next. Moreover, evidence suggests that spring- and summer-fruiting mushrooms may have higher concentrations than those that fruit in the fall,[50] although this observation needs to be confirmed with further samples.

It is distinctly possible that toxin concentrations may be unusually low in the mushrooms fruiting in those regions of the world where A. muscaria and A. pantherina are used as a food source. Nevertheless, eating these mushrooms is a dangerous practice.

A difference of opinion still remains regarding the variations in toxin concentration within the various anatomical parts of an individual fruiting body. People who use this mushroom for food have traditionally peeled young specimens and then boiled them in water, discarding the water before further preparation and eating. They have claimed that the poisons are in the cuticle. In the few studies addressing this question, a conjugate of ibotenic acid and a pigment has been located in the cap's cuticle,[51] but this conjugated toxin represents only a fraction of the specimen's total amount of the toxin, which is present in all portions of the mushroom. Boiling, with subsequent discard of the cooking water, is a good idea, because isoxazole compounds are water soluble and would be partially extracted by the parboiling process.

During drying and storage, a proportion of the ibotenic acid in a mushroom undergoes spontaneous decarboxylation to muscimol. Because this latter compound is considerably more potent than ibotenic acid, drying actually enhances the mushroom's toxicity. It is claimed, but unconfirmed, that undesirable gastrointestinal "complications" are mitigated by this treatment. Whether detoxification or convenience was the reason that the Russian stories usually spoke of the use of dried mushrooms is unclear. However, after prolonged storage, the toxic compounds gradually disappear, with only traces being found after seven years of storage. This characteristic is important for those searching for these agents in herbarium specimens. Quantitative studies become unreliable even after one year.

CLINICAL FEATURES OF ISOXAZOLE POISONING

Onset

Onset of symptoms is almost always between 30 and 120 minutes after mushroom ingestion. In very unusual circumstances, onset is delayed up to six hours.[52]

Signs and Symptoms

The central nervous system manifestations may be preceded by nausea and vomiting. Vomiting is considerably more common in children, although nausea is frequent at all ages. The signs and symptoms of isoxazole poisoning resemble those of alcohol overindulgence, although there are a few distinct differences. Some victims have characterized the intoxicated state as a twilight zone between thinking and dreaming, or between wakefulness and light dozing.

Often the first signs of poisoning are incoordination and ataxia; the patient either is unable to walk or walks with a drunken gait. A rather striking feature, especially in children, is the waxing and waning of activity; they alternate between lethargy and hyperkinetic behavior. The patient may be difficult to arouse for a while, and then he or she becomes almost uncontrollably active, with thrashing about. Muscle fasciculations and twitching frequently develop as the intoxication progresses; then the patient falls into a deep sleep, from which he or she is often difficult to arouse. Muscle cramps, clonic spasms, and tremors are moderately common.

Hallucinations sometimes occur. These experiences are probably better termed illusions, because they are misinterpretations of what the patient is seeing or hearing. A perception that objects have changed size is one of the more frequent manifestations. Most often, surrounding objects appear enlarged, a condition termed macropsia. Alterations in color vision, subjective changes in one's physical abilities, and retrograde amnesia may all occur.

Much of the literature on intentional intoxication refers to manic behavior that develops, if it develops, after the deep sleep. In children, this behavior occurs prior to the onset of the comalike state. Subtle differences in clinical manifestations may distinguish A. muscaria poisoning and A. pantherina poisonings. Amanita pantherina contains compounds such as stizolobic and stizolobinic acids that do not occur in A. muscaria, although their contributions to the intoxication is not clear. In general, the concentrations of toxins are higher in A. pantherina than in A. muscaria. Consequently, poisoning by A. pantherina is usually more severe than that by A. muscaria.

Generalized seizures can occur in adults but are much more frequent in children. In one series, seizures developed in 3 of 10 isoxazole-poisoned children.[22] In the usual case, typical manifestations of either cholinergic or anticholinergic activity are absent, although many cases may manifest subtle mixed findings. Pupils can be either small or dilated, or change in size over time, and bradycardia may develop. In other series, anticholinergic signs were noted, especially with A. pantherina.[53]

Respiratory depression is not generally a feature of the pantherine syndrome in humans. Animals given lethal doses may develop respiratory depression just prior to death. Cardiovascular effects are generally insignificant.

Key Clinical Features of the Inebriation Syndrome

Onset: 30 to 120 minutes

Symptoms peak at 2 to 5 hours

+ Nausea (and vomiting)
+ Confusion or delirium
+ Incoordination and ataxia, dizziness
+ Alternation between lethargy and euphoric and manic behavior
+ Progressive, deep, comalike sleep
+ Hallucinations; visual distortions
+ Muscle fasciculation, with cramps and spasms
+ Generalized seizures (rare)

Duration: 8 to 24 hours

+ Hangover headache possible

PROGNOSIS

In general, the dangers of overtreatment or iatrogenic problems are much greater than any effects the mushrooms are likely to cause. These mushrooms are not lethal, despite their reputation. People who have employed them for suicidal purposes have been most disappointed with the results.

Duration of symptoms depends on the dose, but symptoms usually persist for 8–12 hours. Severely poisoned individuals have been known to sleep for almost a day. Some people awake with a severe, migraine-like headache; others are entirely asymptomatic. With few exceptions, all symptoms have disappeared by 48 hours.

Deaths due to these mushrooms have been reported but are exceedingly rare[21]—at most three in the last 50 years. A fatal outcome usually occurs in patients who already have serious health problems or are very young. One elderly man with underlying heart disease died of isoxazole poisoning. A two-year-old was fed a meal of *Amanita crenulata* (?*gemmata*) by her parents. Six hours later, she became irritable, developed abdominal pain, became listless and rapidly sank into a coma. Following a generalized seizure, she had a cardiorespiratory arrest.[54] In 1994, a teenager died on Vancouver Island after he and his friends washed down a good dose of *A. pantherina* with alcohol. However, the autopsy showed that he actually died from aspiration of his own vomitus.

A famous case involved Count de Vecchia, attaché to the Italian legation in Washington, D.C., who breakfasted on two dozen mushroom caps in the mistaken belief that the *A. muscaria* were *A. caesarea*. His convulsions were so violent that he reportedly broke the bed before lapsing into a terminal coma.[55] His breakfast companion, a Dr. Kelly, had been less gluttonous and had consumed only a dozen caps. Dr. Kelly spent a considerable portion of the day in a profound stupor but was well recovered within 12 hours.

MANAGEMENT OF THE INEBRIATION SYNDROME

Do not overtreat patients poisoned by *A. muscaria* or *A. pantherina*.

In patients presenting within the first one to two hours after mushroom ingestion, induction of emesis is warranted. Syrup of ipecac should be given in the standard dose adjusted for the weight or age of the patient (see Chapter 11). In any patient with a diminished level of consciousness, emetics should be avoided and gastric lavage should be instituted instead. If the patient has already vomited copiously, emesis is unnecessary. In severely lethargic or unconscious patients, gastric lavage can be performed; the airway should be protected by endotracheal intubation. Activated charcoal is frequently added to this regimen and is reasonable, although no controlled studies have been done to demonstrate any specific benefit from this treatment.

Seizures, if they occur, can usually be controlled with standard anticonvulsants. Despite statements that diazepam or phenobarbital may be contraindicated because of experimental data collected from animal models,[45] absolutely no evidence supports this restriction in humans. They can be safely employed in the usual doses in victims of pantherine poisoning.

It is rare for vomiting to be so severe that it leads to significant changes in the hydration state of the patient. However, in those rare cases in which vomiting is excessive or in which the patient cannot be aroused for many hours, intravenous replacement and maintenance fluids may be required.

Many articles, textbooks, and other sources recommend the use of physostigmine in the unlikely event that significant anticholinergic signs appear. Only the most exceptional case would ever warrant its use.

Use of atropine is not indicated. In fact, it could exacerbate the effects of the toxins.

No laboratory tests are needed, except those that might be used to monitor vital signs.

Key Elements of Treatment of the Inebriation Syndrome

- ◆ Avoid overtreatment
- ◆ Emesis or lavage within the first four hours of ingestion
- ◆ Activated charcoal
- ◆ Provide supportive care:
    Hydration, if needed
    Anticonvulsants for seizures

UNANSWERED QUESTIONS ABOUT PANTHERINE SYNDROME

Although *A. muscaria* may be one of the best-studied mushrooms, many aspects of its toxicology are unclear or unknown. The precise pharmacological effects of the isoxazole metabolites are still undetermined. Many questions remain.

1. Why are the clinical effects so different from one collection of mushrooms to the next?

2. Are other compounds responsible for the well-documented historical descriptions, or are these merely the result of variations in the concentration of the toxins?

3. What is the mechanism behind the anticholinergic symptoms experienced by some individuals?

4. What is the clinical role (if any) of muscazone, the derivative of ibotenic acid produced by ultraviolet radiation?

5. What is the anatomic distribution of the toxins in an individual fruiting body? Are these concentrations different at different phases of the life cycle?

6. What is the distribution of toxins and the cause of the symptoms in poisonings by *Amanita gemmata*?

7. What accounts for the differences in the clinical and toxicological features of *A. muscaria* and *A. pantherina* poisonings?

REFERENCES

1. T. Robbins, "*Amanita muscaria*: The Toadstool That Conquered the Universe," *High Times* (1976) December, p. 91.

2. R. Heim, *Les champignons toxiques et hallucinogènes* (Paris: N. Bourbee, 1963).

3. R. E. Schultes and A. Hofmann, *The Botany and Chemistry of Hallucinogens* (Springfield, IL: Thomas, 1973).

4. R. G. Wasson, *Soma: The Divine Mushroom of Immortality*, soft cover ed. (New York: Harcourt Brace Janovich, 1968).

5. R. G. Wasson, "Fly Agaric and Man." In *Ethnopharmacologic Search for Psychoactive Drugs*, ed. D. H. Efron, B. Holmstedt, and N. S. Kline (Washington, D.C.: U.S. Government Printing Office, 1967).

6. V. P. Wasson and R. G. Wasson, *Mushrooms, Russia and History* (New York: Pantheon Books, 1957).

7. J. Allegro, *The Sacred Mushroom and the Cross* (New York: Bantam Books, 1971).

8. T. J. Riedlinger, "Wasson's Alternative Candidates for Soma," *Journal of Psychoactive Drugs* (1993) 25(2):149–156.

9. F. J. von Strahlenburg, *An Historic-Geographic Description of the North and Eastern Parts of Europe and Asia* (Stockholm, 1730).

10. S. P. Krasheninnikov, *Description of the Kamchatka Land* (St. Petersburg, 1755).

11. J. Enderli, *Two Years Among the Chukchi and Koryak* (Gotha, 1903).

12. B. G. von Maydell, *Reisen und Forschungen im Jakutskischen Gebiet Ostsibiriens in den Jahren 1881–1871* (St. Petersburg, 1893). Published in the series *Beiträge zur Kenntniss des russischen Reiches und der angrezenden Länder Asiens*, Vol. 1.

13. C. von Dittmar, *Reisen und Aufenthalt in Kamtschatka in den Jahren 1851–1855* (St. Petersburg, 1900). Published in the series *Beiträge zur Kenntniss des russischen Reiches und der angrezenden Länder Asiens*, Vol. 8, Pt. II, Sec. I, pp. 98–100.

14. A. Erman, *Reise um die Erde durch Nord-Asien und die beiden Oceane in den Jahren 1828, 1829 und 1839 usgeührt* (Berlin, 1833–1848).

15. A. McDonald, "The Present Status of Soma." In *Mushroom Poisoning: Diagnosis and Treatment*, ed. B. H. Rumack and E. Salzman (West Palm Beach, FL: CRC Press, 1978).

16. J. Ott, *Hallucinogenic Plants of North America* (Berkeley, CA: Wingbow, 1976).

17. W. Jocelson, *The Koryak*, Memoir of the American Museum of Natural History (1908).

18. E. Crunwdell, "The Unnatural History of the Fly Agaric," *The Mycologist* (1987) 1(4):178–181.

19. S. Odman, *An Attempt to Explain the Berserk-Raging of Ancient Nordic Warriors* (Stockholm, 1784).

20. D. L. Sayers and E. Rawlins, *The Documents in the Case*, paperback ed. (London: Avon, 1930).

21. J. W. Hotson, "Mushroom Poisoning at Seattle," *Mycologia* (1934) 26:194–195.

22. D. R. Benjamin, "Mushroom Poisoning in Infants and Children: The *Amanita pantherina/muscaria* Group," *Journal of Toxicology and Clinical Toxicology* (1992) 30(1):13–22.

23. R. E. Tullos, "*Amanita crenulata*: History, Taxonomy, Distribution, and Poisonings," *Mycotaxon* (1990) 39:393–405.

24. T. Wieland, "Poisonous Principles of the Genus *Amanita*," *Science* (1968) 159:946–952.

25. D. T. Jenkins, "A Taxonomic and Nomenclatural Study of the Genus *Amanita*, Section *Amanita* for North America," *Bibliotheca Mycologica* (1977) 57:1–26.

26. J. Dearness, "Mushroom Poisoning Due to *Amanita cothurnata*," *Mycologia* (1935) 27:85–86.

27. R. G. Benedict, V. E. Tyler, and L. R. Brady, "Chemotaxonomic Significance of Isoxazole Derivatives in *Amanita* Species," *Lloydia* (1966) 29:333–342.

28. J. A. Beutler, "Chemotaxonomy of *Amanita*: Qualitative and Quantitative Evaluation of the Isoxazoles, Tryptamines, and Cyclopeptides as Chemical Traits," Doctoral thesis, Philadelphia College of Pharmacy and Science, Philadelphia, 1980.

29. C. H. Eugster, "The Chemistry of Muscarine," *Advances in Organic Chemistry* (1960) 2:427–455.

30. D. Stallard and T. E. Edes, "Muscarinic Poisoning from Medicine and Mushrooms: A Puzzling Symptom Complex," *Postgraduate Medicine* (1989) 85(1):341–345.

31. K. Bowden and A. C. Drysdale, "A Novel Constituent of *Amanita muscaria*," *Tetrahedron Letters* (1965) 6(12):727–728.

32. T. Takemoto, Y. Nakajima, and T. Yokobe, "Isolation of a Flycidal Constituent 'Ibotenic Acid' from *Amanita muscaria* and *A. pantherina*," *Yakugaku Zasshi* (1964) 84:1233–1234.

33. T. Takemoto, Y. Nakajima, and T. Yokobe, "Structure of Ibotenic Acid," *Yakugaku Zasshi* (1964) 84:1232–1234.

34. C. H. Eugster, G. F. R. Müller, and R. Good, "Active Principles from *Amanita muscaria*: Ibotenic Acid and Muscazone," *Tetrahedron Letters* (1965) 6(23):1813–1815.

35. Y. Konda, H. Takahashi, and M. Onda, "Structure Elucidation of Pantherine, a Flycidal Alkaloid from *Amanita pantherina* (DC) Fr.," *Chemistry and Pharmaceutical Bulletin* (1985) 33(3):1083–1087.

36. W. S. Chilton, C. P. Hsu, and W. T. Zdybak, "Stizolobic and Stizolobinic Acids in *Amanita pantherina*," *Phytochemistry* (1974) 13:1179–1181.

37. W. S. Chilton and J. Ott, "Toxic Metabolites of *Amanita pantherina*, *A. cothurnata*, *A. muscaria* and Other Species," *Lloydia* (1976) 39:150–157.

38. H. Fritz, A. R. Gagneux, R. Zbinden, and C. H. Eugster, "The Structure of Muscazone," *Tetrahedron Letters* (1965) 6:2075–2076.

39. G. A. R. Johnston, D. R. Curtis, and W. C. deGroat, "Central Actions of Ibotenic Acid and Muscimol," *Biochemical Pharmacology* (1968) 17:2488–2489.

40. H. V. Wheal and G. A. Kerkut, "The Action of Muscimol on the Inhibitory Postsynaptic Membrane of the Crustacean Neuromuscular Junction," *Brain Research* (1976) 109:179–183.

41. R. J. Walker, G. N. Woodruff, and G. A. Kerkut, "The Effect of Ibotenic Acid and Muscimol on Single Neurons of the Snail," *Comparative and General Pharmacology* (1971) 2:186–192.

42. D. R. Curtis, D. Lodge, and H. McLennan, "The Excitation and Depression of Spinal Neurons by Ibotenic Acid," *Journal of Physiology* (1979) 291:19–28.

43. F. V. DeFeudis, L. Ossola, and P. Mandel, "More Muscimol Binding Sites Than GABA Binding Sites in a Particulate Fraction of Rat Brain," *Biochemical Pharmacology* (1979) 28:2687–2689.

44. F. V. DeFeudis, "Binding Studies with Muscimol: Relation to Synaptic γ-minobutyrate Receptors," *Neuroscience* (1980) 5:675–688.

45. A. Scotti de Carolis, F. Lipparini, and V. G. Longo, "Neuropharmacologic Investigations on Muscimol, a Psychotropic Drug Extracted from *Amanita muscaria*," *Psychopharmacologia* (1969) 15:186–195.

46. R. Schwartz, E. Okuno, C. Speciale, C. Kohler, and W. O. J. Whetsell, "Neuronal Degeneration in Animals and Man: The Quinolinic Acid Connection." In *Neurotoxins and Their Pharmacological Implications*, ed. P. Janner (New York: Raven Press, 1987).

47. W. S. Chilton, "Chemistry and Mode of Action of Mushroom Toxins." In *Mushroom Poisoning: Diagnosis and Treatment*, ed. B. H. Rumack and E. Salzman (West Palm Beach, FL: CRC Press, 1978), pp. 87–124.

48. W. Theobald, O. Büch, H. Kunz, P. Krupp, E. G. Stenger, and H. Heimann, [Pharmacological and experimental psychological investigations of two of the components of the fly agaric] (German), *Arzneimittel-Forschung* (1968) 18:311–315.

49. W. S. Chilton, "The Course of an Intentional Poisoning," *McIlvainea* (1975) 2:17.

50. R. G. Benedict, "Mushroom Toxins Other Than *Amanita*." In *Microbial Toxins*, Vol. 8, ed. S. Kadis, A. Ciegler, and S. J. Ajl (New York: Academic Press, 1972), pp. 281–320.

51. P. Catalfomo and C. H. Eugster, "*Amanita muscaria*: Present Understanding of Its Chemistry," *Bulletin of Narcotics (U.N.)* (1970) 22(4):33–41.

52. P. P. Vergeer, "Poisonous Fungi: Mushrooms." In *Fungi Pathogenic for Humans and Animals*, ed. D. H. Howard (New York: Marcel Dekker, 1983), pp. 372–412.

53. D. J. McCormick, A. J. Avbel, and R. B. Gibbons, "Nonlethal Mushroom Poisoning," *Annals of Internal Medicine* (1979) 90(3):332–335.

54. R. Buck, "Mycetism," *New England Journal of Medicine* (1969) 280:1363.

55. O. E. Fischer, "Mushroom Poisoning." In *The Gilled Mushrooms of Michigan and the Great Lakes Region* (Dover reprint), ed. C. H. Kauffman (New York: Dover, 1971).

CHAPTER SEVENTEEN

HALLUCINOGENIC SYNDROME
Poisoning by Tryptamine Derivatives

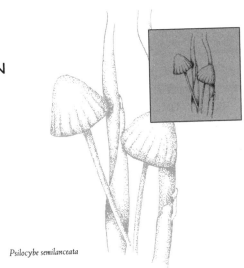

Psilocybe semilanceata

CASE 1

Gathered early that morning from freshly manured ground, more than 20 basidiomes had been ingested by a 34-year-old man. In something of a gourmet breakfast they had been sauteed in garlic and white wine and served heaped over toast. At 8:40 A.M. they were pronounced "some of the most delicious mushrooms that I have had." However, within 40 minutes sensations of detachment from conversation and an inability to take in surroundings became apparent. The man promptly reported to hospital, where at 10 A.M. he described himself as "far away" and incapable of following speech at normal speeds. Although worried about his condition and its possible development, he became elated.

There followed undoubtedly the least pleasant experience of the day when gastric lavage was carried out. In spite of his high spirits he retained sufficient presence of mind to refuse insertion of a second tube when the first became blocked, with the stomach pumping only half complete. By the time he left hospital at about 11:30 A.M., the patient had become frivolous, experiencing some sexual stimulation [not further expounded upon in the report], and a great urge to laugh and smile. After an afternoon's rest in bed, the elation waned in the evening.

This case was reported in Scotland in 1983.[1] The symptoms sound very much like the effects of a *Psilocybe*, although the mushroom involved was not from that genus. During the halcyon days of the hallucinogenic mushroom cult of the 1960s and 1970s, a variety of species in addition to those from the genus *Psilocybe* were publicized as mushrooms with mind-altering properties. Of these, a common lawn-inhabiting species, *Panaeolus foenisecii* (the haymaker's mushroom), was one of the favored species. Studies demonstrated the typical potent tryptamine derivatives such as psilocin or psilocybin in some but not all the tested collections of this species, although closely related but inadequately characterized analogues were always present. It is apparent that certain collections of the *Panaeolus* genus are hallucinogenic and contain psilocybin.[2,3]

CASE 2

They were all eaten by Mrs. Y and myself (a botanist). Peculiar symptoms were perceived in a very short time. Noticed first that I could not collect my thoughts easily, when addressed, nor answer readily. Could not will to arise promptly. Walked a short distance; the time was short, but seemed long drawn out; could think straight, but seemed drowsy; remember little about the walk. Mrs. Y was in about the same condition, according to Mr. Y (her husband did not participate in the experiment). My mind very soon appeared to clear up somewhat and things began to seem funny and rather like intoxication. Walked with Mr. Y. A little later objects took on peculiar bright colors. A field of redtop grass seemed to lie in horizontal stripes of bright red and green, and a peculiar green haze spread itself over all the landscape. At this time Mrs. Y saw nearly everything green but the sky was blue; her white handkerchief appeared green to her, and the tips of her fingers seemed to be like the heads of snakes.

Next, say about a half hour after the eating, both of us had an irresistible impulse to run and jump, which we did freely.

After entering the house, I noticed that the irregular figures on the wall-paper seemed to have creepy and crawling motions, contracting and expanding continually, though not changing their forms; finally they began to project from the wall and grew towards me from it with uncanny motions.

I then had a very disagreeable illusion. Innumerable human faces, of all sorts and sizes, but all hideous, seemed to fill the room and to extend off in multitudes to interminable distances, while many were close to me on all sides. They were all grimacing rapidly and horribly and undergoing contortions, all the time growing more and more hideous.

This is another firsthand report of an hallucinogenic experience. It illustrates the extraordinary effects of tryptamine metabolites from a *Psilocybe* mushroom on the human brain—in this case, the brain of J. H. Sanford.[4]

CASE 3

At the moment she was lying down and experiencing the most glorious visions of color and sounds of music . . . She called later that evening, said that the hallucinations had passed and that she was feeling perfectly normal again, and added that if this were the way that one died of mushroom poisoning, she was all for it.

Gymnopilus spectabilis, known in Japan as the "laughing mushroom," has on occasions been confused with the honey mushroom, *Armillaria mellea*, a desirable edible species. These instances of mistaken identity have shown that *G. spectabilis* can produce hallucinations very similar to those produced by the toxins in the *Psilocybe* group.[5] The preceding account by Walters describes a telephone call he received from a woman, who recounted her experiences after eating a small amount of *G. spectabilis* raw.[6]

Sanford studied many accounts of Japan's "laughing mushrooms."[4] Their history goes back centuries, to a medieval folktale of the eleventh century from a collection called *Konjaku monogatari* (*Tales of Long Ago*).

Long, long ago, some woodcutters from Kyoto went into the mountains and lost their way. Not knowing which way to go, four or five of them were lamenting their condition when they heard a group of people coming from the depth of the mountains. The woodcutters were wondering suspiciously what sort of people it might be when four or five Buddhist nuns came out dancing and singing. Seeing them, the woodcutters became fearful, thinking things like, "Dancing, singing nuns are certainly not human beings but must be demons or goblins." And when the nuns saw the men and started straight towards them, the woodcutters became very frightened and wondered, "How is it that nuns come out of the very depths of the mountains dancing and singing?"

The nuns then said, "Our appearance dancing and singing has no doubt frightened you. But we are simply nuns who live nearby. We came to pick flowers as offerings to Buddha, but after we had all entered the hills together we lost our way and couldn't remember how to get out. Then we came upon some mushrooms, and although we wondered whether we might not be poisoned if we ate them, we were hungry and decided that it was better to pick them than to starve to death. But after we had picked and roasted them we found that they were quite delicious, and thinking, "Aren't these fine!" we ate them. But then as we finished the mushrooms, we found that we couldn't keep from dancing. Even as we were thinking, "How strange!" The woodcutters were no end surprised at this unusual story.

Now the woodcutters were very hungry so they thought, "Better than dying let's ask for some too." And they ate some of the numerous mushrooms that the nuns had picked, whereupon they also were compelled to dance. In that condition the nuns and the woodcutters laughed and danced round and round together. After a while the intoxication seemed to wear off and somehow they all found their separate ways home. After this the mushrooms came to be called "maitake," dancing mushrooms (*mai*, "dance"; *take*, "mushroom").

When we think about it this is a striking story. For even though we still have this kind of mushroom, people who eat them do not dance. Thus this exceedingly strange story has been handed down.

Konjaku monogatari (11th century, Japan)

In his study, Sanford also included a more modern story, which illustrates the characteristics of another of the Japanese hallucinogenic mushrooms called *waraitake*.[4] This story was originally published in 1917 in a newspaper, *The Northcountry News*.

In Ishikawa prefecture, Hagui country, Hinogawa village, Oginotani, one Tsuta (age 40), wife of Oda Yasutaro, and her elder brother were gathering plants on May 11, 1917, at about 2:00 P.M. in a place known locally as Inoya Mulberry field. As they were poking in the dirt on the Oginoshima property of the Ichihoku Sanno Company they found a lot of grey mushrooms that looked a lot like "chestnut mushrooms" growing at the base of a chestnut tree. Mrs. Oda wanted to keep them since they seemed a lucky find but her brother warned her of the dangers of eating mushrooms whose identity was not wholly clear and she finally decided she would throw them out when she got home. However, a neighbor, Mrs. Taniguchi Jutaro (age 35), saw them and said that she had picked some very similar mushrooms at the same spot in March and asked to have a portion now. Mrs. Oda, not wanting to be responsible for a poisoning, refused, but finally gave in under further pressure.

About eight o'clock that evening, Mr. Taniguchi (age 31), Mrs. Taniguchi and Mrs. Taniguchi's brother, Buntsuke (age 41) treated themselves to two bowls of mushroom soup, while the elder Mrs. Taniguchi (age 71) ate one bowl with only two or three mushrooms in it. They had hardly eaten when first Mrs. Taniguchi and then Mr. Taniguchi began to feel odd. Mr. Taniguchi then went next door to ask if someone would fetch a doctor. When he got back home he found his wife dancing around stark naked, playing an imaginary shamisen, and laughing raucously. Even as he stood there amazed at all the uproar, he found that he too was falling into the same crazed state. The older brother also eventually began to dance crazily. The intoxication of Taniguchi's mother was weaker, however, and though she became muddled she never lost complete control of her senses. She did, however, keep repeating the same words over and over and went to every house in the neighborhood apologizing throughout the night for "preparing such a poor meal" and thanking everyone for "putting up with it."

Hokkoku Shimbun (1917)

A number of different Japanese mushrooms are responsible for the effects attributed to the generic "laughing mushroom," including *Gymnopilus spectabilis*, *Panaeolus papilionaceus*, and probably some indigenous *Psilocybe* species, such as *Psilocybe argentipes*.[7]

HISTORICAL BACKGROUND

In sixteenth-century Europe, the effects of alcohol were well known and widely employed as a means to briefly retreat from the harsh realities of living; but apart from sacramental wine, alcohol played no significant role in ceremony. Then, during the Spanish conquest of Central America, stories began to filter back to Europe about the use of inebriants in religious rites. Three substances were mentioned in these reports: a sacred mushroom called teonanacatl, a cactus (peyote), and the seeds of the morning glory flower, named ololiuqui by the local Indians. Most of the reports originated from the clerics who accompanied the Spanish conquistadors to ensure that Christianity was spread, that the soldiers had access to their religion, and that the Church obtained its share of the wealth that was being plucked most unwillingly from the New World. These clerics were also responsible for recording the events as they saw them. And most of them did not like what they saw. In fact, they were horrified. The use of intoxicating plants in rituals were the least of their concerns. They reported on ceremonial human sacrifices, cannibalism, and a variety of other decidedly un-Christian behaviors. A fundamental fear was that the written history of the Indians bore no resemblance whatsoever to the Catholic Church's view of the world or its history.

The systematic and tragic destruction of the great cultures that used the inebriants has been recounted in numerous books. Very few vestiges of their civilizations remain, and apart from the few obscure reports written by the priests who witnessed the disappearance of the peoples, almost all knowledge of the use of intoxicating plants has disappeared.

At the beginning of the twentieth century, interest in these plants was revived. The first hallucinogen to be confirmed was that of the peyote cactus. Because at first there was little evidence of the hallucinogenic effects of the morning glory seeds and the mushrooms, investigators decided that the cactus alone could account for all early reports, or possibly that a nightshade plant was the culprit. Then, just prior to World War II, extracts of the seeds of the morning glory were shown to have a depressive effect on the central nervous system in animal experiments. The great ethnobotanists Schultes and Reko discovered that morning glory plants were still being cultivated and used by local Mexican "doctors" for divination. During these investigations, Schultes and Reko also uncovered the use of hallucinogenic fungi in religious rituals.

The amateur ethnomycologist G. Wasson made a thorough investigation of the practice, and the well-respected mycologist R. Heim, in

Paris, formally identified the mushrooms.[8,9] In an article in *Life* magazine in 1957, Wasson discussed the use and effect of these mushrooms.[10] The words *magic mushroom* became a household term, and this fungus was adopted as a symbol of the counterculture of the 1960s—a device to deal with a world with which the youth were having trouble accepting. It was widely used in the United States during the 1960s and 1970s and became popular in the United Kingdom and elsewhere about a decade later.[11] No other group of mushrooms reflects society, the culture, and its values as clearly as those producing hallucinations.

A complete history of and various ethnomycological musings about magic mushrooms have been gathered together in a book, which is based on papers presented at the Second International Conference on Hallucinogenic Mushrooms held in Port Townsend, Washington, in 1977.[12] It makes fascinating reading for those curious about the topic. A more recent book updates the situation.[13] For a general examination of naturally occurring hallucinogens, the book by Schultes and Hofmann is highly recommended. These scientists are responsible for the pioneering work in the area.[14]

Although the popularity of ingesting hallucinogenic mushrooms has waned, it has by no means disappeared. A recent Danish study confirmed that about 7% of Danish students have had experience with this group of mushrooms.[15] The current incidence of recreational use in the United States remains an underground secret.

The term applied to individuals who have ingested these mushrooms depends on one's vantage point: To physicians, they are patients; to law enforcement officers, they are drug abusers; to parents, they are misguided youth; to toxicologists, they are victims; to scientists, they are subjects; to the individuals themselves, they are just ordinary people seeking a good time through a perfectly reasonable and harmless pastime.

INCIDENCE OF TRYPTAMINE POISONING

There is no reliable information about the number of users of magic mushrooms nor about the incidence of significant adverse reactions requiring medical attention. The largest series of hallucinating patients seen in a hospital included 44 patients admitted over a five-week period.[16] Most presented between three and four hours after the ingestion because of dysphoric symptoms. Forty of the patients had dilated pupils. Twenty-three had experienced nausea and vomiting; whereas less than half had tachycardia, hyperreflexia, or hypertension. Most had distortions of visual perception, parasthesias, and feelings of depersonalization.

Another sizable occurrence followed a "magic mushroom festival" in Cardiff, Wales, where 100 teenagers indulged in a "banquet." Ten patients eventually visited emergency rooms and two were finally admitted. These two patients had each eaten at least 100 "liberty caps" (*Psilocybe semilanceata*).[17]

It seems that humans are not the only species to experiment with states of altered consciousness. A number of our domesticated animals are trying to keep pace. Both dogs and horses are reported to have indulged themselves with these mushrooms, without the assistance of the owners.[18–20] No snacking cats have been noted, but that is no surprise because the cat is an obligate carnivore. Many would say that they are already a little otherworldly and need no further enhancement of that state.

DISTRIBUTION AND HABITATS OF HALLUCINOGENIC MUSHROOMS

Members of the hallucinogenic genera are primarily saprophytic fungi. Some, like *Psilocybe semilanceata*, have a penchant for the rich organic material of dung or manure. Most tend to grow in lawns, pastures, gardens, and parks. They can be found along roadsides, along the edge of forested areas, and in open woods. Some, like *P. baeocystis*, prefer peat moss and wood mulch. Wherever humans have cultivated intensively, one or another species of *Psilocybe* or *Panaeolus* is likely to appear. Heavy mulching with wood chips and bark sometimes produces a considerable flush of species like *P. cyanescens* (color plate 11). Others grow on rotting wood or logs. Some species are very widespread and cosmopolitan; others have a more restricted range.

Representatives of the hallucinogenic genera can be found in every corner of the globe. Particularly rich areas in the United States include the Pacific Northwest and the Gulf Coast. Mycologists have suggested that their transport to Australia, where they are believed to be an introduced species, was by means of the dung of livestock imported from the Cape of Good Hope in 1788.[21] Once in Australia, they may have been spread by dung beetles, also imported from South Africa to deal with the mounting piles of cow chips. Fungi are known to change substrate preference under certain circumstances, so they may have altered their allegiances from cattle dung to kangaroo dung. The islands off the coast of Thailand also host the hallucinogenic mushrooms and those seeking an exotic location for their chemical recreation.[21]

SPECIES CONTAINING HALLUCINOGENIC COMPOUNDS

Many species in two genera and a few species in several other genera produce and accumulate hallucinogenic tryptamine derivatives. The following list, derived from a variety of sources, is incomplete, and debate continues regarding the chemotaxonomy in a number of instances. Interestingly, in Rumack and Salzman's excellent book on mushroom poisoning,[23] three authors who cover the subject of hallucinogenic mushrooms have three different lists.[24–38] Stamet's book covers the cultivation and the identification of the common North American *Psilocybe* species.[39] Guzman's extensive treatise on this genus lists 144 different species worldwide, of which 81 are said to be hallucinogenic.[40]

The compound initially thought to be responsible for the hallucinogenic effects of *Gymnopilus spectabilis* was one of the mushroom's pigments,[41] but the mushroom has been shown to contain psilocybin.[42] Like the variation in so many other mushroom toxins, psilocybin can be found in some collections and not others. Small amounts of this drug have also been identified in *Pluteus salicinus* and *Pluteus nigroviridus*. A number of cases of intoxication in East Germany in 1980 proved to be produced by psilocybin-containing *Inocybe*.[43] Even certain puffballs have been shown to possess the "magic," for example, *Lycoperdon marginatum* and *Lycoperdon mixtecorum*.[14]

Other hallucinogenic compounds have been isolated from mushrooms. Baeocystin and norbaeocystin, identified in *Psilocybe baeocystis*, are two such compounds.[44,45]

TOXICOLOGY AND PHARMACOLOGY OF THE HALLUCINOGENIC MUSHROOMS

It was not long after the clinical and anthropological description of these powerfully hallucinogenic mushrooms before a variety of 4- and 5-hydroxylated *N*-methyltryptamines were isolated from mushrooms. The first was the compound bufotenine, which was identified in *Amanita citrina*. The report came from the Weiland group in Germany, who have devoted their careers to investigations of the *Amanita* genus.[48] Next came Heim's description of psilocybin and its associated

MUSHROOMS CONTAINING TRYPTAMINE DERIVATIVES

Psilocybe species

P. argentipes	P. aztecorum
P. bonetii	P. baeocystis
P. caerulipes	P. caerulescens
P. cambodginiensis	P. coprinifacies
P. cubensis	P. cyanescens (color plate 11)
P. cyanofibrillosus	P. fimetaria
P. mexicana	P. pelliculosa
P. quebecensis	P. semilanceata
P. sempervira	P. serbica
P. strictipes	P. stuntzii (color plate 10)
P. subaeruginosa	P. zapatecorum

Panaeolus species[46]

P. africanus	P. ater
P. cambodginiensis	P. castaneifolius
P. fimcola	P. foenisecii
P. microsporus	P. papillionaceus
P. retrugis	P. sphinctrinus
P. subalteatus	P. tropicalis

Tryptamine derivatives have not been found in all the collections of *Panaeolus* examined.

Conocybe species

C. cyanopus	C. smithii

Inocybe species[47]

I. coelestium	I. haemacta
I. corydalina	I. aeruginascens

Gymnopilus species

G. spectabilis (color plate 12)

Lycoperdon species

L. mixtecorum	L. marginatum

Pluteus genus

P. salicinus	P. nigroviridus

compound psilocin, both isolated from *Psilocybe mexicana*. Colorless psilocybin can be rapidly oxidized to blue products,[49] one of its more characteristic field identification features.

Psilocybin is rapidly dephosphorylated to its phosphate ester, psilocin, by alkaline phosphatase, an enzyme present both in the brush border of the intestinal mucosa and in the liver.[50] The rapid degradation of psilocybin suggests that the compound circulating in the bloodstream is primarily psilocin. In rats, only about one-half of the ingested dose is actually absorbed; the rest is excreted in the feces.[51] Within 30 minutes, the toxin has equilibrated throughout the body, although it accumulates in the liver and the adrenal gland. Blood levels begin to fall after this initial peak and by four hours cannot be detected in the blood or brain. However, the drug that has accumulated in the liver and the adrenal glands persists for up to 48 hours. This residue may account for the prolonged euphoric sensation many people comment on following a dose of these mushrooms.

In rats, circulating psilocin and its metabolites are rapidly excreted in the urine in the first eight hours after ingestion, and by the end of a day, at least two-thirds of the absorbed toxin has been eliminated. The remainder, presumably the fraction in the liver and the adrenal glands, is slowly excreted for up to a week. It is interesting that monoamine oxidase (MAO), the enzyme normally responsible for the metabolism and degradation of the other biogenic amines, has almost no role in the elimination of psilocin.

The usual tryptamine derivatives absorbed from the gastrointestinal tract, such as serotonin, do not normally cross the blood–brain barrier. Psilocin, however, does cross into the brain, probably because of its high lipid solubility, and, once there, it is relatively unaffected by MAO, an immunity allowing its effects to persist.

The pharmacological effects of psilocybin and psilocin are very similar. A dosage effect is evident, with lower doses resulting in a general sense of relaxation and detachment from the environment. This state occurs with as little as 4 mg of psilocybin. As the dose increases to 6–12 mg, perceptual changes develop, with alterations in both space and time sense, and the visual effects become prominent. Only when the dose is much higher do patients start experiencing true hallucinations, which frequently involve distortions of space, time, and color perception. Early in the investigation of these effects, it became evident that they resembled those produced by other hallucinogenic drugs also being investigated at the time, most notably, lysergic acid diethylamide (LSD). Structural similarities to the biogenic amines responsible for neurotransmission in certain neural pathways, such as dopamine and serotonin, were immediately obvious and suggested that the effects of the mushroom toxin were mediated through the same pathways, perhaps by antagonizing serotonin. However, the details of the mechanism of action of psilocin are still unknown.

The compound bufotenine, mentioned earlier, was initially isolated from the toxic secretions of the skin glands of the toad.[52] A dried preparation has been used for many years in the materia medica of China (Ch'an Su). Bufotenine is an isomer of psilocybin and an N,N-dimethyl derivative of serotonin. It is only one of a number of toxic compounds in toad "sweat." The cardiac effects of toad excretions are due to an aglycone and not to bufotenine.[53] This substance, despite its chemical similarity to psilocybin and serotonin, does not have any noticeable effects on the central nervous system of humans. The current "toad-licking" craze reflects the hope that some of these potentially mind-altering compounds will actually produce an effect. My main reason for mentioning it is to complete the circle begun in Chapter 1 connecting mushrooms—toadstools—and toads.

TOXICITY (OR POTENCY, DEPENDING ON ONE'S POINT OF VIEW)

The quantity of psilocybin and psilocin in each fruiting body is variable and partly species dependent.[21] The average thrill seeker has no simple way to measure the concentration of drugs ingested—from as few as 3 to as many as 60 mushrooms may be required to produce the desired effect. Roughly 4–8 mg of psilocybin is present in 20 g of fresh mushrooms or 2 g of dried specimens.[54] As noted in the preceding section, a dose of 12 mg or more of psilocybin produces vivid hallucinations.

Psilocybin is relatively stable, and it retains its activity in dried mushrooms for an appreciable length of time. After a few years, however, it gradually disappears. It is heat stable and water soluble, so steeping dried mushrooms in boiling water to make a decoction or tea is a common way to extract these drugs.

The methods for consuming the mushroom or its extracts are numerous. Some people use the mushrooms fresh; others dry or freeze them for later use. Mushrooms have been powdered and put into capsules for sale and ease of administration. Milkshakes (magic smoothies), tea, soup, stews, omelets, and a simple decoction (magic Kool-Aid) are but a sample of the ways in which the drugs are ingested. Some methods attempt to alleviate the rather acrid taste of the hallucinogenic mushrooms. Packed in honey, they preserve their potency quite well. But probably the least noxious way to eat them is to serve them with chocolate.

CLINICAL FEATURES OF THE HALLUCINOGENIC SYNDROME

Onset

The effects of these mushrooms are noticeable soon after ingestion, usually between 10 and 30 minutes.

Signs and Symptoms

From the clinical standpoint, few specific signs are present other than the unusual behavior of the patient, which may be similar to that produced by many other hallucinogenic agents and common street drugs. Most of the physical effects appear to be due to stimulation of the autonomic nervous system, whereas the motor system is depressed.

Tachycardia is present in less than half the patients, and probably even less than that in older individuals. It has been observed mainly in teenagers who were anxious and having unpleasant experiences.

Dilated pupils are present in over 90% of the patients. Enhancement of the deep tendon reflexes is found in more than half of affected individuals. Other physical signs and symptoms include confusion, headache, vertigo, muscular weakness, and a feeling of numbness.

The early phases of psilocybin intoxication are often characterized by a sense of euphoria—excessive exhilaration with uncontrollable laughter. The initial hallucinations involve visual phenomena in which objects take on particularly vivid colors. This heightened brilliance of colors increases when the eyes are closed. The colors undulate and vibrate, sometimes exploding in pyrotechnical displays, and may be superimposed on dreamlike images that float in and out. Some individuals become very quiet and introspective. Distortions and apparent movement of objects in the immediate environment may occur. The hallucinations developing at this point are extremely variable and are probably most dependent on the mind-set and background of the individual. They can be very pleasant or very frightening, or they can have religious overtones. There is generally a distortion of time, with events seeming to unfold far more slowly than in real time. Misjudgment of distances is common, so walking and climbing are difficult. The eyes and ears appear to be more sensitive to stimulation by both light and sound. I have been unable to find reports on the effects on taste or smell, but changes in tactile sensation also are common.

The effects of psilocybin and psilocin in young children are rather poorly described, because it is difficult to get any sense of what a preverbal child is experiencing. However, evidence suggests that seizures are much more likely to develop in children.[46] Four children who ate *Psilocybe baeocystis* (ages 4, 4, 6, and 9 years) all developed mydriasis and generalized tonic-clonic convulsions.[55]

In children poisoned with these mushrooms, high body temperatures have been noted. Unfortunately, not all the reports in the literature give the patient's temperature. Psilocin is known to be pyrogenic in animals, especially dogs. It is not known whether the convulsions seen in psilocin-poisoned children are related to this febrile reaction or are a direct effect of tryptamine derivatives on the central nervous system. Both may play a role.

Another interesting observation was made in Australia, where a two-year-old child with pica suffered periodic attacks of hysterical behavior accompanied by visual and tactile hallucinations. Her symptoms were confidently attributed to the consumption of *Panaeolus foenisecii* and dirt from her front yard.[56] This supposed poisoning is suspect because the onset of symptoms occurred many hours after the child came inside, timing that is atypical of psilocybin intoxication.[57]

The literature on the clinical effects of psilocybe on adults is huge, because the mushrooms came into vogue just at the time of burgeoning interest in psychotropic agents.[58-62] Numerous studies were done in volunteer subjects, and many comparisons were made between the new agents then being developed, such as LSD.[63] Many of these studies began as very serious investigations into neural mechanisms, with the hope of using some of these compounds in the management of psychiatric illnesses. In some notable institutions, the researchers shifted their focus of interest. Investigations into the purely scientific aspects of the mushrooms, such as dissecting the mechanism of action of the compounds, gave way to the personal use of the mushrooms because of their unusual and interesting psychotropic effects—their legitimacy within the scientific community subsequently declined very rapidly. Soon it was well-nigh impossible to conduct clinical studies, and the mushrooms became part of an underground, but prominent, drug culture.

The following factors influence the effects of the toxins:

♦ The circumstances under which the mushrooms are used, for example, at home, in a group, alone, or as part of a ceremony or ritual. This feature is part of what has been referred to as "the set."
♦ The psychological makeup of the individual, including his or her cultural heritage[64]
♦ Previous experiences with the mushrooms and current expectations
♦ Dose
♦ Method of preparation of the fungi

Each person will have a slightly different psychological experience, but many similarities in the overall nature of the episode are noted. In teenagers, the incidence of unpleasant experiences is much higher than in the slightly older population. In one study in which the mushroom was being regularly consumed by 12- to 24-year-old subjects (mean age 16.5 years), almost half of them had experiences described as frightening and unpleasant.[60]

Hollister administered an average dose of the drug to volunteers and traced the various symptoms through the course of a typical intoxication:[61,62]

◆ 0–30 minutes: Anxiety and tension develop; there may be light-headedness, mild abdominal pain, nausea, weakness, shivering, and muscle aches. The lips may feel numb.
◆ 30–60 minutes: Onset of visual effects and distortions; development of euphoria and an introspective state; increases in auditory acuity, sweating, tearing, and facial flushing; inability to concentrate; depersonalization; incoordination; tremulous speech; feelings of unreality.
◆ 1–2 hours: Visual effects intensify; greater distortion of perception; alteration in time sense; continued euphoria; rumination and introspection.
◆ 2–4 hours: Gradual reduction of symptoms, with complete resolution in the majority of individuals. The subject (victim, patient) is generally completely normal after 4–8 hours. A few people will have headaches or fatigue, and some experience a more subtle sense of well-being for a number of days.

Adolescents and adults differ in the ways they use these mushrooms. Adolescents are much more likely to combine the mushrooms with alcohol, marijuana, and other drugs. In a study of drug-abusing teenagers, 26% reported using "magic" mushrooms. It may be the mixing of agents, the polypharmacy, which accounts for the higher incidence of unpleasant experiences in this age group.[65]

A few individuals have actually attempted to "mainline" psilocybin by heating the mushroom, squeezing out the juice, drawing the extract up in a syringe, and injecting it intravenously.[66] Within 10 minutes, they experience chills, rigors, dyspnea, vomiting, severe generalized myalgia, headache, flank pain, and fever. Mild methemoglobinemia and elevated serum levels of transaminases and bilirubin have been noted in these individuals.[67] These are all symptoms associated with the circulation of a foreign pyrogenic material and are not typical effects of psilocybin. It is remarkable that no one has died from this activity.

Key Clinical Features of the Hallucinogenic Syndrome

Onset
- Rapid, 10–30 minutes

Psychotropic features
- Sense of exhilaration and uncontrollable laughter
- Hallucinations; mostly visual, involving colors and shapes
- Distortion of time sense
- Euphoria, introspective and meditative state

Physical features
- Dilated pupils
- Confusion, vertigo
- Muscular weakness
- Increased deep tendon reflexes

PROGNOSIS

The duration of the effects with the average dose of mushrooms used for "recreational" purposes is generally from four to five hours. Neither headache nor hangover is common, in contrast to the effects caused by *Amanita muscaria* when it is used for similar purposes. Moreover, a sense of peace and serenity that lasts for a number of days is not uncommon. This mood may border on euphoria. In a few patients, flashback phenomena have been described and can continue for a number of months. This feature is probably very rare. Many articles in the literature mention this complication by citing a single case, which was reported by a fourth-year medical student. The report contains insufficient data to make a reasonable assessment of the patient's psychiatric condition.[68]

There is no convincing evidence that the hallucinogens produced by mushrooms are organically addictive. "The inherent danger from the ingestion of wild mushrooms lies not so much in the consumption of psychoactive varieties, but rather in the picking and eating of a toxic species that may resemble a hallucinogenic variety."[21] According to Guzman and colleagues, "field and laboratory studies strongly indicate that psychoactive mushroom use as it normally occurs does not constitute a drug abuse problem or a public health hazard."[69] A California study suggested that "the low frequency and few negative effects of psychoactive mushroom use indicate that abuse does not present a social problem nor is there evidence for predicting the development of a problem."[70] However, a psychological dependence may develop. In addition, a variety of psychiatric disturbances have been associated with their use. Whether this is due to the mushrooms and their toxins per se or to inherently faulty neural connections in the patient is not known.[71]

Deaths due to poisoning by hallucinogenic mushrooms are extraordinarily rare.[72] I found one documented victim—a six-year-old child.[73]

MANAGEMENT OF THE HALLUCINOGENIC SYNDROME

In the overwhelming majority of cases, the individuals using the hallucinogenic mushrooms do not seek any form of treatment. In the usual situation, a patient comes to the hospital only when he or she is having a particularly bad experience with the drug, one that precipitates extreme anxiety or an obvious panic attack. Occasionally children and teenagers are brought by parents or the police, who are concerned about their unusual behavior. In these latter situations, induction of emesis with syrup of ipecac (see page 184) may be considered if the ingestion has been large and very recent. Evidence suggests that gastric lavage may exacerbate the unpleasant effects and should not be used.[16,21,74,75] In one study, the induction of vomiting did not appear to hasten the recovery.[16] If the admission is more than two hours after the ingestion, induction of emesis is useless.

Diazepam (adult dose: 0.2–0.5 mg/kg body weight; child's dose: 0.1–0.5 mg/kg body weight) can be used to control the anxiety and the hallucinations in extreme cases. Chlorpromazine has been used in the past but should be avoided because anticholinergic features are often already present.[76]

Patients should be prevented from harming themselves by providing them with a quiet, safe environment in which to stay while the effects of the chemicals wear off spontaneously. Talking to the patient in a quiet, reassuring manner and calming their fears can be very helpful. Many practitioners feel that the technique of "talking the patient down" is the critical element in successful management.

In children with a high fever, the temperature can be decreased with a tepid sponge bath or acetaminophen.

Seizures, which are much more likely to occur in children, can generally be controlled with intravenous diazepam (0.25–0.4 mg/kg body weight up to 10 mg in a child over 5 years; this dosage can be repeated every 15 minutes). If this treatment fails to control the seizures, then phenobarbital or phenytoin can be employed. Any time repeated doses of anticonvulsant drugs are employed, the patient must be carefully monitored for respiratory depression and may require intubation and ventilatory assistance.

Key Elements of Treatment of the Hallucinogenic Syndrome

- ◆ Provide safe, quiet environment.
- ◆ "Talk patient down" in a reassuring manner.
- ◆ Consider use of an emetic (ipecac) for young children who present within an hour of ingestion of a large amount of mushrooms; otherwise, avoid emesis.
- ◆ Use Diazepam to calm extremely anxious patients or to control seizures.
  Adult dose: 0.2–0.5 mg/kg body weight; may be repeated every 15–30 minutes to at total of 30 mg.
  Juvenile dose: 0.1–0.5 mg/kg body weight; do not exceed 10 mg total.
- ◆ Reduce fever in young children by administering tepid sponge baths or antipyretic agent.

- ◆ Do not use gastric lavage.
- ◆ Do not use anticholinergic agents.

LEGAL STATUS OF THE PSILOCYBIN-CONTAINING MUSHROOMS IN THE UNITED STATES

At the present time, any mushroom or preparation containing psilocybin, psilocin, or bufotenine is classified under the same federal schedule (Schedule I) as marijuana and LSD are.[77] This classification makes it illegal for both sale and possession without a specific permit from the Drug Enforcement Agency, which is in the Division of Alcohol, Tobacco, and Firearms. These compounds, which the Federal government has called drugs, are classified in this manner because there are no established medical uses for them (despite the use of marijuana in the management of certain cancer patients), because of the fear of abuse and because there is no consensus from the medical community for their safe or effective use. During the heyday of the "magic mushroom" years, more arrests were probably made for trespassing than for the actual possession of the mushrooms.

An anomaly in the law does not prohibit the sale or the transportation of mushroom spawn, the mycelia which can be grown at home and even persuaded to fruit under the appropriate conditions. A number of magazines offer this for sale, although the home cultivation and use of the mushrooms is obviously illegal. One late 1960s study of the illicit mushrooms offered for sale on the streets showed that, out of a total of five samples, one was inert and four actually contained LSD, which must have been artificially added for extra effect.[78]

UNANSWERED QUESTIONS ABOUT THE HALLUCINOGENIC SYNDROME

The precise neural mechanisms for the effects of the hallucinogenic mushrooms have not been completely defined. It is still unclear whether compounds such as psilocybin have any role in the management of psychiatric illness or whether they can be helpful in assisting our studies into the mechanisms of neural transmission and understanding our patterns of thinking and perception. We also must determine whether there are any long-term psychiatric complications in a few users of these mushrooms and, if so, what determines them.

The present discussion cites a mere fraction of the material written on the subject. However, the articles cited here will lead the curious reader to the appropriate literature.

REFERENCES

1. A. P. Bennell and R. Watling, "Mushroom Poisonings in Scotland," *Bulletin of the British Mycological Society* (1983) 17(2):104–105.

2. G. M. Ola'h, "Le genre *Panaeolus*," *Revue de Mycologie* (1969), Memoire Hors série 10, Paris.

3. P. A. J.-L. Margot, "Identification Programme for Poisonous and Hallucinogenic Mushrooms of Interest to Forensic Science," Doctoral thesis, University of Strathclyde, Glasgow, Scotland, 1980.

4. J. H. Sanford, "Japan's Laughing Mushrooms," *Economic Botany* (1972) 26:174–177.

5. R. W. Buck, "Psychedelic Effect of *Pholiota spectabilis*," *New England Journal of Medicine* (1967) 276:391–392.

6. M. B. Walters, "*Pholiota spectabilis*, a Hallucinogenic Fungus," *Mycologia* (1965) 57:837–838.

7. M. Musha, A. Ishii, F. Tanaka, and G. Kusano, "Poisoning by Hallucinogenic Mushroom Hikageshibiretake (*Psilocybe argentipes* K. Yokoyama) Indigenous to Japan," *Tohoku Journal of Experimental Medicine* (1986) 148(1):73–78.

8. R. Heim and R. G. Wasson, *Les champignons hallucinogènes du Mexique* (Paris: Editions du Musée Nationale d'Histoire Naturelle, 1958).

9. R. Singer, "Mycological Investigations on Teonanacatl, the Mexican Hallucinogenic Mushroom. I. The History of Teonanacatl, Field Work, and Culture Work," *Mycologia* (1958) 50:239–261.

10. R. G. Wasson, "Seeking the Magic Mushroom," *Life*, 13 May 1957.

11. P. Cooles, "Abuse of the Mushroom *Panaeolus foenisecii*," *British Medical Journal* (1980) 280(6212):446–447.

12. J. Ott and J. Bigwood, *Teonanacatl: Hallucinogenic Mushrooms of North America*, Extracts from the Second International Conference on Hallucinogenic Mushrooms, Port Townsend, WA, 1977 (Seattle: Madrona Publishing, 1978), p. 175.

13. P. E. Furst, *Mushrooms: Psychedelic Fungi* (London: Burke Publishing, 1988).

14. R. E. Schultes and A. Hofmann, *The Botany and Chemistry of Hallucinogens*, 2nd ed. (Springfield, IL: Thomas, 1980).

15. J. F. Lassen, N. F. Lassen, and J. Skov, "Hallucinogenic Mushroom Use by Danish Students: Pattern of Consumption," *Journal of Internal Medicine* (1993) 233(2):111–112.

16. N. R. Peden, S. D. Pringle, and J. Crooks, "The Problem of Psilocybin Mushroom Abuse," *Human Toxicology* (1982) 1(4):417–424.

17. A. D. Harries and V. Evans, "Sequelae of a 'Magic Mushroom Banquet'," *Postgraduate Medical Journal* (1981) 57:571–572.

18. A. P. Kirwin, "'Magic' Mushroom Poisoning in a Dog" (letter), *Veterinary Record* (1990) 126:149.

19. J. Jones, "'Magic' Mushroom Poisoning in a Colt" (letter), *Veterinary Record* (1990) 127:603.

20. P. N. Hyde, "High Horse?" (letter), *Veterinary Record* (1990) 127:554.

21. J. W. Allen, M. D. Merlin, and K. L. Jansen, "An Ethnomycological Review of Psychoactive Agarics in Australia and New Zealand," *Journal of Psychoactive Drugs* (1991) 23(1):39–69.

22. J. W. Allen and M. D. Merlin, "Psychoactive Mushroom Use in Koh Samui and Koh Pha-Ngan, Thailand," *Journal of Ethnopharmacology* (1992) 35(3):205–228.

23. B. H. Rumack and E. Salzman, eds., *Mushroom Poisoning: Diagnosis and Treatment* (West Palm Beach, FL: CRC Press, 1978).

24. R. G. Benedict, L. R. Brady, A. H. Smith, and V. E. Tyler, Jr., "Occurrence of Psilocybin and Psilocin in Certain *Conocybe* and *Psilocybe* Species," *Lloydia* (1962) 25:156–159.

25. R. G. Benedict, L. R. Brady, and V. E. Tyler, "Occurrence of Psilocin in *Psilocybe baeocystis*," *Journal of Pharmaceutical Sciences* (1962) 51:393–394.

26. W. C. Chilton, "Chemistry and Mode of Action of Mushroom Toxins." In *Mushroom Poisoning: Diagnosis and Treatment*, ed. B. H. Rumack and E. Salzman (West Palm Beach, FL: CRC Press, 1978).

27. V. E. Tyler and D. Gröger, "Occurrence of 5-Hydroxytryptamine and 5-Hydroxytryptophane in *Panaeolus sphinctrinus*," *Journal of Pharmaceutical Sciences* (1964) 53:462–463.

28. J. E. Robbers, V. E. Tyler, and G. M. Ola'h, "Additional Evidence Supporting the Occurrence of Psilocybin in *Panaeolus foenisecii*," *Lloydia* (1969) 32:399–400.

29. G. M. Ola'h, "Chemotaxonomic Study on *Panaeolus*: Research on the Presence of Psychotropic Indoles in These Mushrooms," *Comptes Rendus de l'Académie des Sciences, Série D* (1968) 267:1369–1372.

30. P. G. Mantle and E. S. Waight, "Psilocybin in Sporophores of *Psilocybe semilanceata*," *Transactions of the British Mycological Society* (1969) 53:302–304.

31. R. G. Benedict, V. E. Tyler, and R. Watling, "Blueing in *Conocybe, Psilocybe* and *Stropharia* Species and the Detection of Psilocybin," *Lloydia* (1967) 30:150–157.

32. A. Hofmann et al., "Psilocybin and Psilocin, Two Psychoactive Components of the Mexican Intoxicating Mushroom," *Helvetica Chimica Acta* (1959) 42:1557–1572.

33. V. E. Tyler, "Indole Derivatives in Certain North American Mushrooms," *Lloydia* (1961) 24:71–74.

34. J. F. Ammirati and J. A. Traquair, *Poisonous Mushrooms of Northern United States and Canada* (Minneapolis: University of Minnesota Press, 1985).

35. A. Bresinsky and H. Besl, *A Colour Atlas of Poisonous Fungi* (London: Wolfe Publishing, 1990).

36. G. Guzmán and J. Ott, "Description and Chemical Analysis of a New Species of Hallucinogenic *Psilocybe* from the Pacific Northwest," *Mycologia* (1976) 68:1261–1267.

37. T. M. Lee, L. G. West, J. L. McClaughin, L. R. Brady, J. L. Lowe, and A. H. Smith, "Screening for *N*-Methylated Tyramines in Some Higher Fungi," *Lloydia* (1975) 38:450–452.

38. J. Ott, *Hallucinogenic Plants of North America* (Berkeley, CA: Wingbow, 1976).

39. P. Stamets, *Psilocybe Mushrooms and Their Allies* (Seattle: Homestead Book Company, 1978).

40. G. Guzmán, "The Genus *Psilocybe*," *Beihefte zu Nova Hedwigia* (1983) 74:1–439.

41. G. M. Hatfield and L. R. Brady, "Occurrence of bis-Noryangonin in *Gymnopilus spectabilis*," *Journal of Pharmaceutical Sciences* (1969) 58:1298.

42. G. M. Hatfield, L. J. Valdes, and A. H. Smith, "The Occurrence of Psilocybin in *Gymnopilus* Species," *Lloydia* (1978) 41:140–144.

43. G. Drewitz, "Eine halluzinogene Pilzart grünlichverfarbender Risspilz (*Inocybe aeruginascens*)," *Mycologisches Mitteilungsblatt* (1983) 26:11–17.

44. A. Y. Leung and A. G. Paul, "Baeocystin and Norbaeocytin: New Analogs of Psilocybin from *Psilocybe baeocystis*," *Journal of Pharmaceutical Sciences* (1968) 67:1667–1671.

45. A. Y. Leung and A. G. Paul, "Baeocystin, A Mono-methyl Analog of Psilocybin from *Psilocybe baeocystis* Saprophytic Culture," *Journal of Pharmaceutical Sciences* (1967) 56:146.

46. J. Pollock, "Psilocybin Mycetismus with Special Reference to *Panaeolus*," *Journal of Psychedelic Drugs* (1976) 8:42–47.

47. T. Stijve, J. Klán, and T. W. Kuyper, "Occurrence of Psilocybin and Baeocystin in the Genus *Inocybe* (Fr.) Fr.," *Persoonia* (1985) 12(4):469–473.

48. T. Weiland, W. Motzel, and H. Merz, "Amanita Toxins. IX. Presence of Bufotenin in the Yellow *Amanita*," *Justus Liebigs Annalen der Chemie* (1953) 581:10–16.

49. W. Levine, "Formation of Blue Oxidation Product from Psilocybin," *Nature* (1967) 215:1292–1293.

50. A. Horita and L. J. Weber, "Dephosphorylation of Psilocybin to Psilocin by Alkaline Phosphatase," *Proceedings of the Society for Experimental Biology and Medicine* (1961) 106:32–34.

51. F. Kalberer, W. Kreis, and J. Rutschmann, "The Fate of Psilocin in the Rat," *Biochemical Pharmacology* (1962) 11:261–269.

52. H. Wieland, G. Hesse, and H. Mittasch, "Basic Constituents of the Skin Secretions of the Toad," *Chemische Berichte* (1931) 64:2099.

53. V. Deulofeu, "The Chemistry of the Constituents of Toad Venoms," *Progress in the Chemistry of Natural Products* (1948) 5:241.

54. P. Vergeer, "Poisonous Fungi." In *Fungi Pathogenic for Humans and Animals. Part B. Pathogenicity and Detection,* ed. P. A. Lempke (New York: Marcel Dekker, 1983), pp. 373–412.

55. R. G. Benedict, "Mushroom Toxins Other Than *Amanita.*" In *Fungal Toxins,* Vol. 3 of the series *Microbial Toxins: A Comprehensive Treatise,* ed. S. Kadis, A. Ciegler, and S. J. Ajl (New York: Academic Press, 1972), pp. 281–320.

56. R. V. Southcott, "Notes on Some Poisoning and Clinical Effects Following the Ingestion of Australian Fungi," *South Australian Clinician* (1975) 6:441–478.

57. J. W. Allen and M. D. Merlin, "Observations Regarding the Suspected Psychotropic Properties of *Panaeolina foenisecii* Maire." In *Yearbook for Ethnomedicine and the Study of Consciousness,* ed. C. Rätsch (VWB, 1992), pp. 99–115.

58. J. Delay, P. Pichon, and T. Lemperiere, "La psilocybine," *Presse Médicale* (1959) 49:1181.

59. J. Ott, "Recreational Use of Hallucingenic Mushrooms in the United States." In *Mushroom Poisoning: Diagnosis and Treatment,* ed. B. H. Rumack and E. Salzman (West Palm Beach, FL: CRC Press, 1978), pp. 231–243.

60. N. R. Peden, A. F. Bisset, K. E. C. MacAuley, J. Crooks, and A. J. Pelosi, "Clinical Toxicology of 'Magic' Mushroom Ingestion," *Postgraduate Medical Journal* (1981) 57:543–545.

61. L. E. Hollister, "Clinical, Biochemical and Psychologic Effects of Psilocybin," *Archives Internationale de Pharmacodynamie et de Thérapie* (1961) 130:42.

62. L. E. Hollister, J. Prusmack, J. A. Paulsen, and N. Rosenquist, "Comparison of Three Psychotropic Drugs (Psilocybin, JB-329, IT-290) in Volunteer Subjects," *Journal of Nervous and Mental Disease* (1960) 131:428–434.

63. L. E. Hollister and A. M. Hartman, "Mescaline, Lysergic Acid Diethylamide and Psilocybin: Comparison of Clinical Syndromes, Effects on Color Perception and Biochemical Measures," *Comprehensive Psychiatry* (1962) 3:235.

64. A. J. Parashos, "The Psilocybin-induced 'State of Drunkenness' in Normal Volunteers and Schizophrenics," *Behavioural Neuropsychiatry* (1976–7) 8:83–85.

65. R. H. Schwartz and D. E. Smith, "Hallucinogenic Mushrooms," *Clinical Pediatrics* (1988) 27(2):70–73.

66. G. Sivyer and L. Dorrington, "Intravenous Injections of Mushrooms," *Medical Journal of Australia* (1984) 140:182–183.

67. S. C. Curry, "Intravenous Mushroom Poisoning," *Annals of Emergency Medicine* (1985) 14:900–902.

68. C. Benjamin, "Persistent Psychiatric Symptoms After Eating Psilocybin Mushrooms," *British Medical Journal* (1979) 1:1319–1320.

69. G. Guzmán, J. Ott, J. Boydston, and S. M. Pollock, "Psychoactive Mycoflora of Washington, Idaho, Oregon, California and British Columbia," *Mycologia* (1976) 68(6):1267–1271.

70. P. Thompson, M. D. Anglin, W. Emboden, and D. G. Fisher, "Mushroom Use by College Students," *Journal of Drug Education* (1985) 15(2):111–124.

71. C. Hyde, G. Glancy, P. Omerod, D. Hall, and G. S. Taylor, "Abuse of the Indigenous Psilocybin Mushrooms: A New Fashion and Psychiatric Complications," *British Journal of Psychiatry* (1978) 132:602–604.

72. J. W. Allen, "A Private Inquiry into the Circumstances Surrounding the 1972 Death of John Gomilla, Jr., Who Died After Allegedly Consuming Ten Hallucinogenic Mushrooms While Residing in Hawaii," *Journal of Psychoactive Drugs* (1988) 20(4):451–454.

73. E. L. McCawley, R. E. Brummet, and G. W. Dana, "Convulsions from *Psilocybe* Poisoning," *Proceedings of the Western Pharmacological Society* (1962) 5:27–33.

74. J. Francis and V. S. G. Murray, "Review of Inquiries Made to the NPIS Concerning *Psilocybe* Mushroom Ingestions, 1978–1981," *Human Toxicology* (1983) 2:349–352.

75. N. R. Peden and S. D. Pringle, "Hallucinogenic Fungi," *Lancet* (1982) 1(8268):396–397.

76. K. L. R. Jansen, "Magic Mushrooms: A Fast Growing Problem," *Journal of General Practice* (1988) 5:7–10.

77. United States 21 Code of Federal Regulations, 1308,11 (1976).

78. F. E. Cheek, S. Newell, and M. Joffe, "Deceptions in the Illicit Drug Market," *Science* (1970) 167:1276.

CHAPTER EIGHTEEN

MUSCARINE POISONING
PSL or SLUDGE Syndrome

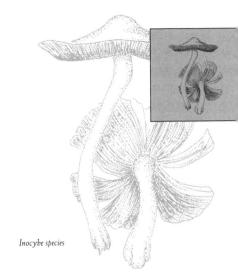

Inocybe species

CASE 1

A woman aged 25 had sudden onset of blurred vision occurring about one hour after having fried fungi for breakfast. Perspiration accompanied by profuse salivation rapidly followed the first symptom. She complained of giddiness and profuse sweating. On examination, within 15 minutes of the onset of the first symptoms, the pupils were not dilated or constricted, but were equal in size and had a sluggish reaction to light. There was profuse sweating and lacrimation. The abdomen was rigid, and generalized tenderness was present. The pulse was slow, about 56, and irregular, the blood pressure 90/50.

Treatment consisted in the immediate administration of 1.3 mg atropine sulfate and gastric lavage until the fluid was returned clear. Two ounces of magnesium sulfate was left in the stomach. Within two hours all symptoms of poisoning had disappeared; convalescence was uneventful.[1] (case report quoted by Lampe[2])

Ever since muscarine was first isolated from *Amanita muscaria* (see Chapter 16), the effects of muscarine—excessive perspiration, salivation, and tearing—have been synonymous with mushroom poisoning. Because of this combination of symptoms, this malady is also called the PSL (perspiration, salivation, lacrimation) syndrome or the SLUDGE (salivation, lacrimation, urination, defecation, gastrointestinal distress, emesis) syndrome. This equation has had a number of undesirable consequences for the management of mushroom poisonings. Atropine, which is a specific antidote for this particular compound, became the standard—almost reflex—therapy for every kind of mushroom poisonings. However, although muscarine is very widespread among both the agarics and the boletes, it reaches clinically significant concentrations in only a small number of species. *Amanita muscaria* is typical of the many

340

mushrooms that contain muscarine only in clinically **insignificant** amounts. The concentration in this particular species is so low that one of the earlier investigators in the field of poisoning used one ton of mushrooms to make his extractions. Moreover, because it was believed that most of the poison was in the cuticle or just beneath it, the investigator had his helpers (no doubt the equivalent of today's graduate students) peel all the mushrooms until late into the night.[3] Nevertheless, for persons poisoned with specimens of the species that do contain large amounts of muscarine, atropine therapy is appropriate.

The widespread presence of muscarine-containing mushrooms in the urban environment makes them prime candidates for the adventurous mycophagist. Although many of them are somewhat small and inconspicuous, they can fruit in such abundance that a meal's worth is easily gathered. They are sometimes mistaken for the fairy ring mushroom (*Marasmius oreades*); in fact, the rings of this excellent edible species sometimes overlap with those of *Clitocybe dealbata* (color plate 15). Moreover, the appearance of some of the toxin-containing species on lawns and in parks increases the likelihood that they will be sampled by curious toddlers. But, despite the ubiquity of the muscarine-containing mushrooms, the actual number of reported muscarine poisonings is very small.

DISTRIBUTION AND HABITATS OF MUSCARINE-CONTAINING MUSHROOMS

Some of the muscarine-containing mushrooms are rather common lawn-dwelling saprophytes, whereas others have a mycorrhizal relationship with the surrounding trees. They are all quite frequent in parks, along roads, on lawns, golf courses, and cemeteries, and in woods.

The genus *Clitocybe* is large, and its numerous species inhabit coniferous forests, hardwood forests, grassy parks, lawns, and gardens. Many *Clitocybe* species fruit in the late fall and early winter. The genus *Inocybe* is even larger, including at least 400 species in North America alone. *Inocybe* species are common in evergreen forests, although some, such as *Inocybe lacera*, fruit in open places and along paths, usually in moss or needles. Although most species fruit in the fall, others fruit throughout the year, whenever the temperature and moisture conditions are propitious. Because neither *Clitocybe* nor *Inocybe* specimens are of culinary interest, most amateurs are unskilled in their identification. Even dedicated professionals have trouble identifying species of these two genera in the field; often a microscope or other laboratory equipment is required for accurate, specific identification.

The genera *Clitocybe* and *Inocybe* epitomize the vast host of hard-to-identify mushrooms that are uniformly disparaged as LBMs (little brown mushrooms). This moniker is routinely employed by mycophagists to describe the small, drab, monotonous mushrooms that lack obvious distinguishing features and range in color from numerous shades of beige to equally numerous shades of brown, gray, and khaki.

Omphalotus species are found in parts of the United States and Europe. Of the toxin-containing species, *Omphalotus olearius* fruits in Europe, and both *O. olearius* and *O. illudens* grow in the eastern United States. *Omphalotus olearius* (jack-o'-lantern mushroom) fruits in large clusters at the base of hardwood stumps, on roots, and on buried wood. Late summer and fall is its usual season; it often fruits right up to the first frost. *Omphalotus olivascens* is found only in the western United States.

MUSHROOMS IMPLICATED IN MUSCARINE POISONINGS

Mushrooms Containing High Concentrations of Muscarine

Two genera—*Inocybe* and *Clitocybe*—have the largest numbers of species in which muscarine is present in clinically significant amounts. The techniques used to identify this compound are complicated, so the reliability of some of the identifications is poor. This caveat especially applies to older investigations that relied on biological assays for muscarine detection. These, although very sensitive, are somewhat nonspecific; acetylcholine, choline, and other analogues are widespread in the higher fungi and also give a positive result in the biological assay. Since the 1960s, newer chemical methods have failed to confirm the presence of muscarine in some species previously thought to contain it. This disparity between the results of chemical assays and bioassays is nicely demonstrated by Chilton.[4] However, the chemical assays are not infallible; some of the chemical techniques cannot distinguish between the active and the inactive forms of muscarine.

Bresinsky's recent book contains a huge table listing the species in which the toxin has been chromatographically identified.[5] However, because of the difficulties in identification, any list of suspects should be viewed as tentative.

Additional toxic compounds have been detected in many muscarine-containing mushrooms.[6] However, their role in the PSL syndrome is not known. It is likely that further toxins will be discovered, because the known toxins do not account for all the reported symptoms. *Inocybe fastigiata*, for example, has been associated with mild liver damage.[7]

Other Species Suspected of Causing Muscarine Poisoning

In addition to the *Inocybe* and *Clitocybe* species listed in the accompanying box, it is also suspected that muscarine is present in *Mycena pura* and some of the boletes, for example, *Boletus calopus* and *Boletus luridus*. These mushrooms and the poisoning they produce are discussed in Chapter 19. *Entoloma rhodopolium*, the cause of numerous poisonings in Japan, probably owes its effects to choline, muscarine, muscaridine, and other toxins.[8]

Species in the genus *Omphalotus* may also contain significant amounts of muscarine, including *O. olearius* (jack-o'-lantern mushroom), *O. olivascens*, *O. subilludens*, and *O. illudens*. The studies on this group were done many years ago and used biological assays. The clinical

COMMON MUSHROOMS CONTAINING MUSCARINE IN CLINICALLY SIGNIFICANT AMOUNTS

Clitocybe species [10–14]

A monograph on this genus has a more complete listing of the muscarine-containing species.[15]

C. angustissima	*C. aurantiacum*
C. candidans	*C. cerussata*
C. dealbata (color plate 15)	*C. dilatata*
C. ericetorum	*C. festiva*
C. gibba	*C. hydogramma*
C. nebularis	*C. rivulosa*
C. suaveolens	*C. trunciola*
C. vermicularis	

Inocybe species [5,16–18]

Muscarine is very widespread throughout this genus, having been found in up to three-quarters of all species surveyed. Common species found in the United States include

I. fastigiata (color plate 14)	*I. patouillardi* (very rare)
I. geophylla	*I. pudica*
I. lacera	*I. sororia*
I. mixtilis	*I. subdestricta*
I. napipes	

characteristics of the poisonings by these mushrooms do not have major muscarinic overtones. In some patients, primarily in Europe, a few symptoms suggested muscarine poisoning.

According to a recent report, *Mycena pura* (var. *rosea*) poisoned a couple who ate a consommé containing approximately 500 g (a little over a pound) of mushrooms. The symptoms were atypically delayed, only commencing 1.5 hours after the meal, and comprised the expected features of sweating, salivation, bradycardia (to below 30 beats per minute), hypotension, diarrhea, and colicky abdominal pain.[9]

Species Containing Muscarine in Insignificant Amounts

Amanita muscaria is the classic example of a mushroom that contains negligible amounts of muscarine. Small amounts of muscarine have also been detected in species belonging to the following genera: *Boletus, Clitopilus, Collybia, Hygrocybe, Hypholoma, Lactarius, Mycena, Paxillus, Entoloma* (=*Rhodophyllus*), *Russula, Tricholoma,* and *Tylopilus*.[10] Dozens of other species undoubtedly contain muscarine, because it is ubiquitous throughout the mushroom world.

TOXICOLOGY AND PHARMACOLOGY OF MUSCARINE

The initial discovery of muscarine dates back to 1869, when Schmeideberg and Koppe isolated from *Amanita muscaria* a compound that they presumed to be the active toxic ingredient in this species. It is a small molecule and proved very resistant to analysis. It has a whole series of isomers, most of which are not physiologically active. The active compound is the L(+) isomer of muscarine. Excellent reviews of the chemistry and toxicology of this compound have been published,[19–23] as have the early studies of its pharmacology.[24,25]

Muscarine binds to some—but not all—of the receptors of the parasympathetic nervous system that are usually stimulated by acetylcholine. Normally, the acetylcholine is rapidly metabolized in the presence of the enzyme acetylcholinesterase, which cleaves and inactivates the molecule. Detachment of the inactive cleavage products frees up the binding site, and another molecule of acetylcholine can then bind to it. Muscarine, however, is not affected by acetylcholinesterase, so its stimulation of the receptor is prolonged and continuous. Those organs and tissues containing abundant muscarinic binding sites are the ones most affected. The primary receptors affected are the postganglionic parasympathetic receptors of smooth muscle and various glands (sweat, tear, bronchial, salivary). Other acetylcholine receptors, such as those at the neuromuscular junctions, are unaffected by muscarine. Atropine competes for the same receptor sites and can dislodge muscarine. However, atropine binding does not stimulate the cell or nerve and consequently reverses the effects of muscarine binding. These interactions are illustrated in the accompanying figure.

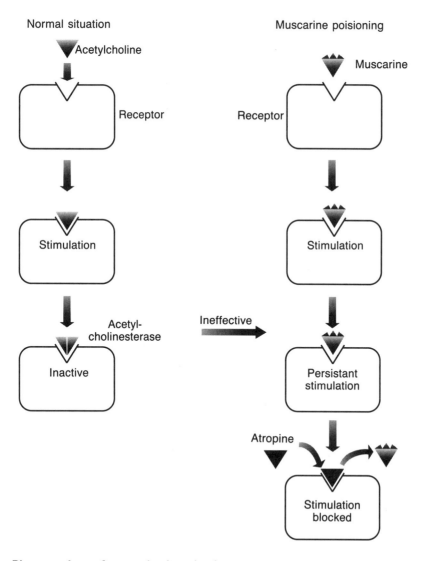

Pharmacology of muscarine intoxication

The muscarine molecule is heat stable and therefore unaffected by cooking. Fortunately, it is rather poorly absorbed from the gastrointestinal tract. In certain animal species, such as the rabbit, it is hardly absorbed at all. However, given intravenously, even in tiny amounts, it has profound effects. For example, as little as 0.01 µg/kg body weight produces hypotension in cats. The cardiovascular system of some animals—for example, the cat—is especially sensitive to the parenteral administration of this agent. Mice are similarly sensitive, showing excessive bronchial secretion, lacrimation, salivation, and defecation prior to their rapid death, often within minutes of administration.

Muscarine does not cross the blood–brain barrier and so cannot be responsible for any of the central nervous system effects occasionally manifested by some of the patients. Any dizziness may be the result of hypotension and decreased cerebral blood flow.

TOXICITY OF MUSCARINE

The lethal dose of muscarine for a human is not precisely known. Estimates range from as high as 180 mg to as low as 40 mg.[5,18]

How much of the muscarine-containing mushrooms does one need to eat to become symptomatic? This question is hard to answer, because of the wide variability in the concentration of muscarine from one species to the next. A man would have to eat the equivalent of his own weight of *Amanita muscaria* before manifesting any muscarinic symptoms—actually, the accompanying indigestion would be a more serious problem. With the small *Inocybe* or *Clitocybe* species, which have high concentrations of active muscarine, 100 g (about 1/4 cup) or less of fresh specimens could produce symptoms. To reiterate: Only the L(+) isomer of muscarine is physiologically active.

CLINICAL FEATURES OF MUSCARINE POISONING

Onset

The effects of muscarine develop rapidly, often within 15 to 30 minutes of eating the mushrooms. Certainly by one hour, almost all muscarine-poisoned victims will manifest symptoms.

Signs and Symptoms

The clinical features of this syndrome result from the stimulation of the parasympathetic nervous system. The extent and the severity of the symptoms depend on the dose. Children may be severely affected because of their small size,[26] although it is rare for them to eat more than a cap or two. Adults usually make a meal out of collected mushrooms and can absorb a considerable quantity of muscarine.

The most frequent manifestation is profuse sweating, accompanied by flushing of the skin. This symptom is usually associated with increased salivation, colicky abdominal pain due to intestinal stimulation, nausea, vomiting, and diarrhea. Excess tear production (lacrimation) can be quite striking. Effects on the eye develop in acutely poisoned victims and include blurred vision resulting from an inability to accommodate and from pupillary constriction (miosis). The cardiovascular manifestations include bradycardia, which may be quite noticeable, and hypotension. A significant decrease in blood pressure is usually seen in only the most severe poisonings.

Other less frequent effects include a painful urge to urinate and difficulty in breathing, which is due to partial bronchoconstriction and excess bronchial secretions. These pulmonary effects produce an asthmatic type of wheezing and pulmonary congestion. There may be a sense of anxiety and dizziness. It has been reported that the urine of patients poisoned with the red-staining *Inocybe* (*I. patouillardi*) turns red after acidification, but this assertion needs further confirmation.

Central nervous system manifestations do not occur, because muscarine does not cross the blood–brain barrier. However, cardiorespiratory failure, which sometimes occurs in the most severely poisoned victims, will eventually lead to cerebral anoxia and coma.

Although this symptom complex is very distinctive, it can on occasion be confused with the panic reaction. In this latter situation, the

Key Clinical Features of Muscarine Poisoning

- Onset is 5–30 minutes after ingestion of mushrooms
- Perspiration
- Salivation
- Lacrimation
- Nausea, vomiting, and diarrhea
- Colicky abdominal pain
- Bradycardia
- Miosis, blurred vision
- Urge to urinate

skin is usually cold and clammy, but excessive tearing and salivation may accompany the nausea and apprehension. During a panic reaction, the pupils are dilated as a result of fright and anxiety, and the pulse rate is always elevated, unlike the slow pulse produced by muscarine.

PROGNOSIS

Although no reports of fatalities have appeared in the American literature, Europe has had a rare case or two. Most fatalities have been associated with *Inocybe patouillardi*, which was also responsible for a mass poisoning in 1963, from which everyone recovered.[27] Supposedly this mushroom does not grow in the United States, although a few collections of it have been made here in the summer.[8]

Without atropine therapy, symptoms may persist for many hours before finally resolving. The duration of the illness is probably dose dependent. Usually the symptoms are so alarming that patients seek immediate medical care. Once atropine therapy has been instituted, most patients recover in less than 30 minutes.

MANAGEMENT OF MUSCARINE POISONING

Remove any residual mushrooms from the stomach. Because patients usually present within the first one to two hours after ingestion, the induction of emesis may be appropriate. Emesis is especially appropriate if the patient presents within 30 minutes after ingestion. The standard dose of syrup of ipecac is recommended (see Chapter 11). If the patient has vomited copiously before presentation, however, induced emesis is unnecessary. Furthermore, when significant bronchospasm and excess bronchial secretions are present, emesis is contraindicated. The physician's judgment of the evolving clinical condition is the key in determining the role or danger of inducing vomiting.

Administer atropine. Muscarine poisoning is the only type of mushroom poisonings that should be treated with atropine. The initial dose for adults is 0.5–1.0 mg intravenously; for children, the initial dose is 0.01–0.3 mg/kg body weight. Similar doses can be given at 10-minute intervals, up to a total dose of 2 mg in adults and 1 mg in children.

The biggest danger is giving too much atropine, thereby producing a whole set of undesirable side effects. Usually miosis, if present, will disappear first. The dose can be titrated on the basis of excessive production of secretions and sweating (bronchial secretion is the more important of these two signs). It is unnecessary to give enough atropine to produce dilated pupils, a sign that formerly was used as the end point of therapy. Doses sufficient to produce this result can cause significant complications, especially in children, who may develop hyperthermia and delirium.[2]

Key Elements of Treatment of Muscarine Poisoning

* Consider inducing emesis to empty the stomach of all mushroom fragments.
* Administer atropine.

* Do not overtreat.

UNANSWERED QUESTIONS ABOUT MUSCARINE POISONING

What role, if any does muscarine play in the poisonings associated with *Mycena pura* and *Omphalotus* species? In most of the reported *Omphalotus* poisonings, muscarinic symptoms have been inconspicuous or absent. However, in a few, increased perspiration and salivation have been commented on. As has been stressed before, such symptoms commonly accompany nausea and vomiting and are not, by themselves, indicative of muscarine.

REFERENCES

1. D. Wilson, "Poisoning by *Inocybe fastigiata*," *British Medical Journal* (1947) ii:297.

2. K. Lampe, "Pharmacology and Therapy of Mushroom Intoxications." In *Mushroom Poisoning: Diagnosis and Treatment*, ed. B. H. Rumack and E. Salzman (West Palm Beach, FL: CRC Press, 1978), pp. 125–169.

3. F. Kögl, H. Duisberg, and J. Erxleben, [Untersuchungen über Pilzgifte. I. Über das Muscarin] (German), *Justus Liebigs Annalen der Chemie* (1931) 489:156–192.

4. W. C. Chilton, "Chemistry and Mode of Action of Mushroom Toxins." In *Mushroom Poisoning: Diagnosis and Treatment*, ed. B. H. Rumack and E. Salzman (West Palm Beach, FL: CRC Press, 1978), pp. 87–124.

5. A. Bresinsky and H. Besl, *A Colour Atlas of Poisonous Fungi* (London: Wolfe Publishing, 1990).

6. G. Sullivan, R. D. Garret, and R. F. Lenehan, "Occurrence of Atromentin and Thelephoric Acid in Cultures of *Clitocybe subilludens*," *Journal of Pharmaceutical Sciences* (1971) 60:1727–1730.

7. L. A. Varese and S. Barbero, "Avvelenamento collettivo da *Inocybe fastigiata*," *Minerva Pediatrica* (1967) 19:337–339.

8. T. Maki, K. Takahashi, and S. Shibata, "Isolation of the Vomiting Principles from the Mushroom *Rhodophyllus rhodopolius*," *Journal of Agricultural and Food Chemistry* (1985) 33:1204–1205.

9. P. Goigoux, "Un cas grave d'intoxication par *Mycena rosea*," *Bulletin de la Fédération de Mycologie de Dauphine Savoie* (1992) 127(Oct):10–11.

10. R. J. Stadelmann, E. Müller, and C. H. Eugster, "Über die Verbreitung der stereomeren Muscarine innerhalb der Ordnung der Agaricales," *Helvetica Chimica Acta* (1976) 59:2432–2436.

11. R. G. Benedict, "Mushroom Toxins Other Than *Amanita*." In *Microbial Toxins*, ed. S. Kadis, A. Ciegler, and S. J. Ajl (New York: Academic Press, 1972), pp. 281–320.

12. G. Lincoff and D. H. Mitchel, *Toxic and Hallucinogenic Mushroom Poisoning* (New York: Van Nostrand Reinhold, 1977).

13. T. J. Duffy and P. P. Vergeer, *California Toxic Fungi*, Toxicology Monograph No. 1, Mycological Society of San Francisco (1977).

14. M. Moser, *Die Röhrlinge und Blätterpilz (Polyporales, Boletales, Agaricales, Russalales)*, published in the series *Kleine Kryptogamenflora* as Vol. IIb/2, ed. G. Fischer (Stuttgart, 1978).

15. H. E. Bigelow, *North American Species of Clitocybe*, Part 1, *Beihafte zu Nova Hedwigia* (1982) 72:1–280.

16. C. H. Eugster, "Isolierung von Muscarine aus *Inocybe patouillardi*," *Helvetica Chimica Acta* (1957) 40:886–887.

17. M. H. Malone, R. C. Robichaud, V. E. J. Tyler, and L. R. Brady, "Relative Muscarinic Potency of Thirty *Inocybe* Species," *Lloydia* (1962) 25:231–237.

18. J. F. Ammirati and J. A. Traquair, *Poisonous Mushrooms of Northern United States and Canada* (Minneapolis: University of Minnesota Press, 1985).

19. S. Wilkinson, "The History and Chemistry of Muscarine," *Quarterly Review of the Chemical Society* (1961) 15:153–171.

20. C. H. Eugster, "Chemie der Wirkstoffe aus dem Fliegpilz (*Amanita muscaria*)." In *Fortschritte der Chemie organischer Naturstoffe*, Vol. 27, ed. L. Zechmeister (Vienna: Springer-Verlag, 1969), pp. 261.

21. F. Jellinek, "The Structure of Muscarine," *Acta Crystallographica* (1957) 10:277–281.

22. K. Bowden and G. A. Mogey, "The Story of Muscarine," *Journal of Pharmacy and Pharmacology* (1969) 10:145.

23. C. H. Eugster, "The Chemistry of Muscarine." In *Advances in Organic Chemistry*, Vol. 2, ed. R. A. Raphael, E. D. Taylor, and H. Wynsberg (New York: Interscience, 1960), pp. 427–455.

24. P. G. Waser, "Chemistry and Pharmacology of Muscarine, Muscarone and Some Related Compounds," *Pharmacology Reviews* (1961) 13:465–515.

25. P. J. Fraser, "Pharmacological Actions or Pure Muscarine Chloride," *British Journal of Pharmacology* (1957) 12:47–52.

26. I. Amitai, O. Peleg, I. Ariel, and N. Binyamini, "Severe Poisoning in a Child by the Mushroom *Inocybe tristis*, Malencon and Bertault," *Israeli Journal of Medical Science* (1982) 18(7):798–801.

27. M. Hermann, "Die Naumberger Massen-Pilzvirgiftung mit dem Ziegelroten Risspilz—*Inocybe patouillardi*—im Juni 1963," *Mycologisches Mitteilungsblatt* (1964) 8:42–44.

CHAPTER NINETEEN

GASTROINTESTINAL SYNDROME

Cholorophyllum molybdites

RING, RING... ring, ring... ring, ring. With a firm nudge, my wife launched me across the bed to the telephone. After a couple of attempts, I found the receiver and lifted it. The radio alarm clock fluoresced 2:30 A.M. "Dr. Benjamin?" a voice inquired.

Being a pathologist, the number of post-midnight calls are fortunately few. "Yes," I responded, trying hard not to wake up too much. I had learned the trick as an intern many years ago. Sometimes I would arrive on the ward the next morning and be quite surprised about the decisions I had made during these interrupted stages of REM sleep.

"Dr. Benjamin, I was given your name by the Poison Control Center . . . beep . . . We have a case of mushroom poisoning in the emergency room and would like some advice . . . beep . . ."

I was now wide awake and knew that there was little chance of getting back to sleep, at least for a few hours. "Tell me the story . . . beep . . .", I said.

"Oh, by the way," said the voice, "you can ignore those beeps. It's just to let you know that this conversation is being recorded. I am calling from Langley Air Force Base, Virginia . . . beep . . . Well, this officer picked some large white mushrooms on the lawn outside one of the buildings and had them for dinner at around 8:30 P.M. Two hours later, he began developing severe cramping abdominal pain. He retched a few times but hasn't vomited. He arrived at around midnight in severe pain, writhing in agony. The pain comes in waves that almost paralyze him by their intensity."

"Any other symptoms?" I inquired, rattling off a list of other things to look for.

"No," she said, "just the severe pain and nausea."

From the intensity of the pain, it sounded like the dreaded *Chlorophyllum molybdites*, although a number of the toxic *Agaricus* species can be equally as bad. I am very poor at the telephone identification of mushrooms. After a few abortive attempts to get her to describe the specimens, I gave her some advice, reassured her that her patient would survive, although would probably not be in any shape for dancing for the next few hours, and told her that I would check again in the morning.

I contacted Dr. Ammirati, who called the next day and coerced a textbook description of *Chlorophyllum* from the attending physician. I made a followup call to hear that all was well and that the pain had finally resolved spontaneously by about five in the morning.

"How ironic," I thought, "that a member of the Central Intelligence Agency had misconstrued a *Chlorophyllum* for an edible *Lepiotas*. No wonder people with lesser observational skills have difficulty distinguishing the two."

D. Benjamin, personal journal (1990)

Under the rubric of the gastrointestinal syndrome are included the reactions to a vast array of mushrooms, all of which have a few features in common. These mushrooms provoke symptoms in many but not all people. The illness is seldom serious, although the victims can be quite miserable. The mushrooms all affect the gastrointestinal tract with various combinations of nausea, vomiting, colicky (cramping) abdominal pain, and diarrhea. The toxins are largely unknown.

The mushrooms responsible for these gastrointestinal upsets are rather poorly studied, from both the clinical and the toxicological standpoint.[1] The only acceptable experimental animal is a human, so we have to rely on the few scattered reports that reach the medical literature and the anecdotes or personal experiences of those who manage cases of mushroom poisoning. Not even impecunious university students—usually enthusiastic volunteers—are willing to help the cause of science in this instance. And, because the poisonings are not life-threatening, they do not generate the interest invested in the other forms of poisoning. Even the mycology of the culprits is not particularly well known, because good mechanisms for collecting the epidemiological and identification information have not been available until recently.

Gastrointestinal syndrome is produced by a very heterogeneous group of mushrooms, and the symptoms produced by the various mushrooms are not caused by related poisons. Although the gastrointestinal tract is the focus of all the distress, subtle differences distinguish the clinical effects of one mushroom from those of the next. This variability reflects significant chemical differences in the toxins. For example, the poisons in *Omphalotus olearius* act primarily as emetic agents, an action suggesting a more direct effect of the absorbed toxins on the vomiting center in the brain. In contrast, *Chlorophyllum molybdites* produces vomiting, diarrhea, and severe abdominal pain. The toxins of this species apparently act largely on the gastrointestinal tract itself.[2]

The diverse secondary compounds in these mushrooms are far from being fully catalogued and even further from being understood. These compounds largely account for the toxic effects and do not appear to play a primary role in the vital metabolic processes required for growth and development of the fungus. However, they may be important in how the fungus communicates with and influences its neighbors and its immediate environment. For those with a bent toward chemistry, a number of reviews of the subject are available.[3-10] Bresinsky uses the following broad categories to catalogue these toxic compounds.

- Terpenoid compounds: This group includes the sesquiterpenes and the triterpenes, which are important toxins in species such as *Omphalotus olearius*.
- Anthroquinones: These compounds are purgative and mutagenic pigments; they are widespread throughout the mushroom world.
- Oxolans: Each oxolan is a five-membered ring containing an oxygen atom; muscarine is an oxolan.
- Nitrogen heterocyclics: This group includes the hallucinogenic tryptamines and isoxazoles.
- Amides and peptides: These small molecules are the most toxic mushroom toxins; the amatoxins are in this group.
- Compounds with N-N bonds: Hydrazines and their relatives are in this group; many are carcinogenic.
- Polysaccharides and other complex or unusual sugars
- Lipids
- Sterols

As noted repeatedly in earlier chapters, there is great variability in the presence and amount of toxin within each mushroom specimen, depending on where in the country it is picked. A few of the mushrooms discussed here are regarded as perfectly edible by some people, yet they cause considerable misery for others. The term *partials* has been used to describe these capricious fungi. Some species express their toxicity when they have not been adequately cooked because a considerable number of their toxins are heat labile. This characteristic accounts for some of the disparity in the field guides in regard to the edibility of selected mushrooms. Many of the reports in the literature are based on single episodes and unsubstantiated references in the older field guides. These warnings have simply been transcribed over the years because new information on which to base a judgment has not been forthcoming. For good reason, latter-day authors are reluctant not to mention that a mushroom might be toxic if some "authority" has pronounced it so in the past. The list provided here covers only those species most likely to be encountered, even though dozens of other fungi have been accused of wrongdoing. For a more complete list, the reader should consult Lincoff and Mitchell.[11]

From the viewpoint of the weekend mycophagist, this assortment of mushrooms, which causes great gastrointestinal distress, is probably the most important group to know about. In a survey of over 40 years of mushroom poisoning, 40% of all notified intoxications were caused by the mushrooms discussed in this chapter.[12] Of course, the number of these reports grossly underrepresents the true number of poisonings, because only a small fraction is ever reported. Many of the species having notoriously grim reputations belong to this assemblage, and some can be rather easily confused with a variety of excellent edible species. For those who enjoy living on the edge and chasing new culinary experiences, one of these mushrooms will be the most likely cause of at least a night or two of woe. But do not be dissuaded from the adventure—if you are in good health, the encounter is unlikely to prove lethal. Conversely, there is no excuse for not positively identifying the mushrooms one plans to eat.

CLINICAL FEATURES AND PROGNOSIS

Although some variation from one mushroom to the next is to be expected, the majority produce symptoms between 15 minutes and two hours after the meal. Nausea, vomiting, colicky abdominal pain, and diarrhea dominate the clinical picture. These symptoms may be accompanied by a sense of anxiety with sweating, cold and clammy extremities, and tachycardia. In young children, the rapid loss of fluids can result in hemodynamic disturbances. Likewise, the elderly are more vulnerable because of the presence of underlying diseases. In the majority of circumstances, the symptoms resolve spontaneously within 8 to 12 hours, but may linger on for a couple of days.

MANAGEMENT OF THE GASTROINTESTINAL SYNDROME

Although the gastrointestinal symptoms discussed here are not caused by a homogeneous group of toxins, the great similarities in the clinical features and the potential complications allow the attending physician to adopt a fairly generic therapeutic approach. Because the majority of toxins are direct gastrointestinal irritants, removing them from the stomach as rapidly and completely as possible is the mainstay of treatment. Rapid resolution of the symptoms generally follows emesis or lavage. Syrup of ipecac is the favored emetic (see page 184), but lavage should be employed in cases in which the patient is not able to protect his or her own airway.

Activated charcoal is usually administered to bind any residual toxin. Intravenous fluids may be required if the fluid loss from vomiting and diarrhea or both has been excessive. This supportive care is especially important in children and elderly patients, that is, those patients most vulnerable to the effects of dehydration. Antispasmodics can be used for the symptomatic relief of the colicky abdominal pain.

Of the mushrooms responsible for the gastrointestinal syndrome, *Chlorophyllum molybdites* accounts for more cases in the United States than any other mushroom and is often associated with greater morbidity.[13–23] For this reason, I discuss it first; the rest follow in alphabetical order.

CHLOROPHYLLUM MOLYBDITES

Chlorophyllum molybdites has been called a host of other names, including green-spored *Lepiota*, Morgan's *Lepiota*, sickening *Lepiota*, *Lepiota morgani*, and sickening parasol. Each one of these names is quite instructive. It is frequently confused with two edible *Lepiota* species, *L. rachodes* (parasol mushroom) and *L. procera* (shaggy parasol mushroom). Indeed, *C. molybdites* was once classified in the *Lepiota* taxon. The most significant difference between it and *Lepiota* species is spore color. *Chlorophyllum molybdites* (illustration at top of page and color plate 16) has green spores. The spores are not bright green and may not be detected by a casual glance at the gill surface, especially in young immature mushrooms. However, a formal spore print reveals a definite green cast.

Chlorophyllum molybdites also has characteristics in common with *Amanita* species: white gills, white or off-white spores, and a ring around the stem. However, *C. molybdites* has no volva, and the surface of the cap usually breaks into fine or coarse scales. The stem is relatively thin in comparison to the large cap, and the appellation of "parasol mushroom" paints an excellent picture of what to expect.

In the United States, *C. molybdites* is widely distributed, although it has not been found in Washington, Oregon, Idaho, and Montana. It is common in California, across the eastern seaboard, and in the Southeast. In general, this is a pantropical mushroom, favoring the warmer climates; it has a worldwide distribution.

In some individuals, the gastrointestinal syndrome, which occurs about one to three hours after the meal, can be very severe, especially the colicky abdominal pain, which can mimic that of a "surgical" abdomen. Symptoms persist for up to six hours, and even longer in a few patients. Nausea, vomiting, and diarrhea complete the picture. The diarrhea may be explosive in nature and become bloody. For children, as mentioned earlier, a risk of dehydration and hypovolemic shock exists if fluid status is not carefully controlled.[17] One childhood fatality has been reported; the child died following the development of seizures about 17 hours after the poisoning. The rare patient may even have gastrointestinal hemorrhage, which in one patient was attributed to the complication of disseminated intravascular coagulation.[24] Some patients become drowsy at the onset of the poisoning and show peripheral cyanosis. Others have reported a sensitivity to both light and sound.

Considerable controversy surrounds the topics of toxicity and edibility of C. molybdites. Many claim that the mushroom is edible, at least for certain people, especially if the mushroom is well cooked.[25] There is fairly widespread agreement that C. molybdites is definitely toxic when eaten raw, an acknowledgment suggesting that at least one of its important toxins is heat labile. However, regional differences in the amount of toxin and differences in individual susceptibilities to its effects must account for some of the disparities in the literature. In one well-documented instance, only one of three people sharing a mushroom meal became ill.[18] This may well be another example of the old adage, "One man's mushroom is another man's poison!"[26]

Only the crudest and most preliminary investigations of the putative toxins have been performed. One toxin may be a heat-labile protein that has a mass of approximately 400,000 daltons and is composed of subunits, each with a mass of 40,000–60,000 daltons. This compound is sensitive to both acid and pepsin and in experimental animals only showed an effect when given by intraperitoneal injection. Whether this compound represents the important toxin in humans is entirely unknown.[16]

AGARICUS SPECIES

The genus Agaricus contains some of the most esculent fungi, such as the meadow mushroom (A. campestris), the supermarket mushroom (A. bisporus = A. brunnescens), the horse mushroom (A. arvensis), and the incomparable prince mushroom (A. augustus). But lurking on lawns and in pastures and woods are hosts of related species not nearly as kind to the digestive system. Distinguishing them is not always easy. Part of the trick is to have a fine sense of smell and to be able to differ-

Agaricus xanthodermus

entiate the odor of anise from that of phenol. The odor of ink must also be avoided. The other part of the equation is to know all the field characteristics, including the color reactions the mushrooms undergo when bruised.

David Arora has aptly named this group of mushrooms the "lose-your-lunch bunch." The most frequent culprits include

A. *xanthodermus* (color plate 18) Common name: yellow stainer
A. *praeclarisquamosus* Common names: flat top, flat-capped
  (=A. *placomyces*,   *Psalliota*
  =A. *meleagris*)
  (color plate 17)
A. *hondensis*[27]
A. *sivicola*[28]
A. *albolutescens*[28]
A. *californicus*
A. *glaber*

Some of these species are rather easily confused with edible species, so poisoning is quite frequent, especially in areas where they are common, such as California. As a general rule, all *Agaricus* species should be approached with considerable circumspection. This genus is not for the careless or the unobservant mycophagist.

One of the toxins in the yellow-staining varieties is presumed to be phenol.[29] Previous suggestions that the sesquiterpenes are the major toxins are probably incorrect. A number of other chemicals have been isolated from *A. xanthodermus*.[30,31] Furthermore, the yellow-staining *Agaricus* species concentrate heavy metals, so it is advisable to limit one's consumption of these varieties. The yellow-staining group includes the horse mushroom (*A. arvensis*).

In Chapter 7, the carcinogenic potential of the genus *Agaricus* is discussed.

ARMILLARIA MELLEA COMPLEX

Around the globe, mushrooms credited with the appellation of *Armillaria mellea* (honey mushroom, boot-lace or shoestring fungus) have a variable reputation, depending on the palate of the mycophagist and the vulnerability of one's gastric mucosa (color plate 20). Charles Badham denounced it as "nauseous and disagreeable," adding that "not to be poisonous is its only recommendation."

> In Vienna it is employed chiefly for making sauces; but we must confess that even in this way, and with a prejudice in favor of Viennese cookery, our experience of it was not satisfactory. It is at best a sorry substitute for the mushroom.
>
> M. C. Cooke and M. J. Berkeley, *Fungi: Their Nature, Influence and Uses* (1888)

Although considered an edible mushroom and voraciously consumed by many people because of its ubiquity and frequency, *Armillaria mellea* causes enough problems for enough people to merit inclusion in this chapter. It is especially troublesome when eaten raw.[32] The toxins, which produce the usual gastrointestinal upset, are not known. An antibiotic (melleolide) has been isolated from the mycelium of this species.[33]

Folklore from different parts of the country suggests various ways to mitigate the toxic effects of the honey mushroom. In the Pacific Northwest, it is said that the mushroom should not be eaten after the first frost; or those mushrooms growing on conifers, rather than hardwoods, should not be collected. In other regions, those growing on eucalyptus are reportedly unsafe. The veracity of all this advice remains unproved.

This taxon is a favorite of some mycologists because of the huge variety of morphological forms the mushroom can take. Endless debates swirl about its classification, with the traditional "lumpers" and "splitters" lined up on either side of the taxonomic arguments.

RED-PORED BOLETES

The few boletes with red pores—rather than yellow, olive, or dull brown—have for many years had a dubious reputation. This bad name can be traced to 1869, when it was stated, "One of the most dangerous fungi is the variety of *Boletus luridus* named by Lenz *Boletus satanas*"[34] (color plate 19). This assertion was made at the time when "muscarine" was being found in a variety of species. Few people have sampled this mushroom, because of the bright blue staining of its bruised tissue and the odor of rotting flesh emanating from it. The dire warnings being circulated and firmly embedded in its name have made it a decidedly unattractive comestible. Experiments on human subjects are few, and some current mycologists doubt its toxicity.[35]

The other red-pored boletes, *Boletus pulcherrimus* (=*eastwoodiae*), *B. luridus*, *B. erythropus*, *B. haematinus*, and *B. subvelutipes*, are all considered to be mildly toxic, although documentation of this assertion is fairly scanty. A few, such as *Boletus erythropus*, are regarded as excellent edible species.

Once again, muscarine has been promoted as the *Boletus* toxin that causes gastrointestinal distress. However, although minute quantities of muscarine can be demonstrated in many red-pored boletes, it is never present in pharmacologically significant amounts.

Currently, little is known about the other toxic ingredients or their purported effects other than that they produce gastrointestinal irritation. A compound called bolesatine has been isolated from *Boletus satanas*. It appears to be a glycoprotein very resistant to any form of

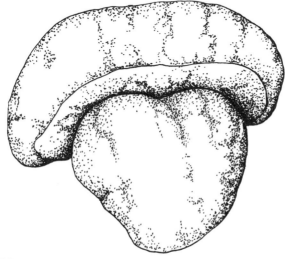

Boletus satanas

proteolysis. It inhibits protein synthesis in cells in vitro.[36] Bolesatine acts on neither DNA nor RNA but appears to inhibit the lengthening of proteins at the level of their assembly on the ribosomes.[37] It is not a classic ribosome-inactivating protein. Its effect can be blocked or inhibited with D-galactose, probably by interfering with bolesatine's binding to the cell wall. It has a molecular weight of 63,000. In addition to its ability to inhibit protein synthesis in the livers and kidneys of mice, in cell-free systems, and in cultured cell lines, bolesatine also agglutinates platelets and red cells, acting like a lectin.[36–39] However, it is difficult to visualize how this activity could play any role in the immediate gastrointestinal toxicity. At this time, it is probably prudent to avoid all red-pored boletes, unless one is willing to become a case report in the medical literature.

In the fall of 1994, a couple developed gastrointestinal distress after eating a meal of *Boletus pulcherrimus*. The wife recovered, but the husband died suddenly. Autopsy demonstrated midgut ischemia and infarction. The pathogenesis of this process is not known. This is the first case of a fatal poisoning by a red-pored bolete of which I am aware.

ENTOLOMA (=RHODOPHYLLUS) SPECIES

Without doubt this is a very poisonous plant, for I once cooked a very small piece of a specimen for luncheon, and was very nearly poisoned to death thereby.

I did not eat a twentieth part of the specimen gathered—I am sure not so much as a quarter of an ounce—and the taste was by no means disagreeable. But mark the result. (It must be borne in mind, too, that though I fell so dangerously ill, I never till the last moment suspected the fungus. Such a confirmed toadstool-eater was I that I laid my symptoms to anything but the true cause.)

About a quarter of an hour after luncheon I left home, and was immediately overtaken by a strange nervous, gloomy, low-spirited feeling quite new to me. Soon a severe headache added its charms to my feelings, and the swimming of the brain commenced, with violent pains in the stomach.

I had now great difficulty to keep upon my legs at all; my senses all appeared leaving me, and every object appeared to be moving with a death-like stillness from side to side, up and down, or round and round.

More dead than alive, I soon returned home, and was horrified to find two others (whom I had invited to partake of my repast) in exactly the same condition as myself. At this moment, and not before, I thought of *Agaricus sinuatus*. These two others had suffered precisely as I had done, and we all three were apparently dying fast. They, however, were attacked by violent vomiting, which I imagine helped their recovery; for after a few

Entoloma sinatum

days of sickness and nausea (with medical assistance) they got well; but it was not so for me; for although I had at first the inclination, I had not the strength left to vomit. During the latter part of the first day I was, however, so continually and fearfully purged, and suffered so much from headache and swimming of the brain, that I really thought that every moment would be my last.

I was very ill for the next four or five days; suffered from loathing and lassitude; fell into a deep sleep, long and troubled; at times found my joints quite stiff; at others, found everything swimming before me; and it was not till a fortnight had elapsed that every bodily derangement had left me.

Charles Badham, *Esculent Funguses of England* (1863)

Potentially toxic species of *Entoloma* include *E. lividum* (=*sinuatum*), *E. rhodopolium, E. mammosum, E. strictius, E. nidrosum, E. vernum,* and *E. pascuum.* Although a number of different species in the genus *Entoloma* are toxic, the one accounting for the majority of serious poisonings is *E. sinuatum* (=*E. lividum,* =*Rhodophyllus sinuatus;* color plate 21). Poisoning by this mushroom is quite common in Europe,[40] accounting for at least 10% of the cases. In Japan, *E. rhodopolium* ranks among the top three species causing mushroom poisoning. It is not clear what edible species these mushrooms are confused with, although members of the genus *Agaricus, Clitopilus prunulus, Leucoagaricus* species, and *Calocybe gambosum* (St. George's mushroom) are all possibilities. Poisoning with *Entoloma* species can be serious, as evidenced by the 70 people treated in hospitals in Geneva alone in 1983.[41] In one of the Swiss victims, the start of the symptoms was delayed for four hours; then, after the usual manifestations of vomiting, diarrhea, and abdominal pain, the patient became unconscious.

Generally, symptoms develop one-half to two hours after the meal and are characterized by vomiting, diarrhea, and headache. These maladies may persist for up to two days. Rare cases of mild liver toxicity have been recorded, as have occasional episodes of psychological disturbance. In these instances, the patients developed a period of depression following an initial phase of euphoria and excitement. Although the depression is generally short-lived, in a few victims it lasted for months.

The most recent information suggests that choline, muscarine, and muscaridine are some of the toxins responsible for the symptoms caused by *E. rhodopolium*.[42]

A purported hemolysin has been identified in some of the toxic *Entoloma* species.[43] It causes diarrhea and death in mice. Its hemolytic activity is destroyed by heating, pepsin, and an acid pH. What role, if any, this toxin plays in human poisoning is currently unknown, although some investigators consider it similar to the bacterial hemolysins produced by organisms such as *Vibrio cholera*, which also have an enterotoxic effect. Interestingly, a number of very potent hemolysins have been identified in a variety of eminently edible mushrooms. Apparently, none of these causes human poisonings, because they either are not absorbed, or are destroyed in the process of cooking, or are inactivated in the gastrointestinal tract. These hemolysins include phallolysin in *Amanita phalloides*, rubescenlysin in the edible *Amanita rubescens* (red blusher),[44] pleurotolysin in the oyster mushroom, flammutoxin in the enoki mushroom (*Flammulina velutipes*),[45,46] and volvatoxin in the common paddy straw mushroom (*Volvariella volvacae*). Administered parenterally to experimental animals, many of these compounds are very potent toxins. Fortunately, this is not the route by which we get our nutrition or our enjoyment.

HEBELOMA CRUSTULINIFORME

Both of the common names of the fungus *Hebeloma crustuliniforme* (poison pie, fairy cake *Hebeloma*) include a culinary insinuation, suggesting that they look good enough to eat (color plate 22). They certainly are not; they cause a rather severe gastrointestinal syndrome.[47] In fact, no *Hebeloma*, of any ilk, should be eaten because several of them give rise to the same misery. This mycorrhizal fungus sometimes grows in large groups or rings. A favorite association is with birch trees, but it occurs under other deciduous trees as well. It may also grow with conifers, especially spruce.

Hebeloma crustuliniforme

The toxins in this group of mushrooms are not known. A variety of compounds have been identified, but their role in human poisoning is currently undetermined.[48–50] In *H. senescens*, unusual sesquiterpenes have been identified.[51] *Hebeloma vinosophyllum* contains a host of neurotoxic triterpene glycosides.[52] Other toxin-containing mushrooms in this group include *H. mesopheum, H. sinapizans*, and *H. spoliatum*.[53]

LACTARIUS SPECIES: THE MILKY CAPS

From their generally assuming a cup or funnel shape, by which the pileus acquires the capability of retaining the rain, the French give them the title of "water drinkers," "eau-boiront," and "Poivres" on account of the pungency the milk possesses in many cases: this pungency, however, has none of the aromatic agreeableness of pepper; it is at first barely perceptible, but presently burns the mouth like Mezeron berries, and in the acrid *rufus* and its dangerous relatives, becomes insupportably violent. This *A. rufus*, our present subject, there is only one opinion about, that it is utterly unfit for the table, and when accidentally taken to the extent of a couple of ounces produced very alarming effects, although timely remedies prevented their proving fatal . . . but though cooking and pickling may remove the evil qualities of the milky agarics by abstracting their acrid juices, let us then rejoice and be thankful that in England we are not under the necessity of trying experiments upon them to appease hunger.

Mrs. T. J. Hussey, *Illustrations of British Mycology* (1847)

Many of the milky caps have a dubious reputation, and a few have resulted in quite impressive poisonings.[54,55] The record must go to *Lactarius helvus,* which, in October 1949, reportedly poisoned 418 people near Leipzig, Germany.[56] They had a very rapid onset of action, with most patients exhibiting the usual gastrointestinal symptoms within 30 minutes. Additional symptoms included a sensation of feeling cold and vertigo.

Lactarius helvus has a strong scent of lovage or celery and has been used in small amounts as a condiment. Such a practice is ill advised.

The edibility and the desirability of the milky caps in general is still controversial. On the one hand, it has been said that all mild-tasting milky caps are edible, but the example of *L. helvus* strongly disputes this dictum. On the other hand, it has also been claimed that all the peppery and pungent milky caps are poisonous, despite their widespread culinary use in eastern Europe. It is probably true that the latter require special preparation, such as pickling or adequate cooking, to make them edible. Many milky caps are first boiled in water (which should be discarded afterward) prior to their final preparation. If not treated in this way, there is little doubt about the symptoms they can provoke.[57] The pungent and peppery components found in *Lactarius* species are largely sesquiterpenes, which tend to be released when the fungus is damaged.[6,58,59] One of these compounds has been given the name isovelleral.[60] Its carbon skeleton bears a striking resemblance to the toxin illudin S, which is found in *Omphalotus illudens* (jack-o'-lantern mushroom). If one samples one of these species by chewing a small portion of a raw specimen and then spitting it out, little effect is noted for a number of seconds; then an intense burning sensation develops in the mouth. Charles Badham's description of *Lactarius piperatus* is illustrative:

> I imagine there are few species in this country more dangerous than this one. So essentially and powerfully acrid is the milk, that if it be allowed to trickle over tender hands it will sting like the contact of nettles; and if a drop is placed on the lips or the tongue, the sensation is like the scalding of boiling water, or the burning of a red-hot iron.

These mushrooms are very irritating to the gastrointestinal tract when eaten raw, and some people even have an unpleasant gastrointestinal syndrome after eating well-cooked or pickled mushrooms. In 1942 a fatality was ascribed to *Lactarius glaucesens;* a 30-month-old child with congenital heart disease died after eating a small amount. Of course, it is difficult to know exactly how much the mushroom contributed to her demise.

In addition to the mushrooms already mentioned, the following *Lactarius* species are all considered to be toxic to various degrees: *L. chryorheus*, *L. pallescens*, *L. pubescens*, *L. representaneus*, *L. scrobiculatus* var. *canadensis*, *L. torminosus*, and *L. uvidus*.

As if this were not enough to condemn many of the milky caps, recent evidence demonstrates mutagenic compounds in some taxa.[61,62] The best-studied species in this regard is *L. necator*, a metabolite of which (necatorin) is actively mutagenic.[63,64]

At present, only those milky caps that exude orange or red latex can still be recommended for the table. These acceptable mushrooms are the delicious milky cap, *L. deliciosus*, and *L. sanguifluus*. There are certainly many other eminently edible species, such as *L. camphoratus* (candy cap), which has an odor of maple syrup. But there is too little information on *L. camphoratus* and other species to make definitive recommendations.

LAETIPORUS SULPHUREUS

The beautiful yellow-orange fungus *Laetiporus sulphureus* (sulphur shelf, chicken-of-the-woods) grows in overlapping shelves on old dead timbers and is one of the delights of the fall woodland (color plate 23). It also appears in late summer; and in the eastern United States and Canada, it may be found growing on oak in the spring. Its common name of chicken-of-the-woods suggests a culinary treat, but most people would only rate it as fair. It has also been associated with significant toxicity. Even our old friend Mrs. Hussey was cautious about this one:

[He] compares the flavour to diluted spirits of vitriol. He says also that the smell, as well as the brilliant yellow colour, are carried off by the spirits of wine. It is strongly purgative, according to the same experimenting authority . . . Nothing can be more beautiful than this Aurora-tinted Fungus; the most dull must be struck with it, the most prejudiced admire it. On the question of utility, which is sure to be asked—that it is not fit for table use we need scarcely state, but it need not therefore be condemned, being probably not more poisonous than medicinal things in general. Whether in that light it be worth attention, we leave to wiser heads.

Mrs. T. J. Hussey, *Illustrations of British Mycology* (1847)

Although generally regarded as edible, *L. sulphureus* has caused distress in a few people, one a child in British Columbia who ate the mushroom raw. This six-year-old girl became disoriented and ataxic about an hour after having eaten a mushroom that was growing on a tree in the garden. She described visual hallucinations with lines, shapes, and bright lights; a rather frightening vision involved a "yellow and orange monster" that floated toward her. She recovered spontaneously, after having her stomach emptied, and was completely well in 20 hours.[65]

Even cooked, this mushroom has been known to cause nausea, vomiting, dizziness, and disorientation.[57] The toxins have not been identified, although *L. sulphureus* is known to contain the phenolethylamine tyramine and its *N*-methyl and *N,N*-dimethyl derivatives (hordenine). These compounds could be responsible for the central nervous system manifestations.[35]

OMPHALOTUS SPECIES

Omphalotus species (jack-o'-lantern mushrooms) poison numerous naive or careless chanterelle pickers because they are the mushrooms most often confused with the edible species.[66-71] In Europe, these mushrooms are known as *O. olearius*; in the United States, *O. illudens* and *O. subilludens* are the local species in the East and *O. olivascens* the species in the West (color plate 25). They are widespread across southeastern Canada, around the Great Lakes, and down the eastern seaboard into the Southeast. At least one of the toxic components isolated from this genus is a sesquiterpene with the trivial name of illudin S. The Japanese relative, *Lampteromyces japonicus* (tsukiyotake, or moon-night mushroom), has also been responsible for many cases of intoxication, because it too contains the toxin illudin S.

These mushrooms all have the property of bioluminescence. They can be found growing on wood and quietly glowing in the dark.

The onset of symptoms is generally between one and three hours. Nausea and vomiting are the most striking features and are associated with abdominal pain, headache, and a sense of exhaustion, weakness, and dizziness. Some patients have increased sweating and salivation, and others complain of a bitter taste in the mouth.

Interestingly, diarrhea is relatively uncommon in patients poisoned by these mushrooms, an observation suggesting that the toxin acts at a central nervous system site rather than as a direct gastrointestinal irritant. However, a more recent series of seven patients casts serious doubt on this notion. Three of the patients developed diarrhea. More worrisome was the fact that the same three also demonstrated a marginal increase in hepatic enzymes.[72]

In another series of 16 patients—campers and counselors poisoned with this mushroom at a summer camp for boys and girls in Maryland in 1987—eight had vomiting, five had diarrhea, and two complained of extreme weakness and sedation. A dose–response relationship was evident, with the sicker patients having eaten larger quantities. A sense of tiredness, weakness, and feeling cold was commented on by another eight. In all these patients, recovery was complete in 18 hours.[71]

It is surprising to hear that the counselors had confused these mushrooms with *Laetiporus sulphureus*, the chicken-of-the-woods, a gill-less, polypore fungus with an entirely different growth habit. Perhaps there really are people who cannot tell the difference between an asparagus shoot and an artichoke.

One large group poisoning in Europe involved 25 patients. It is significant that almost all of the victims were tourists in the area in which they were picking mushrooms.[68] They demonstrated the entire spectrum of symptom—nausea, vomiting, abdominal pain, headache, and fatigue or sleepiness. Some had either perioral paresthesia or tingling at the tips of their fingers. The more severely affected victims also had decreased muscle coordination, visual disturbances, and dysphagia.

Recovery is usually rapid but can be protracted in a few individuals. All patients are better within a week. Rare patients have complained of excessive tiredness for up to a month.

A comparison of the clinical details of the North American and the European patients reveals some distinct differences. For North American victims, the onset was more rapid (15 to 90 minutes), there were fewer, if any, muscarinic effects, and the clinical picture was dominated by nausea (90%), vomiting (73%), and weakness or fatigue (40%). Other symptoms were less common in the North American patients: diarrhea (22%) and sweating (15%).[71,72] The duration of the vomiting was relatively short. The majority of the North American patients were well within 12 hours. It is highly probable that the toxins or their concentrations are different in the various species populating the two continents.

As mentioned earlier, one of the toxins in both the jack-o'-lantern mushroom and its Japanese counterpart is a sesquiterpene, illudin S.[73,74] The illudins are extremely cytotoxic and have even been investigated for their possible role as antitumor agents.[70] Other terpenoids may also be involved.[75] At one time, it was believed that the mushrooms contained either muscarine and/or ergot alkaloids. This has not been confirmed in later studies. The clinical evidence from the cases in the United States does not suggest any contribution from muscarine.

PHOLIOTA SQUARROSA

Those melancholy persons in whom Funguses produce disgust, would feel an instinctive dread of the one now represented: thrusting forth a snaky

Pholiota squarrosa

mass from the stump of a decaying tree, with strange flexous serpentine stems, and bristling scaly coat, it is a decidedly repulsive individual . . . Surely this is a fit associate for witches, bats and owls, as any of its brethren can be, and its "very ancient and fish-like smell," by no means improves the impression received through the sight. It grows at the root of ancient trees in general, though sometimes higher up in the cavities of the decaying trunks. From one central point a tuft of stems proceeds, united at the base, where they are attenuated and compressed in to a very dense mass; as the plant extends, they become variously bent from their direct course of growth, the expansion of the stronger members of the family pushing aside the weaker. The fully matured pileus is from one to five inches across, firm, convex, then expanded, like the boss of a shield (umbo). It is never slimy or glutinous. The colour varies, from tawny to bright yellow, and the whole plant partakes of the same hue. The pileus retains the remains of the veil, in the form of tufted dark scales, which are recurved, giving it a bristling appearance.

Mrs. T. J. Hussey, *Illustrations of British Mycology* (1847)

With this rather gruesome, albeit very accurate, description of this widespread mushroom, it may seem to be an unlikely constituent for the stew pot, but it is on occasions. *Pholiota squarrosa* is listed as edible in a number of the standard field guides (color plate 27). However, a small percentage of individuals have a mild to moderate adverse gastrointestinal reaction to this species.[76] A few other species in the same group have also been indicted, including *P. limonella* and *P. highlandensis.*

RUSSULA SPECIES

It gives no warning by its scent or by any other external circumstances of its deleterious quality; if the ignoramus should be tempted to taste, for a few moments all appears harmless, for it is tardily acrid: but it fully makes up for the delay, as the tortured investigator, with burning lips and fauces, and tearful eyes, seeks in vain for alleviation. If not swallowed, however, the effect shortly subsides; it is an agaric unlikely to prove fatal, because the acrimony is not lost in cooking, and they who could eat enough to harm themselves seriously, must have fire-proof palates.

Mrs. T. J. Hussey, *Illustrations of British Mycology* (1847)

Russula emetica (the sickener, the emetic *Russula*) is perhaps the prototype of mushrooms with a burning peppery, pungent taste (color plate 26). In general, members of the genus *Russula*, despite being some of the most prevalent mushrooms in temperate woodlands and forests, have only a limited culinary role. The Russians pickle or marinate *Russula brevipes*, and a few people in the United States eat *Russula cyanoxantha*. The rest are largely overturned in the hope of finding something truly edible or are actively destroyed by the frustrated mushroom picker—the motto "better kicked than picked" has been well applied to this genus.

The peppery or bitter-tasting *Russula* species can cause severe gastrointestinal upset if eaten raw. This rarely occurs, because of these most unpleasant properties. However, it is not uncommon for amateur mycologists to chew on a small portion of a cap as part of the identification process. Even though this practice is generally harmless—assuming, of course, that the piece of mushroom is discarded and not swallowed—vomiting and diarrhea developed in a mycologist who sampled a large number of specimens during one foray. His repeated sampling was apparently sufficient to produce the distress.

The cause of the gastrointestinal symptoms produced by *Russula* species has not been fully documented. It is most likely sesquiterpenes, which are similar to those found in the genus *Lactarius*, to which the genus *Russula* is closely allied. Sesquiterpenes have been identified in a number of different *Russula* species, for example, *Russula sardonia.*[77] It has long been believed that muscarine plays a role in the toxicity of *Russula* species, because some of the early biological assays and chromatographic studies demonstrated its presence in collections. However, this belief has been contradicted by other studies. At present, muscarine's contribution, if any, is unclear and probably negligible.

SUILLUS SPECIES

The slippery jacks, in all their guises, are a common food source for many eastern Europeans, both in their native homelands and in their adopted country, the United States. Almost all mushroom field guides list them as edible, although some mention that they may have a cathartic effect on some individuals. Perhaps that is why they are eaten—for their natural laxative activity, because the effect seems to be more frequent than advertised. Lincoff[11] recommends removal of the pellicle prior to eating this mushroom. There is at least one case documenting the usefulness of this precaution.

The sacrificial lamb in this instance was a 31-year-old male who had eaten *Suillus luteus* on many occasions,[78] always peeling the glutinous cuticle off the surface of the cap. These mushrooms had never given him any problems. On one occasion, he came upon a collection of young, firm specimens in such excellent condition that he decided not to peel them, because the caps, while still glutinous, did not resemble slug slime, as they are wont to do. He developed watery diarrhea 15 minutes after eating a meal. Recurrent episodes continued over the next two days—each time he ate a meal of the mushrooms. This was not associated with any other symptoms such a nausea, vomiting, or pain. Because he did not draw any connection between the mushrooms and the diarrhea at the time, he dried some for later consumption. A few months later, when he used the stored mushrooms, the diarrhea instantly recurred. At this point, he became suspicious of the mushrooms and was astute enough to continue the experiment. He ate some broth in which he had cooked the mushrooms and found it had no effect. When he ate a couple of slices of the mushroom, however, the diarrhea returned. From this single, not particularly well controlled experiment, we can conclude that the toxin, whatever it is, is nonvolatile, heat stable, not readily extracted from the mushrooms by boiling, and can withstand drying.

TRICHOLOMA PARDINUM AND RELATIVES

Tricholoma pardinum (tiger or spotted *Tricholoma*) is a very common cause of poisoning in Europe. For a time in Switzerland, it accounted for over 20% of the intoxications.[12] Although nothing is known about the toxin in this mushroom, it produces both vomiting and diarrhea, which can be severe.[79,80] Certain animals are equally sensitive to this mushroom. Its potency is illustrated by the cat and seven people who became violently ill after sharing a dish containing only a couple of mushroom caps.[81,82] The symptoms develop one-quarter to two hours after the meal and can be accompanied by dizziness and malaise. As with many of the gastrointestinal syndromes, some patients exhibit signs of anxiety, including sweating. Rapid recovery is the norm, although patients may not feel quite well for four to six days.

A number of other *Tricholoma* species have also been accused of causing illness, for example, *T. venenatum* (which is a close relative of *T. pardinum*), *T. pessundatum*, *T. inamoenum*, and *T. zelleri*.

RARELY EATEN TOXIC MUSHROOMS

The following list identifies rarely eaten mushrooms that reportedly are responsible for a variety of different symptoms, some of which are not gastrointestinal in character. However, they are included here for completeness. The references provide additional clinical details, although in many cases the medical features are rather poorly described and are based on only a few case reports.

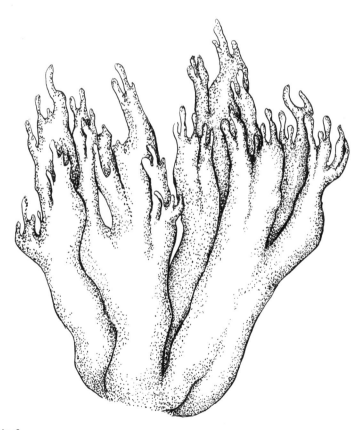

Ramaria formosa

Formal and common names	Symptoms
Boletus pulcherrimus[83]	Muscarinic effects
Clathrus cancellatus[84] (=*C. columnatus*)	Convulsions, dysarthria, coma
Collybia dryophila[85]	Gastrointestinal upset
Collybia acervata	Gastrointestinal upset
Gomphus floccosus[27,86,87] (woolly chanterelle)	Nausea, vomiting, diarrhea (CNS depression, muscle weakness in rats). Onset can be delayed. May be due to norcaperatic acid.
Gomphus kauffmanii[88]	Same as for *G. floccosus*
Mycena pura[89,90] (lilac mycena)	Muscarinic symptoms
Paxillus involutus[91–93] (roll-rim fungus)	Gastrointestinal symptoms; immune hemolytic anemia (see Chapter 20)
Phaeolepiota aurea[94,95]	Vomiting, diarrhea, and colicky abdominal pain
Ramaria formosa[35] (yellow-tipped coral fungus)	Gastrointestinal syndrome
Ramaria pallida[81] (Bauchweh-Koralle, colic coral)	Gastrointestinal syndrome
Ramaria flavobrunnescens[96]	Death of livestock in South America
Scleroderma cepa[97,98] and *S. aurantium* (thick-skinned pigskin, common earth-ball)	Abdominal pain, nausea, generalized tingling sensation; marked spasms, rigidity
Stropharia coronilla[99]	Malaise, headache, generalized aching, ataxia, dizziness, vomiting, hallucinations, confusion
Tricholomopsis platyphylla[100,101]	Colicky abdominal pain, vomiting and/or diarrhea, muscle cramps and spasms
Tricholoma irinum[102] *Tricholoma sulphurum*[103] (sulfurous or gas agaric)	Nausea, vomiting, neurological symptoms

REFERENCES

1. R. G. Benedict, "Mushroom Toxins Other Than *Amanita*." In *Microbial Toxins*, Vol. 8, ed. S. Kadis, A. Ciegler, and S. J. Ajl (New York: Academic Press, 1972), pp. 281–320.

2. K. Lampe, "Pharmacology and Therapy of Mushroom Intoxications." In *Mushroom Poisoning: Diagnosis and Treatment*, ed. B. H. Rumack and E. Salzman (West Palm Beach, FL: CRC Press, Inc., 1978), pp. 125–169.

3. M. K. Wassef, "Fungal Lipids," *Advances in Lipid Research* (1977) 15:159–232.

4. A. Bresinsky, "Chemotaxonomie der Pilze." In *Beiträge zur Biologie der niederen Pflanzen*, ed. W. Frey, H. Hurka, and F. Oberwinkler (Stuttgart and New York: Gustav Fischer, 1977).

5. W. A. Ayer and L. M. Browne, "Terpinoid Metabolites of Mushrooms and Related Basidiomycetes," *Tetrahedron* (1981) 37:2199–2248.

6. E. Barreto-Bergter and P. A. J. Gorin, "Structural Chemistry of Polysaccharides from Fungi and Lichens," *Advances in Carbohydrate Chemistry and Biochemistry* (1983) 41:61–103.

7. S. Shibata, S. Natori, and S. Udagawa, *List of Fungal Products* (Springfield, IL: Thomas, 1964), p. 170.

8. W. Steglich, "Pigments of Higher Fungi (Macromycetes)." In *Pigments in Plants*, 2nd ed., ed. F.-C. Czygan (Stuttgart and New York: Gustav Fischer, 1980), pp. 393–412.

9. W. B. Turner, *Fungal Metabolites* (New York: Academic Press, 1971), p. 446.

10. W. B. Turner and D. Aldridge, *Fungal Metabolites II* (London: Academic Press, 1983), p. 631.

11. G. H. Lincoff and D. H. Mitchel, *Toxic and Hallucinogenic Mushroom Poisoning* (New York: Van Nostrand Reinhold, 1977).

12. A. E. Alder, "Pilzvergiftungen in der Schweiz während 40 Jahren," *Schweizerische Zeitschrift für Pilzkunde* (1960) 38:65–73.

13. D. A. Reid and A. Eicker, "A Comprehensive Account of *Chlorophyllum molybdites*," *Botanical Bulletin of Academia Sinica (Taipei)* (1991) 32(4):317–334.

14. C. O. Smith, "*Lepiota morgani* in Southern California," *Mycologia* (1936) 28:86.

15. P. W. Graff, "The Green-spored *Lepiota*," *Mycologia* (1927) 19:322–326.

16. F. I. Eilers and L. R. Nelson, "Characterization and Partial Purification of the Toxin of *Lepiota morganii*," *Toxicon* (1974) 12:557–563.

17. P. H. Stenklyft and W. L. Augenstein, "*Chlorophyllum molybdites*—Severe Mushroom Poisoning in a Child," *Journal of Toxicology and Clinical Toxicology* (1990) 28:156–159.

18. D. Blayney, E. Rosenkranz, and A. Zettner, "Mushroom Poisoning from *Chlorophyllum molybdites*," *Western Journal of Medicine* (1980) 132(1):74–77.

19. C. W. Smith, "Mushroom Poisoning by *Chlorophyllum molybdites* in Hawaii," *Hawaii Medical Journal* (1980) 39(1):13–14.

20. G. R. B. MacCarter, "Mushroom Poisoning in Rhodesia with a Report of a Case of Poisoning due to *Lepiota morgani*," *Central Africa Journal of Medicine* (1959) 5(8):412.

21. G. R. Whitaker and J. F. Box, "*Chlorophyllum molybdites* Mushroom Poisoning: A Case Report and Review of the Syndrome," *Journal of the Arkansas Medical Society* (1985) 82:220–222.

22. U. K. Chestnut, "Poisonous Properties of the Green-spored *Lepiota*," *Asa Gray Bulletin* (1900) 8:87–93.

23. P. F. Lehmann and U. Khazan, "Mushroom Poisoning by *Chlorophyllum molybdites* in the Midwest United States. Cases and a Review of the Syndrome," *Mycopathologia* (1992) 118(1):3–13.

24. D. Levitan, J. I. Macy, and J. Weissman, "Mechanism of Gastrointestinal Hemorrhage in a Case of Mushroom Poisoning by *Chlorophyllum molybdites*," *Toxicon* (1981) 19(1):179–180.

25. T. Young, "Poisonings by *Chlorophyllum molybdites* in Australia," *The Mycologist* (1989) 3(1):11–12.

26. D. P. Rogers, "Identification of Edible and Poisonous Mushrooms," *Transactions of the New York Academy of Sciences* (1957) 19:545–547.

27. A. H. Smith, *Mushrooms in Their Natural Habitats* (Portland, OR: Sawyer's, 1949).

28. M. McKenny and D. Stuntz, *The Savory Wild Mushroom* (revised by J. Ammirati) (Seattle: University of Washington Press, 1990).

29. M. Gill and R. J. Strauch, "Constituents of *Agaricus xanthodermus* Genevier: The First Naturally Endogenous Azo Compound and Toxic Phenolic Metabolites," *Zeitschrift Naturforschung* (1984) 39c:1027–1029.

30. K. Dornberger, W. Ihn, W. Schade, D. Tresselt, A. Zureck, and L. Radics, "Antibiotics from Basidiomycetes. Evidence for the Occurrence of the 4-Hydroxy-benzenediazonium Ion in the Extracts of *Agaricus xanthodermus* (Agaricales)," *Tetrahedron Letters* (1986) 27:559–560.

31. S. Hilbig, W. Andries, W. Steglich, and T. Anke, "The Chemistry and Antibiotic Activity of the Toadstool *Agaricus xanthoderma* (Agaricales)," *Angewandte Chemie, International Edition in English* (1985) 24:1063–1065.

32. J. Kubicka and J. Veselsky "Über die bedingt toxische Wirkung einiger Speisepilze," *Südwestdeutsches Pilzrundschau* (1981) 17(1):12–14.

33. S. L. Midland, R. R. Izak, R. M. Wing, A. I. Zaki, D. E. Munnecke, and J. J. Sims, "Melleolide, a New Antibiotic from *Armillaria mellea*," *Tetrahedron Letters* (1982) 23:2515–2518.

34. O. Schmeideberg and R. Koppe, *Das Muskarin, das giftige Alkaloid des Fliegenpilzes (Agaricus muscarius)* (Leipzig: Vogel, 1869), p. 111.

35. A. Bresinsky and H. Besl, *A Colour Atlas of Poisonous Fungi* (London: Wolfe Publishing, 1990).

36. O. Kretz, E. E. Creppy, and G. Dirheimer, "Disposition of the Toxic Protein Bolesatine in Rats: Its Resistance to Proteolytic Enzymes," *Xenobiotica* (1991) 21(1):65–73.

37. O. Kretz, E. E. Creppy, and G. Dirheimer, "Characterization of Bolesatine, a Toxic Protein from the Mushroom *Boletus satanas* Lenz and Its Effect on Kidney Cells," *Toxicology* (1991) 66(2):213–224.

38. O. Kretz, E. E. Creppy, Y. Boulanger, and G. Dirheimer, "Purification and Some Properties of Bolesatine, a Protein Inhibiting in vitro Protein Synthesis, from the Mushroom *Boletus satanas*," *Lenz (Boletaceae) Archives of Toxicology Supplement* (1989) 13:422–427.

39. O. Kretz, L. Barbieri, E. E. Creppy, and G. Dirheimer, "Inhibition of Protein Synthesis in Liver and Kidney of Mice by Bolesatine: Mechanistic Approaches to the Mode of Action at the Molecular Level," *Toxicology* (1992) 73(3):297–304.

40. A. E. Alder, "Erkennung und Behandlung der Pilzvergiftung," *Deutsche Medizinische Wochenschrift* (1961) 86(23):1121–1127.

41. J.-R. Chapuis, "Jahresbericht des Verbandstoxikologen für das Jahr 1983," *Schweizerische Zeitschrift für Pilzkunde* (1984) 62:196–197.

42. T Maki, K. Takahashi, and S. Shibata, "Isolation of the Vomiting Principles from the Mushroom *Rhodophyllus rhodopolius*," *Journal of Agricultural and Food Chemistry* (1985) 33:1204–1205.

43. K. Suzuki, T. Une, M. Yamazaki, and T. Takeda, "Purification and Some Properties of a Hemolysin from the Poisonous Mushroom *Rhodophyllus rhodopolius*," *Toxicon* (1990) 28(9):1019–1028.

44. K. P. Odenthal, R. Seeger, R. Braatz, E. Petzinger, H. Moshaf, and D. C. Schmitz, "Damage in vitro to Various Organs and Tissues by Rubescenslysin from the Edible Mushroom *Amanita rubescens*," *Toxicon* (1982) 20(4):765–781.

45. J. Y. Lin, Y. J. Lin, C. C. Chen, H. L. Wu, G. Y. Shi, and T. W. Jeng, "Cardiotoxic Protein from Edible Mushrooms," *Nature* (1974) 252:235–237.

46. J. Y. Lin, "Studies on a New Cardiotoxin Isolated from the Edible Mushroom *Volvariella volvaceae* and *Flammulina velutipes*," *Toxicon* (1975) 13(2):107–111.

47. H. W. Price, "Mushroom Poisoning due to *Hebeloma crustuliniforme*," *American Journal of Diseases of Childhood* (1927) 34:441–442.

48. H. Fujimoto, K. Haginiwa, and K. Suzuki, "Survey of the Toxic *Hebeloma* and Isolation of the Toxic Principle of *H. vinosophyllum*," *Transactions of the Mycological Society of Japan* (1982) 23:405–412.

49. H. Fujimoto, K. Haginiwa, K. Suzuki, and M. Yamazaki, "New Toxic Metabolites from a Mushroom, *Hebeloma vinosophyllum*. II. Isolation and Structures of Hebevinosides VII–XI," *Chemistry and Pharmaceutical Bulletin* (1987) 35:2254–2260.

50. M. De Bernardi, G. Fronza, G. Gianotti, G. Mellerio, G. Vidari, and P. Vita-Finzi, "New Cytotoxic Triterpene from *Hebeloma* Species (Basidiomycetes)," *Tetrahedron Letters* (1983) 24:1635–1638.

51. M. Bocchi, L. Garrlaschelli, G. Vidari, and G. Mellerio, "New Farnesane Sesquiterpenes from *Hebeloma senescens*," *Journal of Natural Products (Lloydia)* (1992) 55(4):428–431.

52. H. Fujimoto, K. Maeda, and M. Yamzaki, "New Toxic Metabolites from a Mushroom, *Hebeloma vinosophyllum*: III. Isolation and Structures of Three New Glycosides, Hebevinosides XII, XIII, and XIV, and Productivity at Three Growth Stages of the Mushroom," *Chemistry and Pharmaceutical Bulletin (Tokyo)* (1991) 39(8):1958–1961.

53. H. Fujimoto, Y. Takano, and M. Yamazaki, "Isolation, Identification and Pharmacological Studies on Three Toxic Metabolites from a Mushroom, *Hebeloma spoliatum*," *Chemistry and Pharmaceutical Bulletin (Tokyo)* (1992) 40(4):869–872.

54. H. Goldman, "Über Vergiftigen mit dem Giftpilz *Agaricus torminosus*," *Wiener Klinische Wochenschrift* (1901) 14:279.

55. V. K. Charles, "Mushroom Poisoning Caused by *Lactaria glaucesens*," *Mycologia* (1942) 34:112–113.

56. G. Klemm, "Beobachtungen über den Verlauf einer Massenvergiftung mit dem Bruchkreizker *Lactarius helvus* Fries," *Mycologisches Mitteilungsblatt* (1961) 5:1–4.

57. D. N. Pegler and R. Watling, "British Toxic Fungi," *Bulletin of the British Mycological Society* (1982) 16:66–75.

58. M. De Bernardi, M. A. Girometta, G. Mellerio, G. Vidari, and P. Vita-Finzi, "Observation on Russulaceae Chemotaxonomy," *Micologia Italiana* (1982) 11:25–37.

59. O. Sterner, R. Bergman, L. Kesler, J. Nilsson, J. Oluwadiya, and B. Wickberg, "Velutinal Esters of *Lactarius vellereus* and *L. necator*. The Preparation of Free Velutinal," *Tetrahedron Letters* (1983) 24:1415–1418.

60. P. H. List and H. Hackenberg, "Die scharf schmeckenden Stoffe von *Lactarius vellereus* Fries," *Zeitschrift für Pilzkunde* (1973) 39:97–102.

61. O. Sterner et al., "Mutagens in Larger Fungi. I. Forty-eight Species Screened for Mutagenic Activity in the *Salmonella*/Microsome Assay," *Mutation Research* (1982) 101:269–281.

62. J. Knuutinen and A. von Wright, "The Mutagenicity of *Lactarius* Mushrooms," *Mutation Research* (1982) 103:115–118.

63. O. Sterner, R. Bergman, C. Franzén, E. Kesler, and L. Nilsson, "The Mutagenicity of Commercial Pickled *Lactarius necator* in the *Salmonella* Assay," *Mutation Research* (1982) 104:233–237.

64. A. von Wright and T. Suortti, "Preliminary Characterization of the Mutagenic Properties of 'Necatorin,' a Strongly Mutagenic Compound of the Mushroom *Lactarius necator,*" *Mutation Research* (1983) 121(2):103–106.

65. R. E. Appleton, J. E. Jan, and P. D. Kroeger, "*Laetiporus sulphureus* Causing Visual Hallucinations and Ataxia in a Child," *Canadian Medical Association Journal* (1988) 139(1):48–49.

66. S. T. Carey, "*Clitocybe illudens*: Its Cultivation, Chemistry and Classification," *Mycologia* (1974) 66:951.

67. Z. Maretic, "Poisoning by the Mushroom *Clitocybe olearia* Maire," *Toxicon* (1967) 4:263–267.

68. Z. Maretic, F. E. Russel, and V. Golobic. "Twenty-five Cases of Poisoning by the Mushroom *Pleurotus olearius,*" *Toxicon* (1975) 13:379–381.

69. F. J. Seaver, "Poisoning with *Clitocybe illudens,*" *Mycologia* (1939) 31:110.

70. T. C. McMorris, M. J. Kelner, W. Wang, S. Moon, and R. Taetle, "On the Mechanism of Toxicity of Illudins: The Role of Glutathione," *Chemical Research in Toxicology* (1990) 3:574–579.

71. A. L. French and L. K. Garrettson, "Poisoning with the North American Jack-o'-Lantern Mushroom, *Omphalotus illudens,*" *Journal of Toxicology and Clinical Toxicology* (1988) 26(1–2):81–88.

72. T. L. Vanden Hoek, T. Erickson, D. Hryhorczuk, and K. Narasimhan, "Jack-o'-Lantern Mushroom Poisoning," *Annals of Emergency Medicine* (1991) 20(5):559–561.

73. K. Nakanishi, M. Ohashi, M. Tada, and Y. Yamada, "Illudin S (Lampterol)," *Tetrahedron* (1965) 21:1231–1246.

74. T. C. McMorris and M. Anchel, "The Structure of the Basidiomycete Metabolites Illudin S and Illudin M," *Journal of the American Chemical Society* (1963) 85:831–832.

75. W. A. Ayer and L. M. Browne, "Terpenoid Metabolites of Mushrooms and Related Basidiomycetes," *Tetrahedron* (1981) 37:2199–2248.

76. R. L. Shaffer, "Poisoning by *Pholiota squarrosa,*" *Mycologia* (1965) 57:318–319.

77. D. Andina et al., "Sesquiterpenes from *Russula sardonia,*" *Phytochemistry* (1980) 19:93–97.

78. M. H. Prager and R. D. Goos, "A Case of Poisoning from *Suillus luteus,*" *Mycopathologia* (1984) 85:175–176.

79. G. Dittrich, "Über Vergiftigen durch Pilze der Gattungen und Tricholoma," *Berichte der Deutschen Botanischen Gesellschaft* (1918) 36:456–459.

80. F. Begenat, "Die 'grundfalschen' Nebelkappen," *Südwestdeutsche Pilzrundschau* (1974) 10(1):5–6.

81. A. E. Alder, "Die Pilzvergiftungen in der Schweiz in Jahre 1948," *Schweizerische Zeitschrift für Pilzkunde* (1950) 28(8):122–132.

82. A. E. Alder, "Neuere Beobachtungen und Erfahrungen bei Pilzvergiftungen," *Schweizerische Zeitschrift für Pilzkunde* (1950) 28(1):6–10.

83. J. R. Kienholz, "A Poisonous *Boletus* from Oregon," *Mycologia* (1934) 26:275–276.

84. W. G. Farlow, "Poisonous Action of *Clathrus columnatus,*" *Botanical Gazette* (1890) 15:45–46.

85. H. Hendrickson, "Poisonings by Edible Mushrooms," *Spore Prints (Bulletin of the Puget Sound Mycological Society)* (1985) January issue, p. 4.

86. J. Miyata, V. E. Tyler, L. R. Brady, and M. H. Malone, "The Occurrence of Norcaperatic Acid in *Cantharellus floccosus,*" *Lloydia* (1966) 29:43.

87. E. D. Henry and G. Sullivan, "Phytochemical Evaluation of Some Cantherelloid Fungi," *Journal of Pharmaceutical Sciences* (1969) 58:1497–1500.

88. R. A. Carrano and M. H. Malone, "Pharmacologic Study of Norcaperatic and Agaric Acids," *Journal of Pharmaceutical Sciences* (1967) 56:1611–1614.

89. J. Kubicka and J. Veselky, *"Mycena rosea,"* (Bull.) ex Sacc. et Dalla Costa ist giftig. *Česká Mykologie* (1978) 32:167–168.

90. G. J. Krieglsteiner and H. Schwöbel, *"Mycena diosma* spec. nov. und der *Mycena pura* Formenkreis in Mitteleuropa," *Zeitschrift für Mykologie* (1982) 48:25–34.

91. J. Kubicka and J. Veselsky, "Die Schädlichkeit des Kahlen Kremplings–*Paxillus involutus* (Batsch ex Fr.) Fr.," *Mycologisches Mitteilungsblatt* (1975) 19:1–5.

92. S. Cochet, "Notes sur Divers cas d'empoisonnements Mortels par le *Paxillus involutus* (Batsch) Fr.," *Bulletin de la Société Mycologique de France* (1974) 90:66–67.

93. S. Grzymala, "Vergiftungen durch *Paxillus involutus* (Batsch)," *Zeitschrift für Pilzkunde* (1958) 24:19–21.

94. V. L. Wells and P. E. Kempton, "Studies on the Fleshy Fungi of Alaska: I," *Lloydia* (1967) 30:258–268.

95. V. L. Wells and P. E. Kempton, *"Togaria aurea in Alaska,"* *Mycologia* (1965) 57:316–318.

96. O. Fidalgo and M. E. P. K. Fidalgo, "A Poisonous *Ramaria* from Southern Brazil," *Rickia Arquivos de Botanica do Estado de Sao Paulo Serie Cryptogamica* (1970) 5:71–91; *Biological Abstracts* (1972) 54:6564.

97. J. A. Stevenson and R. C. Benjamin, *"Scleroderma* Poisoning," *Mycologia* (1961) 53(4):438–439.

98. M. W. Beug, "A Case of *Scleroderma* Poisoning in the Pacific Northwest," *McIlvainea* (1984) 6(2):33.

99. H. W. Thomas, D. H. Mitchell, and B. H. Rumack, "Poisoning from the Mushroom *Stropharia coronilla* (Bull. ex Fr.) Quel.," *Journal of the Arkansas Medical Society* (1977) 73:311–312.

100. R. D. Goos and C. R. Schoop, "A Case of Mushroom Poisoning Caused by *Tricolomopsis platyphylla,"* *Mycologia* (1980) 72(2):433–435.

101. R. D. Goos, "Another Case of Mushroom Poisoning Involving *Tricholomopsis platyphylla,"* *Mycologia* (1984) 76:350–351.

102. A. H. Smith, *"Tricholoma irinum* in Michigan," *Michigan Botany* (1962) 1:51.

103. J. Veselský and J. Dvorák, "Über den Verlauf einer Vergiftung durch den Schwefelgelben Ritterling—*Tricholoma sulphureum* (Bull. ex Fr.) Kumm.," *Česká Mykologie* (1981) 35:114–115.

CHAPTER TWENTY

MISCELLANY
OF TOXINS

A VARIETY OF MUSHROOMS with unique toxicity produce syndromes not easily classified into any of the other groups. Some of these are no more than curiosities for the average mycophagist or physician, because neither is likely to encounter them very often. However, they afford us a glimpse of some of the unusual compounds mushrooms synthesize, they teach us about many aspects of human biology, and perhaps they will help us uncover substances that are potentially exploitable in biomedical science for therapy or research. As one might expect, none of these species have been well studied. Their toxicity is a fruitful area for future investigation.

AMANITA SMITHIANA
(=AMANITA SOLITARIA):
RENAL FAILURE

Although many of the older field guides suggest that *Amanita smithiana* contains isoxazole derivatives, the most recent evidence does not support that contention. This white *Amanita* can readily be misidentified by the neophyte matsutake (*Tricholoma magnivalere*) hunter (color plate 29). Although it bears a superficial resemblance to matsutake, it has a long tapering stem and lacks the distinctive aroma of the matsutake, two traits that should be sufficient to distinguish it from the edible look-alike. Five reasonably well documented poisonings have been ascribed to *A. smithiana*. Two of the five patients developed renal failure, and a third, both renal and hepatic failure.[1] The first two recovered normal renal performance, but the third never regained normal kidney function, although the liver did recover. Orellanine was claimed to be

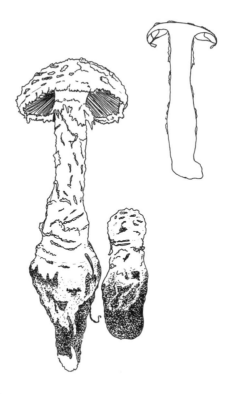

Amanita smithiana

present in one patient despite the lack of a history of *Cortinarius* inges-tion. This observation has not been confirmed or validated. Whether or not *A. smithiana*-induced renal failure is related to orellanine still remains to be determined. It seems most unlikely in light of studies demonstrating a different spectrum of activity against three renal epithelial cell lines.[2]

In three of the *A. smithiana* patients, symptoms of vomiting, abdominal pain, and diarrhea developed between 4 and 11 hours after mushroom ingestion. It is entirely probable that *A. smithiana* contains a renal epithe-lial toxin in addition to other compounds responsible for a gastro-intestinal syndrome. At this time, evidence is accumulating that the toxin or toxins may be similar to those described in *Amanita abrupta*—pentynoic acid and allenic norleucine (2-amino-4,5-hexadienoic acid). The structures of the toxins are unusual in nature. Although studies are still incomplete, they may eventually represent a new class of poisons in the mushroom toxin realm. Although poisoning by these mushrooms is quite unusual, there is great interest in the novel putative compounds.

AURICULARIA AURICULA: EASY BRUISING AND EXCESSIVE BLEEDING

Auricularia auricula (black tree fungus, Jew's ear fungus, tree ear, wood's ear, Judas's ear fungus) is one of the oldest recorded fungi in the West's materia medica (color plate 28). It was recommended for use in soothing sore throats. Its common name arises from the Biblical story of Judas; after leaving the Last Supper, it was said that he hanged himself in a beech tree. Where he had died, a fungus grew in the shape reminiscent of an ear.

This mushroom has a firm, jellylike consistency and has never been much favored in Western cuisine. In China, however, where it is called mo-er, it is a very popular fungus and is used in a variety of dishes. Although it has little intrinsic flavor, it adds an interesting texture to the dish and carries all the other flavors well. Of course, according to the Chinese, it also has medicinal benefits. Some people call it the black Chinese fungus. It is extensively cultivated on logs, and large quantities are imported from other countries to satisfy the demand. When dried, it has a rather unprepossessing appearance, being a small, dark brown to black, irregular and brittle structure without any specific form. It looks like a black, dysmorphic potato chip. When rehydrated in warm water, it swells immensely, still maintaining its flat, irregular, formless shape, but becoming rubbery in the process. It is usually not used in intact pieces but is cut into thin slices or strips. Common dishes containing this fungus are hot and sour soup and moo-shu pork. These are generally banquet-style dishes, and this mushroom is not eaten on a daily basis by the population at large.

In the late 1970s and early 1980s, there was a change in the Chinese restaurant scene in the United States from the all-pervasive Cantonese-style restaurants to those serving the foods of Szechwan, Hunan, Peking, Shanghai, and elsewhere. As this transition progressed, a few patients sought treatment because of small, blotchy hemorrhages in the skin. This condition has subsequently been dubbed the Szechwan restaurant syndrome or Szechwan purpura.

A Minnesotan hematologist astutely connected the hemorrhagic condition to the black tree fungus in their meals.[3] Dr. Hammerschmidt had been working on platelets at the time of his discovery. One day, during an experiment in which he was using his own blood, he noted that his platelets did not respond as anticipated. He knew that he was not taking any medicines, such as aspirin, which could affect platelet function. On the morning of the experiment he had also suffered a nosebleed and he had been unable to stop the bleeding associated with a shaving nick for an unduly long time. The night before, he had had generous quantities of ma-po dofu, a spicy bean curd dish containing generous amounts of both *Auricularia auricula* and sar quort, a water chestnut substitute. He found that extracts of both foods inhibited platelets in the test tube but that only the fungus was active when taken orally.[4]

Further work has suggested that the effect is partly due to adenosine and changes in the cyclic AMP that is involved in platelet aggregation.[5] Other inhibitory compounds are almost certainly present but have yet to be identified. The abnormality in platelet function lasts for a few days before returning to normal. The activity of different collections of this fungus is variable, some showing almost no activity at all. The likelihood of anyone developing spontaneous bleeding is very low, unless they are truly gluttonous eaters. However, women may note that their menstrual bleeding is a little heavier after a meal generously laced with *A. auricula*.

The traditional role for *A. auricula* in China is interesting. It has been recommended as a longevity tonic, and at least in some regions of the country it is used to treat blood clotting and inflammation of the veins in the legs after delivery (postpartum thrombophlebitis).[4] Conversely, it is also employed in a number of conditions in which bleeding is a problem. It is not known whether its antiinflammatory effect is more beneficial than its effects on platelets, although this is certainly the case for drugs like aspirin. Ginger, another staple of Oriental cuisine, has also been shown to have platelet-inhibiting properties.[6]

HYPHOLOMA FASCICULARE (=NAEMATOLOMA FASCICULARE)

This mushroom grows in large clusters from a common base on dead or rotting wood and is a ubiquitous sight in any woodland setting. Its common name is the sulfur tuft (color plate 31). It has always had a shady reputation despite attempts by the likes of Charles McIlvaine to reform it for the table.[7] For many years, mushroom field guides vacillated about its toxic potential. It was commonly known that most specimens were bitter in flavor, which in itself dissuaded people from eating it. In the past two decades, however, it has become evident that *H. fasciculare* is definitely a toxic species. Seldom is it picked for itself, but it may be inadvertently plucked off its host wood because it has been mistaken for the honey mushroom (*Armillaria mellea*) or the edible *Naematoloma capnoides*.

Although no *H. fasciculare* poisonings have been recorded in the United States, many examples from both Europe and Japan have been reported.[8] A characteristic feature of this poisoning is the long latent period—from 5 to 10 hours—prior to the onset of nausea, vomiting, diarrhea, proteinuria, and possible collapse. Impaired vision and paralysis also have been described in connection with this poisoning.[9] The symptoms gradually improve over the ensuing days. One recorded fatality, caused by a fulminant hepatitislike disorder, resembled amatoxin poisoning.[10] The autopsy demonstrated acute hepatic injury and fatty infiltration of the kidneys, myocardium, and cerebral ganglion cells. However, this patient had eaten a mixture of mushrooms, including specimens of *Leccinum* and *Russula* species, and perhaps others, thus making it difficult to place all the blame on the sulfur tuft.

Compounds that inhibit calmodulin have been found in *H. fasciculare*. Calmodulin is a protein that plays a key role in cellular homeostasis and cell signaling. The inhibitory compounds have been called fasciculic acids.[11,12] A number of other metabolites also have been described, but it is unlikely that they play a role in human poisoning.[13-15]

PANAEOLUS FOENISECII

Panaeolus (=*Panaeoline*, =*Psathyrella*) *foenisecii* is called haymaker's mushroom and mower's mushroom (color plate 13). It is a common mushroom on lawns and golf courses and in gardens, parks, cemeteries, and pastures. It is often evanescent, standing erect in the morning and wilted or gone by the end of the day. Some collections of this species contain hallucinogenic compounds, but their presence is variable from one collection to the next. This mushroom is reportedly toxic for children who sample it in the raw state.[16-18] In all reported cases, the toxicity is compatible with a psychotropic compound. However, at present, its mechanism of action is not known. Some mycologists doubt the toxicity of this particular mushroom and suggest that it has been confused with other members of the genus *Panaeolus*, some of which are known to contain psilocybin.[19]

PAXILLUS INVOLUTUS AND THE PAXILLUS SYNDROME: IMMUNE HEMOLYTIC ANEMIA

The syndrome of an acute hemolytic anemia associated with mushroom ingestion has only been recognized in the past few decades and is quite rare.[20-24] As far as I can ascertain, no confirmed cases have been reported in the literature from North America, probably because this mushroom is not a common article of diet for the average American mycophagist. If such poisonings have occurred, they were probably unrecognized and diagnosed as idiopathic immune hemolytic anemia. There was a couple in Vancouver, Washington, who became ill very soon after eating these mushrooms. The husband, who had consumed the larger quantity, developed severe chest and lower back pains, associated with hypotension. He had a hemolytic anemia and was in intensive care for almost a week. He also suffered some permanent loss of vision ascribed to central retinal necrosis. His wife had only a couple of caps but also developed low back pain and was mildly anemic. Her recovery was rapid (Jan Lindgren, Chairperson, Toxicology Committee, Oregon Mycological Society, personal communication).

The mushroom responsible, *Paxillus involutus* (called brown roll-rim, inrolled *Paxillus*, roll-rim fungus), is a common urban mushroom. It is found on cultivated lawns, in parks and cemeteries, on golf courses, and in coniferous and deciduous forests. In fact, it is hard to miss this mushroom in the fall (color plate 30).

In addition to the disastrous and potentially fatal hemolytic complication, *Paxillus involutus* is also known to be a gastrointestinal irritant, especially when eaten raw or incompletely cooked. Nevertheless, it has had a long culinary history in eastern Europe, where it is employed in a number of ways in the kitchen. Pickling and salting are two ways in which it is prepared. So popular is this fungus in countries like Poland that the gastrointestinal symptoms it causes are reported to be the third most common type of poisoning there.

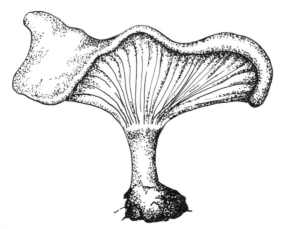

Paxillus involutus

Hemolytic anemia generally develops in individuals who have eaten *Paxillus involutus* for many years with no ill effect. For reasons presently unclear, a few people produce IgG antibodies to an unidentified antigen in the mushroom. During the course of a subsequent meal, antigen–antibody complexes form, agglutination occurs, complement is fixed, and the red blood cells undergo intravascular hemolysis. The onset of the symptoms is rapid, developing within two hours of the mushroom meal. The initial symptoms include vomiting and diarrhea, abdominal pain, and collapse with hypotension. A rapidly developing anemia, with a rise in indirect bilirubin and free hemoglobin (if the hemolysis is massive), a fall in the level of haptoglobin, and hemoglobinuria are all part of the syndrome. The usual renal complications may follow, with kidney failure and renal pain. An agglutination test has been devised by Schmidt and is described in Bresinsky and Besl.[25]

Some poisonings have proved fatal. A 49-year-old male died 3.5 days after his last mushroom meal in protracted, unremitting shock.[21] The autopsy showed widespread intravascular coagulation and extensive fat emboli in the lungs.

Treatment consists of gastric lavage and the management of the acute hemolytic crisis. Because shock may be life threatening, it must be managed with aggressive standard measures. Plasmapheresis may be useful if rapidly instituted.[26] Corticosteroids are generally used as well.[27] Peritoneal or hemodialysis may be required for the kidney failure.

Because of this syndrome and the gastrointestinal symptoms associated with *P. involutus*, firm, sometimes strident, recommendations to avoid eating it have been broadcast.[23] There is no reason to court unnecessary risk. Furthermore, recent evidence has shown that this mushroom contains compounds that directly damage chromosomes, thus raising the specter of mutagenicity and carcinogenicity.[28]

Hemolytic crises have occurred after ingestion of other mushrooms, including *Suillus luteus* and *Amanita vaginata*.[29,30] Some of these may be due to the presence of hemolysins, which are normally inactivated during cooking and not absorbed. However, for the rare individual who eats these mushrooms raw or partially cooked and has the ability to absorb the toxin, this risk (albeit very small) does exist. This form of hemolysis is clearly different from the immune-mediated hemolysis provoked by *Paxillus involutus*.

TWO OTHER TOXINS

We have only begun to explore the many chemicals manufactured by the higher fungi. Our knowledge is rudimentary and has naturally focused on those that do us obvious harm on a regular basis. We have not been tempted to eat the majority of mushrooms—indeed many mushrooms are still waiting to be described, let alone cooked and eaten. Within this vast kingdom, numerous compounds with powerful physiological effects are surely present. I shall mention two that have come to light, one relatively recently.

Acromelic Acid

This fascinating compound has been isolated from *Clitocybe acromelalga*.[31] Ingestion of this acromelic acid-containing mushroom produces sharp pains with marked, reddish swelling of the hands and feet (erythromelalgia) about a week later. This chemical is closely related to L-kainic acid and is similar to domoic acid, the compound that is found in some shellfish and is responsible for producing a devastating neurological disorder. Acromelic acid also bears a structural relationship to the neurotransmitter glutamate and has been shown to be a powerful glutamate agonist at the neuromuscular junction.

Hydrocyanic Acid

Various cyanogens are common throughout the plant and fungal kingdoms. In mushrooms, hydrocyanic acid is found in a variety of genera, but because it is so volatile and usually present in minute concentrations, it never poses much of a health hazard. However, it is responsible for the delightful oil of almond odor in mushrooms such as *Marasmius oreades* (the fairy ring mushroom). It is also present in other edible mushrooms such as the blewitt (*Lepista nuda*), the gypsy mushroom (*Rozites caperata*), *Phaelepiota aurea*, and angel wings (*Pleurocybella porrigens*).[32–34]

REFERENCES

1. R. E. Tullos and J. E. Lindgren, "*Amanita smithiana*—Taxonomy, Distribution and Poisonings," *Mycotaxon* (1992) 45(Oct-Dec):373–387.

2. V. Pelizzari, E. Feifal, M. M. Rohrmoser, G. Gstraunthaler, and M. Moser, "Partial Purification of a Toxic Component of *Amanita smithiana*," *Mycologia* (1994) 86(4):555–560.

3. D. E. Hammerschmidt, "Szechwan Purpura," *New England Journal of Medicine* (1980) 302:1191–1193.

4. D. E. Hammerschmidt, "Chinese Diet and Traditional Materia Medica: Effects on Platelet Function and Atherogenesis." In *Plants in Indigenous Medicine and Diet: Biobehavioral Approaches*, ed. N. L. Etkin (New York: Redgrave Publishing, 1986), p. 171.

5. A. N. Makheja and J. M. Bailey, "Identification of the Anti-Platelet Substance in the Chinese Black Ear Fungus" (letter), *New England Journal of Medicine* (1981) 304:175.

6. C. R. Dorso, R. I. Levin, A. Eldor, E. A. Jaffe, and B. B. Weksler, "Chinese Food and Platelets," *New England Journal of Medicine* (1980) 303:756–757.

7. C. McIlvaine and R. K. MacAdam, *One Thousand American Fungi*, rev. ed. (Indianapolis, IN: Bowen-Merrill). Reprinted with nomenclatural changes by R. L Schaffer, ed. (New York: Dover, 1973).

8. B. P. Wasiljkow, "Die Vergiftungfälle des büscheligen Schwefelkopfes, *Hypholoma fasciculare*," *Schweizerische Zeitschrift für Pilzkunde* (1963) 41:117–121.

9. W. S. Chilton, "Poisonous Mushrooms," *Pacific Search* (1972) March issue.

10. J. Herbich, K. Lohwag, and R. Rotter, "Tödliche Vergiftung mit dem grünblättrigen Schwefelkopf," *Archives of Toxicology* (1966) 21:310–320.

11. I. Kubo, A. Matsumoto, M. Kozuka, and W. F. Wood, "Calmodulin Inhibitors from the Bitter Mushroom *Naematoloma fasciculare* (Fr.) Karst (Strophariaceae) and Absolute Configuration of Fasciculols," *Chemistry and Pharmaceutical Bulletin (Tokyo)* (1985) 33(9):3281–3285.

12. A. Takahashi, G. Kusano, T. Ohta, Y. Ohizumi, and S. Nozoe, "Fasciculic Acids A, B, and C as Calmodulin Antagonists from the Mushroom *Naematoloma fasciculare*," *Chemistry and Pharmaceutical Bulletin (Tokyo)* (1989) 37(12):3247–3250.

13. M. Ikeda, H. Watanabe, A. Hayakawa, K. Sato, T. Sassa, and Y. Miura, "Structures of Fasciculol B and Its Depsipeptide, New Biologically Active Substances from *Nematoloma fasciculare*," *Agricultural and Biological Chemistry* (1977) 41:1543–1545.

14. M. Ikeda, M. Sato, T. Izawa, T. Sassa, and Y. Miura, "Isolation and Structure of Fasciculol A, a New Plant Growth Inhibitor from *Nematoloma fasciculare*," *Agricultural and Biological Chemistry* (1977) 41:1539–1541.

15. M. Ikeda, K. Niwa, T. Tohyama, T. Sassa, and Y. Miura, "Structures of Fasciculol C and Its Depsipeptides, New Biologically Active Substances from *Nematoloma fasciculare*," *Agricultural and Biological Chemistry* (1977) 41:1803–1805.

16. M. Holden, "A Possible Case of Poisoning by *Panaeolina foenisecii*," *Bulletin of the British Mycological Society* (1965) 25:9.

17. R. V. Southcott, "Notes on Some Poisoning and Clinical Effects Following the Ingestion of Australian Fungi," *South Australian Clinics* (1975) 6:441–478.

18. R. Watling, "Studies in the Genera *Lacrymaria* and *Panaeolus*," *Notes of the Royal Botanic Gardens of Edinburgh* (1979) 37:369.

19. J. W. Allen and M. D. Merlin, "Observations Regarding the Suspected Psychotropic Properties of *Panaeolina foenisecii* Maire." In *Yearbook for Ethnomedicine and the Study of Consciousness*, ed. C. Rätsch (VWB, 1992), pp. 99–115.

20. J. Schmidt, A. Hartmann, A. Würstlin, and H. Deicher, "Akutes Nierenversagen durch immunhämolytische Anämie nach Genuss des Kahlen Kremplings (*Paxillus involutus*)," *Deutsche Medizinische Wochenschrift* (1971) 69:1188–1191.

21. M. Winkelmann, F. Borchard, W. Strangel, and B. Grabensee, "Tödlich verlaufene immunhämolytische Anämie nach Genuss des Kahlen Kremplings (*Paxillus involutus*)" [Fatal Immunohemolytic Anemia after Eating the Mushroom *Paxillus involutus*], *Deutsche Medizinische Wochenschrift* (1982) 107:1190–1194.

22. H. Lefevre, "Immunhämolytische Anämie nach Genuss des Kahlen Kremplings (*Paxillus involutus*)," *Deutsche Medizinische Wochenschrift* (1982) 107:1374.

23. R. Flammer, "Hämolyse bei Pilzvergiftungen: Fakten und Hypothesen," *Schwiezerische Medizinische Wochenschrift* (1983) 113:1555–1561.

24. H. Deicher and W. Strangel, "Akute immunhämolytische Anämie nach Genuss des Kahlen Kremplings (*Paxillus involutus*)," *Verhandlungen der Deutschen Gesellschaft für Innere Medizin* (1977) 83:1606.

25. A. Bresinsky and H. Besl, *A Colour Atlas of Poisonous Fungi* (London: Wolfe Publishing, 1990).

26. M. Winkelmann, W. Stangel, I. Schedel, and B. Grabensee, "Severe Hemolysis Caused by Antibodies Against the Mushroom *Paxillus involutus* and Its Therapy by Plasma Exchange," *Klinische Wochenschrift* (1986) 64(19):935–938.

27. L. L. Olesen, [Poisoning with the Brown Roll-rim Mushroom, *Paxillus involutus*] (Danish), *Ugeskrift for Laeger* (1991) 153(6):445.

28. J. Gilot-Delhalle, J. Moutschen, and M. Moutschen-Dahmen, "Chromosome-breaking Activity of Extracts of the Mushroom *Paxillus involutus* Fries ex Batsch," *Experientia* (1991) 47(3):282–284.

29. D. M. Simons, *The Mushroom Toxins from a Chemical Point of View*, Syllabus, Toxic Mushroom Workshop (Bronx: New York Botanical Garden, 1978).

30. P. Vergeer, "Poisonous Fungi." In *Fungi Pathogenic for Humans and Animals. Part B. Pathogenicity and Detection*, ed. P. A. Lempke (New York: Marcel Dekker, 1983), p. 373.

31. H. Shinozaki, M. Ishida, and T. Okamoto, "Acromelic Acid, a Novel Excitatory Amino Acid from a Poisonous Mushroom: Effects on the Crayfish Neuromuscular Junction," *Brain Research* (1986) 399(2):395–398.

32. P. Heinemann, "Observations sur les Basidiomycètes à acide cyanohydrique II," *Lejeunia* (1949) 13:99–100.

33. P. Heinemann, "Observations sur les Basidiomycètes à acide cyanohydrique," *Bull. Trimest. Soc. Mycol. Fr.* (1942) 58:99–104.

34. L. Göttl, "Blausäurebildende Basidiomyzeten. Hat Cyanogenese einen taxonomischen Wert?" *Zeitschrift für Pilzkunde* (1976) 42:185–194.

EPILOGUE

THE FASCINATION and intrigue of mushrooms is reflected in the passion with which so many people react to their presence on the table or in the woods. In mycophobic North America and the British Isles, they are, at best, ignored, their fleeting presence during the time of fruiting barely a reminder that within the soil or the roots or the branches are organisms busily contributing to the essential fabric of every ecosystem. At worst, they are scorned and trampled underfoot. In these countries, they still evoke a sense of danger and fear. Here mycophagists are still regarded with suspicion or considered to be quaint eccentrics near the fringe of acceptable society. School biology courses scarcely mention their role in nature. Even university-level botany and biology course give fungi short shrift. Yeasts are regarded simply as tools in molecular biology. Taxonomic mycologists are poorly supported despite the fact that many members of the fungal kingdom have yet to be identified, catalogued, and studied.

To those familiar with the beauty and the pleasure of mushrooms, however, the fear and loathing is replaced by a sense of reverence, if not awe. This appreciation reaches its zenith in the great Slavic cultures of eastern Europe and the peoples across the Asian continent. There and elsewhere, in Europe and Africa, mushrooms contribute to the joy of living, providing nourishment for both the body and soul. There is more than nutrition involved in this nourishment—the delightful flavors and textures blend almost imperceptibly into a magic, enhancing the pleasures of daily living. And there are even those who have used the mushrooms to explore different realities.

A few years ago, the venerable doyenne of the TV kitchen, Julia Child, berated a group of nutritionists for their cold, calculating approach to food. One of her important contributions was to note that food, in the eyes of many dietitians, physicians, and the lay public, had become generally classified into two categories, medicine or poison. She rightly pointed out that food and eating were far more than calories, nutrients, fat, and danger. Mushrooms epitomize this attitude in the most basic way. Widely regarded as having "no food value," they are viewed with great distrust by the majority because of the very real danger of poisoning. At the same time, others are attempting to market the medicinal properties of mushrooms. The reality of mushrooms in the diet is far from both these extremes. They offer a panoply of tastes and textures and flavors. They can and do enhance meals. They are eagerly sought in the market and in the field by all those who have learned to embrace the joy and the pleasure of the table and the fellowship of the meal. Hunting them in the woods recalls the old days of our foraging forebears and the thrill of discovery.

But the fungi did not evolve either for our pleasure or health or for our destruction. Each one of the myriad species evolved to fill a local niche in the ecosystem, part of the shimmering and ever-changing fabric of Earth. We still know very little of how this system is put together, how it functions, or what the consequences will be once it is significantly perturbed by our behavior. Fungi, and the mushrooms they spawn at the time of their reproduction, are integral to the success of life on Earth. Part of our obligation, as a sapient species who happens to be sharing the planet with them, is to appreciate and understand them.

CHEMISTRY OF MUSHROOM TOXINS AND METHODS OF ANALYSIS

T HIS APPENDIX is included for the toxicologists and mycologists who need a procedure for the detection of a specific toxin in a mushroom specimen and for clinical laboratory scientists who are required to detect and quantitate the poisons and possibly their metabolites in the human body. In the latter situation, only the detection of amanitins and orellanine has been found to be clinically relevant.

When possible, the assays should be performed on fresh or recently dried samples. Studies have shown that a number of compounds disappear during storage, to the point of being undetectable. Within each group, a number of procedures are listed. Some are older, more tedious, and not always specific; the method selected will depend on the resources available to the chemist.

CYCLOPEPTIDES: AMATOXINS

The amatoxins make up a family of cyclopeptides.[1]

Amatoxins

	R^1	R^2	R^3	R^4	R^5
α-Amanitin	CH$_2$OH	OH	NH$_2$	OH	OH
β-Amanitin	CH$_2$OH	OH	OH	OH	OH
γ-Amanitin	CH$_3$	OH	NH$_2$	OH	OH
ε-Amanitin	CH$_3$	OH	OH	OH	OH
Amanin	CH$_2$OH	OH	OH	H	OH
Amaninamide	CH$_2$OH	OH	NH$_2$	H	OH
Amanulin	CH$_3$	H	NH$_2$	OH	OH
Amanullinic acid	CH$_3$	H	OH	OH	OH
Proamanullin	CH$_3$	H	NH$_2$	OH	H

- Wieland (Meixner) newspaper test for the presence of amatoxins in a mushroom.[2-4] A full description of this test appears on pages 191–192. Detection limit: 0.02 mg/mL of toxin
- High-performance liquid chromatography. Detection limit: 50 ng of amanitin[5-10]
- Radioimmunoassay[11]
- Thin-layer chromatography. Detection limit: 0.05 μg[12-14]

COPRINE

- Thin-layer chromatography[15-17]

Coprine

GYROMITRIN AND ITS HYDRAZINE METABOLITES

Gyromitrin

N-methyl-N-formylhydrazine

H$_2$N—NHCH$_3$

Monomethylhydrazine

- ◆ Gas–liquid chromatography.[18,19] Limitation: This method may not account for the association of the MFH with higher molecular weight compounds
- ◆ Thin-layer chromatography[18,20–22]
- ◆ Spectroscopic methods;[23] fluorimetry has a sensitivity of 0.2 ng of toxin[24]

PSILOCYBIN AND PSILOCIN

Psilocybin

Psilocin

- ◆ General[25]
- ◆ Thin-layer chromatography[26]
- ◆ Thin-layer chromatography and mass spectroscopy[27]
- ◆ High-performance liquid chromatography[28–31]
- ◆ Gas–liquid chromatography and mass spectroscopy[32]
- ◆ High-performance liquid chromatography and mass spectroscopy[33]

ISOXAZOLE DERIVATIVES

Ibotenic acid

Muscimol

Muscazone

- ◆ Amino acid analysis[34]
- ◆ Gas–liquid chromatography[35]
- ◆ High-performance liquid chromatography[36]
- ◆ High-performance thin-layer chromatography[22]

MUSCARINE

All the procedures for the detection of muscarine consist of a large number of steps. None are trivial and most are tedious. There is no clinical indication for measuring muscarine in patient's body fluids. The older biological methods should not be used. Although they are very sensitive, they are also rather nonspecific.

Muscarine

+ Gas chromatography[37]
+ Thin-layer chromatography[22]
+ High-performance liquid chromatography[38]

ORELLANINE

Although the controversy about the nature of the toxin still exists, a number of assays are available for the detection of the putative poisonous compounds.

Orellanine

+ Thin-layer chromatography with UV fluorescence.[39–42] This test is useful for rapid detection
+ High-pressure liquid chromatography[43,44]

REFERENCES

1. K. Lampe and M. A. McCann, "Differential Diagnosis of Poisoning by North American Mushrooms with Particular Emphasis on *Amanita phalloides*-like Intoxication," *Annals of Emergency Medicine* (1987) 36(9):956–962.

2. J. A. Beutler and P. Vergeer, "Amatoxins in American Mushrooms: Evaluation of the Meixner Test," *Mycologia* (1980) 72(6):1142–1149.

3. A. Meixner, "Amatoxin-Nachweis in Pilzen," *Zeitschrift für Mykologie* (1979) 45:137–139.

4. P. P. Vergeer, "The Meixner Test Evaluated," *McIlvainea* (1986) 7(2):61–70.

5. W. Rieck and D. Platt, "High-performance Liquid Chromatographic Method for the Determination of α-Amanitin and Phalloidin in Human Plasma Using the Column

Switching Technique and Its Application in Suspected Cases of Poisoning by the Green Species of Amanita Mushroom (Amanita phalloides)," Journal of Chromatography (1988) 425(1):121–134.

6. F. Jehl, C. Gallion, P. Birkel, A. Jaeger, F. Flesch, and R. Minck, "Determination of α-Amanitin and β-Amanitin in Human Biological Fluids by High Performance Liquid Chromatography," Analytical Biochemistry (1985) 149(1):35–42.

7. G. Caccialanza, C. Gandini, and R. Ponci, "Direct Simultaneous Determination of α-Amanitin, β-Amanitin and Phalloidine by High Performance Liquid Chromatography," Journal of Pharmaceutical and Biomedical Analysis (1985) 3:179–183.

8. L. Pastorello et al., "Determination of α-Amanitin by High-performance Liquid Chromatography," Journal of Chromatography (1982) 233:398–403.

9. F. Belliardo and G. Massano, "Determination of α-Amanitin in Serum by HPLC," Journal of Liquid Chromatography (1983) 6:551–558.

10. F. C. Enjalbert, C. Gallion, F. Jehl, H. Monteil, and H. Faulstich, "Simultaneous Assay for Amatoxins and Phallotoxins in Amanita phalloides Fr. by High-performance Liquid Chromatography," Journal of Chromatography (1992) 598:227–236.

11. H. Faulstich, "New Aspects of Amanita Poisoning," Klinische Wochenschrift (1979) 57:1143–1152.

12. T. Stijve and R. Seeger, [Determination of α-, β-, and γ-amanitin by high performance thin layer chromatography in Amanita phalloides (Vaill. ex Fr.) Secr. from various origins] (German) Zeitschrift für Naturforschung (1979) 34c:1133–1138.

13. C. Andary, F. Enjalbert, G. Privat, and B. Mandrou, "Dosage des amatoxines par spectrophotométrie directe sur chromatogramme chez Amanita phalloides Fries (Basidiomycetes)," Journal of Chromatography (1977) 132:525–532.

14. T. Stijve [High performance thin layer chromatography for the toxic principles of some poisonous mushrooms] (German), Mitteilungen aus dem Gebiete der Lebensmitteluntersuchung und Hygiene (1981) 72:44–54.

15. M. Gillio-Tos, S. A. Previtera, and A. Vimercati, "Separation of Some Aromatic Amines by Thin-layer Chromatography," Journal of Chromatography (1964) 13:571–572.

16. G. M. Hatfield and J. P. Schaumberg, "Isolation and Structural Studies of Coprine, The Disulfiram-like Constituent of Coprinus atramentarius," Lloydia (1975) 38:489–496.

17. P. Lindberg, R. Bergman, and B. Wickberg, "Isolation and Structure of Coprine, the in vivo Aldehyde Dehydrogenase Inhibitor of Coprinus atramentarius: Synthesis of Coprine and Related Cyclopropanone Derivatives," Journal of the Chemical Society, Perkin Transactions (1977) 1:684–691.

18. T. Stijve, "Ethylidene Gyromitrin and N-Methyl-N-formylhydrazine in Commercially Available Dried False Morels, Gyromitra esculenta Fr. ex Pers.," Trav. Chim. Alim. Hyg. (1978) 69:492–504.

19. H. Pyysalo and A. Niskanen, "On the Occurrence of N-Methyl-N-formylhydrazones in Fresh and Processed False Morel, Gyromitra esculenta," Agricultural and Food Chemistry (1977) 25:644–647.

20. C. Andary and G. Privat, "Variations of Monomethylhydrazine Content in Gyromitra esculenta," Mycologia (1985) 77(2):259–264.

21. W. S. Chilton, "Chemistry and Mode of Action of Mushroom Toxins." In Mushroom Poisoning: Diagnosis and Treatment, ed. B. H. Rumack and E. Salzman (West Palm Beach, FL: CRC Press, 1978), pp. 87–124.

22. T. Stijve, [High-performance thin-layer chromatographic determination of the toxic principles of certain poisonous mushrooms] (German), Mitteilungen aus dem Gebiete der Lebensmitteluntersuchung und Hygiene (1981) 72:44–54.

23. G. A. McClusky, R. G. Cooks, and A. M. Knevel, "Direct Analysis of Mushroom Constituents by Mass Spectrometry," *Tetrahedron Letters* (1978) 19:4471–4474.

24. C. Andary, G. Privat, and M. F. Bourrier, "Microdosage spectrofluorimétrique sur couches minces de la monométhylhydrazine chez *Gyromitra esculenta*," *Journal of Chromatography* (1984) 287:419–424.

25. P. B. Baker and G. F. Phillips, "The Forensic Analysis of Drugs of Abuse," *The Analyst* (1983) 108:777–807.

26. T. Stijve, C. Hischebhuber, and D. Ashley, "Occurrence of 5-Hydroxylated Indole Derivatives in *Panaeolina foenisecii* (Fries) Kuehner from Various Origins," *Zeitschrift für Mykologie* (1984) 50:361–368.

27. J. D. Doms and P. F. Lott, "Forensic Chromatography," *Trends in Analytical Chemistry* (1982) 1:105–110.

28. A. L. Christiansen and K. E. Rasmussen, "Screening of Hallucinogenic Mushrooms with High Performance Liquid Chromatography and Multiple Detectors," *Journal of Chromatography* (1983) 270:293–300.

29. M. W. Beug and J. Bigwood, "Quantitative Analysis of Psilocybin and Psilocin in *Psilocybe baeocystis* (Singer and Smith) by High-performance Liquid Chromatography and Thin-layer Chromatography," *Journal of Chromatography* (1982) 207:379–385.

30. M. W. Beug and J. Bigwood, "Psilocybin and Psilocin Levels in Twenty Species from Seven Genera of Wild Mushrooms in the Pacific Northwest, USA," *Journal of Ethnopharmacology* (1982) 5:271–285.

31. S. Borner and R. Brenneisen, "Determination of Tryptamine Derivatives in Hallucinogenic Mushrooms Using High-performance Liquid Chromatography with Photodiode Array Detection," *Journal of Chromatography* (1987) 408:402–408.

32. D. B. Repke, D. T. Leslie, D. M. Mandell, and N. G. Kish, "GLC-Mass Spectral Analysis of Psilocin and Psilocybin," *Journal of Pharmaceutical Sciences* (1977) 66:734–744.

33. M. Wurst, M. Semerdzieva, and J. Vokoun, "Analysis of Psychotropic Compounds in Fungi of the Genus *Psilocybe* by Reversed-phase High-performance Liquid Chromatography," *Journal of Chromatography* (1984) 286:229–235.

34. M. G. Gore and P. M. Jordan, "Microbore Single-column Analysis of Pharmacologically Active Alkaloids from the Fly-agaric Mushroom, *Amanita muscaria*," *Journal of Chromatography* (1982) 243:323–328.

35. D. B. Repke, D. T. Leslie, and N. G. Kish, "GLC-Mass Spectral Analysis of Fungal Metabolites," *Journal of Pharmaceutical Sciences* (1978) 67:485–487.

36. U. Lund, "Estimation of Muscimol and Ibotenic Acid in *Amanita muscaria* Using High-performance Liquid Chromatography," *Archiv for Pharmaci og Chemi, Scientific Edition* (1979) 7:115–118.

37. L. V. Cunningham, "Microanalytical Determination of Muscarine in *Amanita muscaria* by Gas Chromatography, Doctoral thesis, University of Connecticut, 1975.

38. S. E. Unger, A. Vincze, R. G. Cook, R. Chrisman, and L. D. Rothman, "Identification of Quaternary Alkaloids in Mushroom by Chromatography/Secondary Ion Mass Spectrometry," *Analytical Chemistry* (1981) 53:976–981.

39. T. Schumaker and K. Høiland, "Mushroom Poisoning Caused by Species of the Genus *Cortinarius* Fries," *Archives of Toxicology* (1983) 53:87–106.

40. H. Kürnsteiner and M. Moser, "Isolation of a Lethal Toxin from *Cortinarius orellanus* Fr.," *Mycopathologia* (1981) 74:65–72.

41. B. Caddy, C. B. M. Kidd, J. Robertson, I. R. Tebbett, W. J. Tilstone, and R. Watling, "*Cortinarius speciosissimus* Toxins—A Preliminary Report," *Experientia* (1982) 38:1439–1440.

42. S. Rapior, N. Delpech, C. Andary, and G. Huchard, "Intoxication by *Cortinarius orellanus*: Detection and Assay of Orellanine in Biological Fluids and Renal Biopsies," *Mycopathologia* (1989) 108(3):155–161.

43. I. R. Tebbett and B. Caddy, "Analysis of *Cortinarius* Toxins by Reversed-phase High-performance Liquid Chromatography," *Journal of Chromatography* (1984) 283:417–420.

44. I. R. Tebbett and B. Caddy, "Analysis of *Cortinarius* Mushrooms by High-pressure Liquid Chromatography," *Journal of Chromatography* (1983) 268:535–538.

NATIONAL AND REGIONAL FIELD GUIDES FOR MUSHROOM IDENTIFICATION

T HE FOLLOWING bibliography lists some of the more common and widely available field guides for the United States and Canada. Many other countries have excellent guides, and anyone traveling to a new area is encouraged to employ a local or regional guide.

GENERAL FIELD GUIDES

R. and E. Watling, *A Literature Guide for Identifying Mushrooms* (Eureka, CA: Mad River Press, 1980).

This book is an outstanding reference for anyone looking for monographs on almost any family or genus of mushrooms. In addition, it lists field guides for most countries. It is an annotated bibliography and the first place to look for a major literature reference. The drawback is that it only covers publications up to 1980. We can only hope that it will be updated in the next decade.

A. Bresinsky and H. Besl, *A Colour Atlas of Poisonous Fungi*. Translated by N. G. Bisset (London: Wolfe Publishing, 1990).

A superb color atlas with very good descriptions of the mushrooms and excellent discussions of their chemistry and toxicology. This book should be in every poison center. Its focus is on European mushrooms, but it is still very valuable in the United States.

J. Ammirati, J. A. Traquair, and P. A. Horgen, *Poisonous Mushrooms of the Northern United States and Canada* (Minneapolis: University of Minnesota Press, 1985).

The most extensive listing and detailed descriptions of the poisonous species in North America.

J. W. Groves, *Edible and Poisonous Mushrooms of Canada*, Publication No. 112 (Ottawa, Ontario, Canada: Research Branch, Canadian Department of Agriculture, 1962).

A. H. Smith and N. Smith Weber, *The Mushroom Hunter's Field Guide*, 3rd ed. (Ann Arbor: University of Michigan Press, 1980), 316 pp.
This guide is one of the North American classics. It was first published in 1958 and undergoes revision occasionally. About 200 of the more common mushrooms can be identified with the aid of this book, but many of them may not grow in your neck of the woods. More recent guides have a larger number of species.

O. K. Miller, *Mushrooms of North America* (New York: Dutton, 1978).
One of the classic compendiums of North American mushrooms.

G. Lincoff, *The Audubon Society Field Guide to North American Mushrooms* (New York, Knopf, 1981), 926 pp.
A popular guide covering 756 species. Some mushroom hunters find this book difficult to use because of the arrangement of text and photographs. The latter are outstanding, however.

N. J. Turner and A. F. Szczawinski, *Common Poisonous Plants and Mushrooms of North America* (Portland, OR: Timber Press, 1991).

R. Phillips, *Mushrooms of North America* (Toronto, Ontario, Canada: Little Brown, 1991), 319 pp.
The largest photographic collection of North American species (over 1000) in one book. Its large format confines it to the coffee table.

D. W. Fischer and A. E. Bissette, *Edible Wild Mushrooms of North America: A Field to Kitchen Guide* (Austin: University of Texas Press, 1992), 254 pp.

K. and V. McKnight, *A Field Guide to Mushrooms* (Boston: Houghton Mifflin, 1987), 429 pp.

A. Bessette and W. J. Sundberg, *Mushroom—A Quick Reference Guide to Mushrooms of North America*, Macmillan Field Guide Series (New York: Macmillan, 1987), 171 pp.

REGIONAL FIELD GUIDES

West

D. Arora, *Mushrooms Demystified* (Berkeley, CA: Ten Speed Press, 1989), 959 pp.

This guide has become the standard field guide for the entire West Coast. It is easy to use, humorous, comprehensive, and entertaining. In the latest editions, fungi from other regions have been included, so, although it emphasizes western fungi, I recommend it for all amateur mycologists, anywhere. A hip-pocket companion guide, *All That the Rain Promises and More* (Berkeley, CA: Ten Speed Press, 1991) is equally valuable.

D. Biek, *The Mushrooms of Northern California* (Redding, CA: Spore Prints, 1984).

B. Guild, *The Alaskan Mushroom Hunter's Guide* (Seattle: Alaska Northwest Books, 1977).

M. McKenny and D. Stuntz, *The Savory Wild Mushroom*. Revised by J. Ammirati. (Seattle: University of Washington Press, 1990).
A classic guide that is very useful for the identification of most of the common local species. The color pictures that were added in the last revision are excellent.

H. Parker, *Alaska's Mushrooms: A Practical Guide* (Seattle: Alaska Northwest Books, 1994).

R. T. Orr and D. B. Orr, *Mushrooms of Western North America* (Berkeley, CA: University of California Press, 1979).

H. M. E. Schalkwijk-Barensen, *Mushrooms of Western Canada* (Edmonton, Saskatchewan, Canada: Lone Pine Publishing, 1991), 414 pp.

A. H. Smith, *A Field Guide to Western Mushrooms* (Ann Arbor: University of Michigan Press, 1975), 280 pp.

E. Tylukti, *Mushrooms of Idaho and the Pacific Northwest: Discomycetes* (Moscow: University of Idaho Press, 1994).

E. Tylukti, *Mushrooms of Idaho and the Pacific Northwest: Nongilled Hymenomycetes* (Moscow: University of Idaho Press, 1987).

E. Tylukti, *Mushrooms of Idaho and the Pacific Northwest: Nongilled Basidiomycetes* (Moscow: University of Idaho Press, 1985).

East

A. E. Bessette, *Mushrooms of the Adirondacks: A Field Guide* (Utica, NY: North Country Books, 1988), 145 pp.

L. R. Hesler, *Mushrooms of the Great Smokies* (Knoxville: University of Tennessee Press, 1960).

South and Southwest

S. and V. Meltzer, *Texas Mushrooms* (Austin: University of Texas Press, 1992), 350 pp.

J. S. States, *Mushrooms and Truffles of the Southwest* (Tucson: University of Arizona Press, 1990), 280 pp.

N. Smith Weber and A. H. Smith, *A Field Guide to Southern Mushrooms* (Ann Arbor: University of Michigan Press, 1985), 326 pp.

Midwest

D. M. Huffman, L. H. Tiffany, and G. Knaphus, *Mushrooms and Other Fungi of the Midcontinental United States* (Ames: Iowa State University Press, 1986).

B. Horn, R. Kay, and D. Abel, *A Guide to Kansas Mushrooms* (Kansas City: Kansas University Press, 1993).

MYCOLOGICAL ASSOCIATIONS AND CONSULTANTS

MYCOLOGICAL SOCIETIES and clubs have members who can provide identification assistance. In addition to these organizations, universities and community colleges frequently have resident mycologists. The POISINDEX also provides a current list of consultants: a local or regional poison center should be able to provide this information. Unfortunately, not all of the mycological societies have permanent homes. The addresses given here may change in the future. Colleges other than those cited also can provide assistance; those listed here are known to have a special interest in fungi.

ALASKA

Alaska Mycological Society
P. O. Box 2526
Homer, AK 99603-2526

Glacier Bay Mycological Society
P. O. Box 65
Gustavus, AK 99826-0065

Southeastern Alaskan Mycological
Society
P. O. Box 956
Sitka, AK 99835-0956

ARKANSAS

Arkansas Mycological Society
5115 S. Main Street
Pine Bluff, AR 71601-7452

CALIFORNIA

Fungus Federation of Santa Cruz
1305 E. Cliff Drive
Santa Cruz, CA 95062-3722

Humboldt Bay Mycological Society
P. O. Box 4419
Arcata, CA 95521-1419

Los Angeles Mycological Society
5151 State University Drive
Los Angeles, CA 99032-4221

Mount Shasta Mycological Society
623 Pony Trail
Mount Shasta, CA 96067-9769

Mycological Society of San Francisco
P. O. Box 882163
San Francisco, CA 94188-2163

Humboldt State College
Department of Biological Sciences
Arcata, CA 95521
(707) 826-4841

San Francisco State College
Department of Biology
1600 Hollaway Avenue
San Francisco, CA 94132
(415) 469-2439

COLORADO

Colorado Mycological Society
P. O. Box 9621
Denver, CO 80209-0621

Pikes Peak Mycological Society
P. O. Box 1961
Colorado Springs, CO 80901-1961

Rocky Mountain Poison Center
Denver General Hospital
West 8th and Cherokee Street
(303) 629-1123
Colorado Watts line (800) 332-3073

CONNECTICUT

Connecticut Valley Mycological Society
10 Lounsbury Road
Trumbull, CT 06611-4429

Nutmeg Mycological Society
191 Mile Creek Road
Old Lyme, CT 06371-1719

FLORIDA

University of Florida
Department of Botany
Gainsville, FL 32611
(904) 392-2158

Florida State University
Department of Biological Sciences
Tallahassee, FL 32306

GEORGIA

Centers for Disease Control
Infectious Disease Section
Atlanta, GA 30333
(404) 329-3311

IDAHO

North Idaho Mycological Association
10543A Friar Avenue
Hayden Lake, ID 83835-8572

Southern Idaho Mycological
Association
P. O. Box 843
Boise, ID 83701-0843

Palouse Mycological Association
1499 Hawthorne Drive, Apt. 102
Moscow, ID 83843-9325

University of Idaho
Department of Biological Sciences
Moscow, ID 83843
(208) 885-6349

ILLINOIS

Illinois Mycological Association
13535 Longview Drive
Lockport, IL 60441-9440

Illinois Mycological Society
1183 Scott Avenue
Winnetka, IL 60039

Field Museum of Natural History
Roosevelt Road and Lakeshore Drive
Chicago, IL 60605
(312) 992-9410

Southern Illinois University
Department of Botany
Carbondale, IL 62901
(618) 536-2331

Eastern Illinois University
Department of Botany
Charleston, IL 61920
(217) 581-3525

IOWA

Prairie States Mushroom Club
310 Central Drive
Pella, IA 50219-1901

KANSAS

Kaw Valley Mycological Society
601 Mississippi Street
Lawrence, KS 66044-2349

Pittsburg State University
Department of Biology
Pittsburg, KS 66762
(316) 231-8076

University of Kansas
Department of Botany
Lawrence, KS 66045

KENTUCKY

University of Kentucky
School of Biological Science
Lexington, KY 40506
(606) 266-5148

LOUISIANA

Gulf States Mycological Society
414 Kent Avenue
Metairie, LA 70001-4324

MAINE

Maine Mycological Society
RR 1 Box 1920
Litchfield, ME 04350-9736

MARYLAND

Lower East Shore Mushroom Club
14011 Cooley Road
Princess Anne,MD 21853-3247

Mycological Association of Washington
1820 Mount Ephraim Road
Adamstown, MD 21710-8500

Mycology Laboratory, USDA, ARS, NE Region
Agricultural Research Center
Rm 313, Bioscience 011A
Beltsville, MD 20705

National Fungus Collections
Plant Industry Station
Beltsville, MD 20705

National Clearing House for Poison Control Centers
5600 Fishers Lane, Room 1345
Rockville, MD 20857
(301) 443-6260

MASSACHUSETTS

Berkshire Mycological Society
Pleasant Valley Sanctuary
Lenox, MA 02140

Boston Mycological Club
21 1/2 Inman Street
Cambridge,MA 02139-2406

University of Massachusetts
Department of Botany
Amherst, MA 01003
(413) 545-2235

Harvard University
Farlow Herbarium
20 Divinity Avenue
Cambridge, MA 02138
(617) 495-2368

College of Holy Cross
Biology Department
Worcester, MA 01610
(617) 793-2656 or (617) 793-3417

MICHIGAN

Albion Mushroom Club
Whitehouse Nature Center
Albion, MI 49224

Chippewa Nature Center
400 S. Badour Road
Midland, MI 48640-8660

Michigan Mushroom Hunters Club
2297 19th Street
Wyandotte, MI 48192-4125

North American Mycological Association
3556 Oakwood Street
Ann Arbor, MI 48104-5213

West Michigan Mycological Society
923 E. Ludington Avenue
Ludington, MI 49341-2437

Michigan State University
Plant Biology Laboratory
East Lansing, MI 48823
(517) 355-0107

Central Michigan University
Department of Biology
Mt. Pleasant, MI 48859
(517) 774-3279

University of Michigan
University Herbarium
Ann Arbor, MI 48109
(313) 764-2406

Blodgett Regional Poison Center
1840 Wealthy S.E.
Grand Rapids, MI 49506
(616) 774-7855

MINNESOTA

Minnesota Mycological Society
4424 Judson Lane
Edina, MN 55435-1621

University of Minnesota
Plant Pathology Building #304
St. Paul, MI 55101

MISSOURI

Missouri Mycological Society
2888 Ossenfort Road
Glencoe, MO 63038-1716

MONTANA

Southwest Montana Mycological
Association
7315 12th Avenue
Bozeman, MT 59715-4211

NEW HAMPSHIRE

Monadnock Mushroomers Unlimited
P. O. Box 6296
Keene, NH 03431-6296

Montshire Mycological Club
P. O. Box 59
Sunapee, NH 03782-0059

New Hampshire Mycological Society
84 Cannongate III
Nashua, NH 03063-1948

NEW JERSEY

New Jersey Mycological Association
19 Oak Avenue
Denville, NJ 07834-2223

NEW MEXICO

New Mexico Mycological Society
1511 Marble Avenue N.W.
Albuquerque, NM 87104-1347

NEW YORK

Central New York Mycological Society
343 Randolph Street
Syracuse, NY 13205-2357

COMA
99 Old Mill River Road
Pound Ridge, NY 10576-1833

Long Island Mycological Club
34 Heights Road
Northport, NY 11768-2648

Mid-Hudson Mycological Association
1846 Route 32
Modena, NY 12548

Mid-York Mycological Society
P. O. Box 164
Clinton, NY 13323-0164

New York Mycological Society
140 W. 13th Streetl
New York, NY 10011-7802

Rochester Area Mycological Society
54 Roosevelt Road
Rochester, NY 14618

Susquehanna Valley Mycological
Society
1106 Rodman Road
Endicott, NY 13760-1248

New York Botanic Gardens
Bronx, New York, NY 10458
(212) 220-8612

Cornell University
Department of Plant Pathology
Ithaca, NY
(607) 273-0508

Syracuse University
College of Forestry
Syracuse, NY 13210

Buffalo Museum of Science
Research Library, Humboldt Parkway
Buffalo, NY 14211
(800) 633-6088

NORTH CAROLINA

Asheville Mushroom Club
51 Kentwood Lane
Pisgah Forest, NC 28768-9511

Blue Ridge Mushroom Club
P. O. Box 1107
North Wilkeboro, NC 28659-1107

Cape Fear Mycological Society
8132 Rock Creek Road N.E.
Leland, NC 28451

Triangle Area Mushroom Club
P. O. Box 61061
Durham, NC 27715-1061

University of North Carolina
Botany Library
(919) 933-3783

Western Carolina Poison Center
509 Biltmore Avenue
Asheville, NC 28801
(704) 255-4490 or (704) 255-4032

OHIO

Ohio Mushroom Society
9632 Omega Court
Mentor, OH 44060

OREGON

Lincoln County Mycological Society
6504 SW Inlet Avenue
Lincoln City, OR 97367-1140

Mount Mazama Mushroom Association
817 Garfield Street
Medford, OR 97501-4465

North American Truffling Society
P. O. Box 296
Corvallis, OR 97399-0296

Oregon Coast Mycological Society
P. O. Box 1590
Florence, OR 97439-0103

Oregon Mycological Society
22427 S. Horner Road
Estacada, OR 97023

Willamette Valley Mushroom Society
1454 Manzanita Street N.E.
Salem, OR 97303-1915

Oregon State University
Department of Botany and Plant
Pathology
Corvallis, OR 97331
(503) 754-3451

PENNSYLVANIA

Mellon Institute
Department of Biological Sciences
Carnegie-Mellon University
Pittsburgh, PA 15213

RHODE ISLAND

University of Rhode Island
Department of Botany
Kingston, RI 02881
(401) 792-2161

TENNESSEE

University of Tennessee
Department of Botany
Knoxville, TN 37916
(615) 974-2256

TEXAS

Texas Mycological Society
7445 Dillon Street
Houston, TX 77061-2721

UTAH

Utah State University
Biology Department UMC53
Logan, UT 84322

VERMONT

Vermont Mycology Club
P. O. Box 792
Burlington, VT 05402-0792

VIRGINIA

Virginia Polytechnic Institute and State
University
Blackburg, VA 24061
(703) 961-6407 or (703) 961-6765

WASHINGTON

Kitsap Peninsula Mycological Sociiety
P. O. Box 265
Bremerton, WA 98310-0054

Northwest Mushroomers Association
831 Mason Street
Bellingham, WA 98225-5715

Olympic Mountain Mycological Society
P. O. Box 720
Forks, WA 98331-0720

Pacific Northwest Key Council
1943 S.E. Locust Avenue
Portland, OR 97412-4826

Puget Sound Mycological Society
U. W. Urban Horticulture #GF-15
Seattle, WA 98195-0001

Snohomish County Mycological
Society
P. O. Box 2822
Everett, WA 98203-0822

South Sound Mushroom Club
6439 32nd Avenue N.W.
Olympia, WA 98502-9519

Spokane Mushroom Club
P. O. Box 2791
Spokane, WA 99220-2791

Tacoma Mushroom Society
P. O. Box 99577
Tacoma, WA 98499-0577

Tri-Cities Mycological Society
1628 West Clark Street
Pasco, WA 99301

Twin Harbors Mushroom Club
Route 2 Box 193
Hoquiam, WA 98550

Wenatchee Valley Mushroom Society
P. O. Box 296
Monitor, WA 98836-0296

University of Washington
Department of Botany
Seattle, WA 98105

Evergreen State College
Department of Biology
Olympia, WA 98502

Western Washington University
Department of Biology
Bellingham, WA 98225

WISCONSIN

NW Wisconsin Mycological Society
RR 03 Box 17
Frederic, WI 54837-9700

Parkside Mycological Club
5219 85th Street
Kenosha, WI 53142-2209

Wisconsin Mycological Society
800 W. Watts Street, Room 614
Milwaukee, WI 53233-1404

Center for Forest Mycology Research
FPL, P. O. Box 5130
Madison, WI 53705
(608) 264-5634

Milwaukee Public Museum
Section for Botany
800 W. Wells Street
Milwaukee, WI 53233
(414) 278-2771

CANADA

Chibougamau Mycological Club
804 5e rue
Chibougamau, Québec G8P 1V4

Edmonton Mycological Club
7416 182nd Street
Edmonton, Alberta T5T 2G7

Cercle des Mycologues de Montréal
4101 rue Sherbrooke Est
Montréal, Québec H1X 2B2

Mycological Society of Toronto
2 Deepwood Crescent
North York, Ontario M3C 1N8

Vancouver Mycological Society
403 Third Street
New Westminster, British Columbia
V3L 2S1

University of British Columbia
Department of Botany and Biology
Vancouver, British Columbia V6T 2B1
(604) 228-3346

University of Toronto
Department of Botany
Toronto, Ontario, M5S 1A1

University of Manitoba
Department of Botany
Winnepeg, Manitoba

Cercle des Mycologues de Québec
Pavillon Comtois
Université de Laval
Ste.-Foy, Québec G1K 7P4

Cercle des Mycologues de Rimouski
University of Quebec
Rimouski, Quebec

Cercle des Mycologues de Saguenay
438 rue Perrault
Chicoutimi, Québec G7J 3Y9

MUSHROOM COOKBOOKS

T HE CONCEPT of a specialty cookbook devoted solely to the preparation of mushrooms may come as a surprise to those who are not dedicated mycophagists. How much more surprised are they to discover that one can easily devote a bookshelf to this genre alone. Most of these books have appeared in the last decade. For many years, the sole significant contribution to the field of fungal gastronomy was a book published by the members of the Puget Sound Mycological Society. It remains a wonderful addition to any library, although it is somewhat dated and is unfortunately out of print. But if one is able to find a copy in a second-hand book store, it is still worth collecting.

This bibliography is designed to influence cookbook purchases of both the neophyte and the more experienced mushroom chef. All these cookbooks have appealing features, although the approaches and styles are as disparate as the writers who produced them. Of course, for cookbook addicts like myself, no less than the entire collection will do.

I have divided the books into two groups: books written by one or two authors and collections of recipes published by local mushroom clubs. The former reflect the prejudices and palate of one or two individuals, whereas the latter are generally an eclectic assemblage of recipes.

COOKBOOKS BY ONE OR TWO AUTHORS

Jane Grigson, *The Mushroom Feast* (New York, Knopf, 1979), 305 pp.

This book is a must for any collection, as are the rest of her writings. She is one of the finest English food writers, in the tradition of Elizabeth David. Her experience is firmly based on the European continent and is heavily influenced by all the time she has vacationed or lived in France. The book is garnished with delightful anecdotes and a liberal sprinkling of history. The recipes are frequently inspirational as well as practical—they all work well. Following the introductory general recipes, she explores the role of mushrooms with fish, game, poultry, and meat. The fish selection is perhaps the largest anywhere. The recipes vary from the simple to the ambitious. An entire chapter is devoted to Chinese and Japanese dishes. Truffles are well covered— it is unfortunate that few of us have the opportunity to explore them in depth.

Joe Czarnecki, *Joe's Book of Mushroom Cookery* (New York: Atheneum, 1986), 340 pp.

Here is a unique book, the result of one family's two-generation dedication to the cooking of wild mushrooms. The Czarneckis have operated the only wild mushroom restaurant in the United States, in Reading, Pennsylvania. On the basis of the author's name, one might expect a strong eastern European influence in the style of preparation. Indeed, an underlying Polish theme bubbles its way through the recipes. A superb section describes the ways to extract the maximum flavor from any wild mushroom. It is a chapter that all new chefs should be required to read, because the emphasis is on taste—not on texture or presentation or some of the other ancillary attributes of food. The recipes all work extremely well. One can usually visualize the taste from the description, and one is seldom disappointed. The book's only weakness is its concentration on the most common wild mushrooms, such as *Boletus edulis*, chanterelles, and morels. Only one brief chapter, entitled "Some Recipes for Very Wild Mushrooms," is devoted to the less common species. There are a few books that range further afield and explore the more "exotic" mushrooms that so often fill the basket in the fall. Unfortunately, this is not one of them.

Antonio Carluccio, *A Passion for Mushrooms* (Boston: Salem House, 1989), 192 pp.

For sheer Italian exuberance and gorgeous photography, few books surpass this one. A frequent guest on British television, Antonio has the ability to communicate an infectious enthusiasm that could start a mycophagy epidemic. Preparations are generally rather simple, in the style of the best Italian cuisine, relying on the flavors and textures of the natural ingredients. Few gimmicks are employed, and the essential nature of the mushrooms are highlighted. My one reservation about this book concerns his recommendations for eating raw mushrooms. Although it may be safe for the two species for which he makes this recommendation, it is a habit that poison experts are trying to eliminate.

Margaret Leibenstein, *The Edible Mushroom; A Gourmet Cook's Guide* (New York: Ballantine Books, 1986), 205 pp.

This small, rather unpretentious book is a little gem. It contains some of the most innovative, tasty, and successful recipes. This is the book to use if you want to do something special with the hard-won harvest. The combination of chanterelles and ginger is both unique and satisfying. Other memorable meals include the *funghi trifolati* (mushrooms braised in olive oil), a bolete and goat cheese tart, a wild mushroom and pheasant casserole, and an incredible *wild mushroom and eggs Gennargentu,* which has its origins in Sardinia. One cannot classify the recipes into any particular style or cuisine—it borrows from the entire world, yet blends them all into new creations. This book is not for the gastronomically timid, but anyone having a passing knowledge with a whisk and a chef's knife will find it a treasure. I hope that one day it will reappear in an expanded edition.

Arleen Ranis Bessette and Alan E. Bessette, *Taming the Wild Mushroom—A Culinary Guide to Market Foraging* (Austin: University of Texas Press, 1993), 113 pp.

This recent addition to the culinary scene is designed to make the newcomer to mushroom cookery feel comfortable, while providing an interesting introduction to some of the wonderful possibilities of mushroom dishes. Nice photographs illustrate how a particular dish might look prior to serving. The food styling may not win any awards, but such photographs are often a helpful guide for the average home cook. There are a number of interesting recipes, including a good selection of ways to prepare the wonderful matsutake (Japanese pine mushroom) and an intriguing mushroom mayonnaise sauce. Recipes are organized by mushroom species.

Hope Miller, *Hope's Mushroom Cookbook* (Eureka, CA: Mad River Press, 1994), 220 pp.

The already crowded field of mushroom chefs has been joined by Hope Miller, who is the wife of a mycologist and has spent as many years collecting recipes as her husband has spent collecting and identifying mushrooms. This spiral-bound book has the look of a Junior League publication from a small town and the feel of a mushroom club's collective culinary wisdom. The recipes are arranged by meal courses rather than by mushroom species, and the extensive listing of possible substitutions suggests that each recipe is not necessarily designed to make optimal use of the ingredients. The recipes are relatively simple—this is straightforward, down-home cooking. What it lacks in sophistication, however, it makes up in sheer volume. The more than 300 recipes will keep the dedicated mycophagist busily cooking and eating for quite a few years.

MYCOLOGICAL SOCIETY COOKBOOKS

Puget Sound Mycological Society, *Wild Mushroom Recipes* (Seattle, WA: Pacific Search Press, 1980), 178 pp.

This classic book still contains the widest selection of recipes for the greatest number of species. It originated in 1969 as *Oft Told Mushroom Recipes* and has been reprinted three times. The influence of the Asian community in the Pacific Northwest is reflected by the pages on matsutake and the use of soy and monosodium glutamate (MSG). Other groups of mushrooms are equally well covered. In fact, one would be hard pressed not to find a recipe for just about any genus one is likely to stumble across. Some of the recipes are of the "open-a-can" variety, typical of the eclectic tastes and preferences of club members with very different culinary backgrounds. Almost all are very straightforward and simple to execute.

Mycological Society of San Francisco, *Wild About Mushrooms*, ed. Louise Freedman (Berkeley, CA: Harris Publishing, 1987), 239 pp.

This well-edited book mirrors the slightly more sophisticated tastes of a great city and is moderately influenced by the California-style cuisine that blossomed in the 1980s. There is good variety and an excellent collection of recipes, from the basic to the more difficult. Substitutions for each species are listed at the start of each recipe, greatly broadening the possibilities.

Oregon Mycological Society, *Wild Mushroom Cookery* (Portland, OR: Murty Printing-Publishing Services, 1987), 203 pp.

A more down-home version of the society mushroom cookbook was created by the members of the Oregon club, in a spiral-bound, entirely charming production. It is a recreation of an even earlier publication produced in 1965. The influence of Maggie Rogers, one of the editors, in the historical tidbits and other items of mycological interest adds further zest to this book. The other editors, Mike Wells and Roxanne and Dick Piekenbrock, also did a fine job in distilling many interesting recipes and mushroom lore into this valuable book.

INDEX

Boldface numbers indicate the most detailed discussion of the topic.

Acromelic acid, 385

Activated charcoal, **185**, 225, 355

Aflatoxin, 59

Africa, 14–16

Agarick, 19, 42–44, 80, 98

Agaricus albolutescens, 357

Agaricus arvensis, 123, 356, 358

Agaricus augustus, 123, 168, 356

Agaricus bisporus, 63, 64, 71, 72, 87, 91, 116, 117, 119, 120, 146
    cancer and, 127–128
    nutrition and, 64–65

Agaricus californicus, 357

Agaricus campestris, 104, 114, 123, 124, 144, 224, 356

Agaricus glaber, 357

Agaricus hondensis, 104, 357

Agaricus praeclarisquamosus, 357; color plate 17
    spores of, 194

Agaricus silvicola, 104, 357

Agaricus species, 83, 103, 123, 125, **356–358**

Agaricus xanthodermus, 104, 357, 358; color plate 18

Agaritine, 127–128

Agrippina, 33

Alcohol dehydrogenase, 286–287

Alcohol, coprine, 182, 283–284, **286–289**, 290–292

Alcohol, effect on prognosis of amatoxin poisoning, 215–216

Alcohol intoxication, 109, 115, 313

Aleuria aurantica, 146

Alexius, Czar of Russia, 35

Alkaline phosphatase, 219

Allegro, John, 296

Allergic contact dermatitis, 120

Allergic reaction, 109, 116–122, 160

Amadou, 46

Amaninamide, 208

Amanita, 15, 39, 58, 102, 104, 105, 139, 355

Amanita abrupta, 233–234, 380

Amanita bisporigera, 105, 195, 202, 205

Amanita caesarea, 32, 33, 104, 166, 313

Amanita calyptroderma, 104, 158, 166

Amanita citrina, 41, 325

Amanita cothurnata, 304

Amanita crenulata, 304–305, 312

Amanita frostiana, 304

Amanita gemmata, 105, 142, 314

Amanita hygroscopia, 202

Amanita magnivelaris, 202

Amanita mappa, 41

Amanita muscaria, 16, 38, 48, 104, 117, 122, 139, 141, 142, 155, 168, 169, 177, **295–314**, 332, 340, 344, 346; color plate 8
    distribution and habitat of, 305
    in English literature, 302–303
    as inebriant, 297–301

Amanita muscaria (continued)
poisoning by (*see* Isoxazole poisoning)
in religion, 296–297
spores of, 194
in warfare, 301
Amanita ocreata, 192, 200, 202, 228
distribution and habitat of, 205
Amanita pantherina, 40, 139, 142, 155,
168, 172, 177, 295, 301,
303–304, 306, 311, 313;
color plate 9
distribution and habitat of, 305
poisoning by (*see* Isoxazole poisoning)
Amanita phalloides, 16, 21, 33–34, 38,
41, 101, 104, 110, 116,
138, 139, 148, 152, 154,
155, 157, 158, 166, 168,
169, 192, 199, 201, 202,
207, 213, 214, 224, 234,
247, 362; color plates 1
and 2
distribution and habitat of, 167,
204–205
poisoning by
case history, 198–199
history in United States, 154
incidence, 155, 200
spores of, 194
Amanita pseudoporphyria, 234
Amanita rubescens, 104, 124, 147, 303,
362
Amanita smithiana, 177, 179, 180, 234,
244, 245, **378–379**; color
plate 29
Amanita solitaria (see *Amanita smithiana*)
Amanita strobiliformis, 304, 307
Amanita suballiacea, 202
Amanita tenufolia, 202
Amanita vaginata, 384
Amanita verna, 105, 192, 200, 202, 205
Amanita virosa, 21, 104, 105, 155,
192, 200, 202, 205, 208,
224; color plate 3
Amanita, identification of, 190
Amanitins (*see* Amatoxins)

Amatoxin poisoning
case history, 198–199
clinical features, 179, 216–218
historical treatment, 212–216
incidence, 199–200
laboratory findings, 219–220,
222–223
pathology, 232–233
prognosis
factors influencing, 215, 220–
223
historical, 215
long-term outcome, 231– 232
treatment, 223–229
Bastien's method, 213
controversial, 230–231
Amatoxins, 208, 382
in breast milk, 211
chemical identification of
(Meixner test), 191–192
in edible mushrooms, 147
immunotherapy for, 212
pharmacokinetics of, 209–210
pregnancy and, 210–211
renal excretion of, 210, 225– 226
species containing, 202
distribution and habitats of,
203–205
toxicity of, 211–212
toxicology of, 209–210
Amber, 29, 31
Amino acids in mushrooms, 65– 66
Anatomy of mushroom, **58–61**
development, 56–57
fruiting, 58, 158, 167
microscopic structure, 55
reproduction, 55
Animals, mushroom poisoning of,
148–149, 244, 285, 324
Antabuse, 182, 283, 284, 286, 288.
See also Coprine poisoning.
Antanamide, 209–210
Antibiotics, 84–85
Anticancer (antineoplastic) activity,
82–84, 93–94
Anticonvulsants, 187, 188
Antimicrobial activity, 84–87

Antispasmodics, 189
Antiviral activity, 86–87, 94
Anxiety, management of, 188
Aphrodisiacs, 8, 43, 49–50
Apicius, 32
Armillaria, 54
Armillaria mellea complex, 102, 104, 140, 146, 155, 320, **358**, 382; color plate 20
Arora, D., 25, 104, 357
Arsenic, 125
Arthrobotrys, 53
Ascobolus, 54
Ascomycetes, 55, 56, 119
Asthma, 119
Ataxia, 311–312, 365
Athabascans, 17
Atropine, 44, 173, 341, 345, **348–349**
Auricularia auricula, 26, 43, 47,63, 65, 72, 83, 177, **380– 381**; color plate 28
Australia, 23, 26, 153, 324, 330

Bacterial food poisoning, 109, 111–113
Badham, Charles, 12, 13, 69–70, 361, 364
Babar, 12
Basidiomycetes, 55, 56–57, 119
Bastien's treatment for amatoxin poisoning, 213
Battarra, 31
Beserkers, 301
Bible, 30
Biliary drainage, 225
Blackfoot, 19
Blood pressure, 87, 187
Blyton, Enid, 11
Bolestatine, 359–360
Boletus (bolete), 20, 31, 32, 39, 121, **359**
Boletus calopus, 343
Boletus edulis, 20, 21, 43, 44, 63, 68, 70, 71, 83, 86, 125, 145, 146
Boletus erythropus, 359
Boletus haematinus, 359

Boletus luridus, 284, 291, 343, 359
Boletus pulcherrimus, 359, 360, 372
Boletus satanas, 359; color plate 19 spores of, 194
Boletus subvelutipes, 359
Botulism, 112–113
Brazil, 23
Bufotenine, 192, 325, **328**
Burial monuments, 17

Cadmium, 123–124
Calmette-Guerin bacillus (BCG), 83
Calmodulin, 382
Calocybe gambosum, 361
Calocybe indica, 17
Caloric content of mushrooms, 64–65
Calvacin, 84
Calvatia gigantea, 17, 82, 86
Cancer, 127–130, 272–273
Cantharellus cibarius, 15, 20, 21, 63, 68, 69, 71, 104, 108, 113, 121, 129, 243
Carbohydrate content of mushrooms, 66–67
Cardiac arrhythmias, 218, 290
Carol, Lewis, 295, 302
Cassava, 140, 143
Caterpillars, 24
Cathartics, 185–186, 225
Cautery, 45–46
Cell wall, 51, 67, 72
Cèpe (see *Boletus edulis*)
Cesium, 125–126
Chanterelle (see *Cantharellus cibarius*)
Charles VI, Holy Roman Emperor, 35
Chemical contaminants, 113–115
Chernobyl, 36, 125–126, 157
Chestnut blight, 54
Chicken dung, 38, 39–40
Children, mushroom poisoning in, **149**, 163–164, 174, 222, 272, 311, 329, 330, 333, 347, 354, 356
China, 24–25, 42, 78–80, 88, 94, 380–381

Chitin, 51, 67, 72
Chlorophyllum molybdites, 105, 122,
   155, 167, 193, 351, 353,
   355–356; color plate 16
Chlorpromazine, 188, 333
Cholesterol, 87, 91, 92–93
Chronic hepatitis, 232
Cicero, 32
Cimetidine, 231
Clathrus cancellatus, 372
Claudius I, Emperor of Rome, 33,
   34
Claviceps purpurea, 59
Clement VII, Pope, 35
Clitocybe, 101, 105, 306, 341–343
Clitocybe acromelalga, 385
Clitocybe angustissima, 343
Clitocybe aurantiacum, 343
Clitocybe candidans, 343
Clitocybe cerussata, 343
Clitocybe clavipes, 284, 286, 289
Clitocybe dealbata, 341, 343; color
   plate 15
Clitocybe dilatata, 343
Clitocybe ericetorum, 343
Clitocybe festiva, 343
Clitocybe gibba, 343
Clitocybe hydogramma, 343
Clitocybe nebularis, 343
Clitocybe rivulosa, 343
Clitocybe suaveolens, 343
Clitocybe trunciola, 343
Clitocybe vermicularis, 343
Clitopilus prunulus, 361
Clostridium botulinum, 113, 126
Clusius, 31
Coagulopathy
   in amatoxin poisoning, 218, 220,
      222–223
   with *Auricularia auricula*, 177, 381
Collybia acervata, 372
Collybia dryophila, 372
Conocybe, 105, 165, 202, 206
Conocybe cyanopus, 326
Conocybe filaris, 202, 206
Conocybe rugosa, 202

Conocybe smithii, 326
Coprine, 182
   toxicology of, 286–287
Coprine poisoning, 182, 283–294
   case history, 382, 291–292
   clinical features, 182, 288–289
   management, 290
   mushrooms associated with, 284
   prognosis, 289–290
Coprinus, 54, 102
Coprinus atramentarius, 104, 115, 182,
   283, 284, 285, **286**, 287,
   290; color plate 7
Coprinus comatus, 63, 86, 104, 124,
   146, 284, 285,
Coprinus erethistes, 284
Coprinus fuscescens, 284
Coprinus insignis, 284
Coprinus micaceus, 284, 285
Coprinus quadrifidus, 284
Coprinus variegatus, 284
Cordyceps sinensis, 24, 53, 77
Coriolus (=*Trametes*) *versicolor*, 83
Coriolus (*see* Trametes)
Cortinarins, **247–249**, 258
Cortinarius, 104, 126, 148, 243,
   244, 245, 259, 380
   distribution and habitat, 245–
      246
Cortinarius atrovirens, 246
Cortinarius callisteus, 247
Cortinarius gentilis, 246
Cortinarius orellanoides, 246, 247, 259
Cortinarius orellanus, 169, 244, 245,
   247, 258, 259
   spores of, 194
Cortinarius rainerensis, 246
Cortinarius speciosissimus, 21, 167, 244,
   246, 247, 259
Cortinarius splendens, 246, **258**, 259
Cortinarius venenosus, 246
Cortinarius violaceus, 246, 247
Corynebacterium parvum, 83
Craterellus cornucopoides, 84
Culpeper, Nicholas, 44–45
Cyclopeptides (*see* Amatoxins)

Cytochrome c, 214, 231
Cyttaria darwinia, 23

Dactyella, 53
Dakota, 18
Darwin, Charles, 7, 23
de Brunhoff, Jean, 12
de Jacourt, Louis, 5
de Sales, Saint Francis, 2
Diazepam, 188, 333–334
Dioscorides, 38, 43
Diphilus, 31
Disciotis, 267
Disseminated intravascular coagula-
    tion, 218
Disulfiram, 284. See also Antabuse.
Dogwood anthracnose, 54
Doyle, Arthur Conan, 10
Drosophila melanogaster, 148, 208
Dutch elm disease, 54

Echo power, 19
Edibilty characteristics of mush-
    rooms, 3, 137–138
Elaphomyces granulatus, 43, 49
Elderly, 168
Elyot, Thomas,   2
Emesis, 183–185, 313
Enterohepatic circulation, 209–210,
    214, 227
Entoloma, 105, **360–362**
Entoloma lividum (=sinuatum), 153, 361;
    color plate 21
Entoloma mammosum, 361
Entoloma nidrosum, 361
Entoloma pascuum, 361
Entoloma rhodopolium, 154, 343, 361,
    362
Entoloma strictius, 361
Entoloma vernum, 361
Euripedes, 30

Fasciculic acids, 382
Fat content of mushrooms, 66
Fiber content of mushrooms, 66–67
Finland, 20–21

Fistulina hepatica, 69
Flammulina velutipes, 73, 83, 84, 104,
    113, 143, 362
Flavor enhancer, 242–243
Fly agaric (see Amanita muscaria)
Folic acid, 276
Fomes fomentarius, 42–43, 45–46, 80
Fomes officinalis, 19, 42–43, 45–46
Forced diuresis, 186, 225–226
Fossil fungi, 29
Fowl's dung, 38–40
Free hemaglobin, 179, 180
Fungicides (see Chemical contami-
    nants)

Galen, 39, 40, 43, 44
Galerina, 104, 105, 165, 205–206
    identification of, 190–191
Galerina autumnalis, 192, 201, 202;
    color plate 4
    distribution and habitat of, 205
    spores of, 194
Galerina badipes, 202
Galerina beithii, 202
Galerina fasciculata, 202
Galerina marginata, 202, 205
Galerina sulcipes, 202, 234
Galerina unicolor, 202
Galerina venenata, 111, 164, 201, 202
    distribution and habitat of, 205
Gamma amino butyric acid, 270,
    307–308
Ganoderma applanatum, 83
Ganoderma lucidum, 24, 26, 87, **88–91**,
    126
Ganoderma sinense, 79
Gastric lavage, 183, 184–185
Gastrointestinal syndrome, 183,
    351–372
    clincal features, 354
    management, 354–355
Geastrum, 18, 19
Gérard, F., 38
Gerard, John, 5, 6, 47
Glutamate, 66, 117, 306, 385
Glycogen, 51

Glycopyrrolate, 189
Gomphus floccosus, 103, 372
Gomphus kauffmanii, 372
Gould, Stephen Jay, 13
Greeks, 30, 42
Grete Herball, 5
Grey squirrel, 148
Grifola frondosa, 26, 84, 87
Grifola umbellata, 84
Gymnopilus spectabilis, 104, 320, 321,
    325; color plate 12
Gyromitra, 56, 72, 103, 105, 122,
    143, 166, 189, 268
Gyromitra ambigua, 266, 267
Gyromitra californica, 266, 267
Gyromitra esculenta, 21, 105, 140, 144,
    155, 264, 265, 266, 268,
    272, 279; color plate 6
  cancer risk from, 128–129,
  spores of, 194
Gyromitra fastigiata, 266, 267
Gyromitra gigas, 105, 266
Gyromitra montana, 267
Gyromitra sphaerospora, 266, 267
Gyromitrin (hydrazines)
  toxicity of, 271–272
  toxicology of, 268–270
Gyromitrin poisoning, 179, 264–282
  clinical features, 179, 273– 275
  incidence, 265–266
  management, 276–277
  mushrooms responsible for, 266–
    267
    distribution and habitats of,
    268

Hallucinogenic syndrome, 182,
    318–335
  case histories, 318, 319–321
  clinical features, 182, 329–332
  history, 322–323
  incidence, 323–324
  management, 333–334
  mushrooms responsible for, 325–
    326
  prognosis, 332–334

Haptoglobin, 179, 180, 384
Hay, W. D., 11
Heavy-metal contamination, 36,
    122–125, 358
Hebeloma, 105
Hebeloma crustuliniforme, 362–363;
    color plate 22
Hebeloma mesopheum, 363
Hebeloma senescens, 363
Hebeloma sinapizans, 363
Hebeloma spoliatum, 363
Hebeloma vinosophyllum, 363
Helvella, 267
Hemadsorption, 186–187, 230,
    256
Hemolysins, 72, 145, 213, 362
Hemolysis/hemolytic anemia, 146,
    173, 177, 180, 274–276,
    372, 382–384
Hepatic failure, 173, 178–180, 212,
    217–218, 274–275, 378. See
    also Amatoxin poisoning;
    Gyromitrin poisoning.
Herbicides (see Chemical contami-
    nants)
Hippocrates, 30, 45, 75
Honey mushroom (see Armillaria
    mellea)
Hooker, J. D., 45
Horace, 101
Hussey, Mrs. J. T., 47, 68, 102, 363,
    365, 367–368
Hydna, 69
Hydrazines, 128. See also Gyromitrin.
Hydrocyanic acid, 144, 385
Hyperbaric oxygen therapy, 230
Hypersensitivity pneumanitis, 119–
    120
Hypholoma (=Naematoloma) fasciculare,
    55, 102, 381–382; color
    plate 31
Hypocalcemia, 219
Hypogeus fungi, 48, 148. See also
    Truffles.
Hypoglycemia, 217, 220, 221, 223,
    228

Hypomyces lactifluorum, 146
Hypotensive affect of mushrooms, 87

Ibotenic acid (*see* Isoxazoles)
Ibotenic acid/muscimol, 182
Idiosyncratic reaction, 116–117
Illudins, 364, 366–367
Immigrants, 165–166, 198, 199
Immune potentiation, 83, 90–91, 93–94
India, 16–17
Indian bread, 18, 24
Inebriation syndrome (*see* Isoxazole poisoning)
Inky caps (*see* Coprinus)
Inocybe, 105, 147, 165, 341–343
Inocybe aeruginascens, 326
Inocybe coelestium, 326
Inocybe corydalina, 326
Inocybe fastigiata, 342, 343; color plate 14
   spores of, 194
Inocybe geophylla, 343
Inocybe haemacta, 326
Inocybe lacera, 341, 343
Inocybe mixtilis, 343
Inocybe napipes, 343
Inocybe patouillardi, 152, 343, 347
Inocybe pudica, 343
Inocybe sororia, 343
Inocybe subdestricta, 343
Insecticides (*see* Chemical contaminants)
Interstitial nephritis, 253, 256–257, 259
Intestinal obstruction, 35, 109, 122
Intestinal perforation, 218
Irish potato blight, 54
Iroquois, 17
Isovelleral, 364
Isoxazole poisoning
   clinical features, 182, 310–312
   incidence, 303–304
   management, 313–314
   prognosis, 312–313

Isoxazoles
   toxicity of, 309–310
   toxicology of, 305–309

Japan, 25, 88, 94, 127, 154, 233, 306
Jovian, Emperor of Rome, 34–35

Kamchatka, 297–298, 299, 301
Kidney (*see* Renal failure)
Kootenay, 17
Koryak myths, 301
Kuehneromyces mutablis, 105
Kutkin, 214–215

L.B.M.'s (little brown mushrooms), 165, 342
Laccaria amethystina, 124
Lactarius, 20, 103, 126, 138, 146, **363–365,** 369
Lactarius camphoratus, 365
Lactarius chryorheus, 365
Lactarius deliciosus, 20, 365
Lactarius glaucesens, 364
Lactarius helvus, 364
Lactarius necator, 129
Lactarius pallescens, 365
Lactarius piperatus, 47, 364
Lactarius pubescens, 365
Lactarius representaneus, 365
Lactarius rufus, color plate 24
Lactarius sanguifluus, 365
Lactarius scorbiculatus var. *canadensis,* 365
Lactarius torminosus, 38, 47, 364, 365
Lactarius uvidus, 365
Laetiporus sulphureus, 146, **365– 366,** 367; color plate 23
Lampe, K., 171
Lampteromyces japonicus, 154, 366
Laughing mushroom (*see Gymnopilus spectabilis*)
Lawrence, D. H., 10
Laxative, 117
Lead, 123–125
Lectins, 72, 91,117, 143, 360

Lentinus (=Lentulina) edodes, 26, 63, 73, 83, 86, 87, **91–94**
Lentinus (=Neolentinus) lepideus, 83
Lepiota, 103, 201
    identification of, 191
    species of containing amatoxins, 202–203; color plate 5
Lepiota brunneo-incarnata, 202
Lepiota castanae, 201, 202, 203
Lepiota excoriata, 165
Lepiota helveola, 194, 202
Lepiota heteri, 202, 203
Lepiota josserandi, 165, 202, 203
Lepiota naucinoides, 145
Lepiota procera, 105
Lepiota rachodes, 105, 124
Lepiota scobinella,   202
Lepiota subincarnata, 202
Lepista nuda, 124, 146, 385
Leucoagaricus naucinus, 105, 146
Lichen, 30
Life cycle, 56–57
Lincoff, G., 103, 172, 174, 264, 354, 370
Ling chi (see Ganoderma lucidum)
Lipids, 87, 91
Liver dysfunction
    with Amanita smithiana, 378
    in gyromitrin poisoning, 274
    See also Amatoxin poisoning.
Liver transplantation, 228–229
Longfellow, Henry W., 9
Look-alikes, 103–106
Lycoperdon marginatum, 325, 326
Lycoperdon mixtecorum, 325, 326
Lycoperdon perlatum, 124
Lycoperdon species, 18, 105
Lycoperdonosis, 119
Lysergic acid diethylamide (LSD), 300, 327, 330, 335

Magic mushroom, 3, 323, 331
Magnesium sulfate, 186
Mainlining mushroom extracts, 331
Maitake (see Grifola frondosa)
Manna, 30

Marasmius oreades, 12, 102, 105, 144, 341, 385
Martial, 31–32
Mass poisonings, 24, 156–158, 200, 203, 234, 244, 258, 364, 367
Matsutake (see Tricholoma magnivalere)
Meixner test for amatoxins, 191–192
Melzer's reagent, 190
Meperidine, 187
Mercury, 123, 124, 125
Mesenteric venous thrombosis, 218
Methemaglobin, 179, 180, 274, 276, 331
Metroclopramide, 188
Midazolam, 188
Miwok, 18
Monomethylhydrazine (see Gyromitrin)
Monosodium glutamate (MSG), 117
Morchella (morel), 16, 56, 67, 70, 71, 72, 73, 105, 143, 146
    adverse effects of, 144, 278, 291
Morchella esculenta, 63
Morchella hortensis, 83
Morphine, 187
Moxabustion, 45–46
Munchausen's syndrome, 121
Muscarine, 30, 147, 305–306, 340, 362
    toxicity of, 346
    toxicology of, 344–346
Muscarine poisoning, 181, 340–349
    case history, 340
    clinical features, 181, 346–348
    management, 348–349
    mushrooms responsible for, 341–344
    prognosis, 348
Muscimol (see Isoxazoles)
Muscle fasciculation, 275, 311–312
Mushroom
    anatomy, 55, **58–61**, 158, 167
    habitats, 51–54, 138–139
    identification, 189–195
    nutritional value, 62–68

types, 59
See also specific genera and species.
Mycelium, 55
Mycena pura, 343, 344, 349, 372
Mycetism, 60
Mycetoxicosis, 60
Mycophiles/mycophilia, 3, 22, 24, 199
Mycophobes, 3, **4–14**
Mycorrhiza, 25, **52**, 54, 73, 167, 202, 204, 305, 362
Mycoviruses, 86–87, 94
Myths for detecting poisonous mushrooms, 100

N-acetylcystine, 231
Naematoloma capnoides, 381
Naematoloma fasciculare (see *Hypholoma fasciculare*)
Nausea, management of, 188–189
Navajo, 18
Naxalk, 17
Nematode, 25, 86
Nicander, 36, 39
North American Mycological Association, 106, 121, 136, 145, 155–156, 169, 265
North American native peoples, 17–19
Nothofagus, 23
Nutritional value of mushrooms
  amino acids, 65–66
  caloric content, 64–65
  carbohydrate, 66–67
  fat content, 66
  fiber, 66–67
  minerals, 67
  protein, 65
  water content, 64

Ojibway, 297
Omaha, 17
Ompholotus, 342, 349, **366–367**
Ompholotus illudens, 342, 343, 364, 366
Ompholotus olearius, 104, 155, 342, 343, 352, 353, 366

Ompholotus olivascens, 342, 343, 366; color plate 25
Ompholotus subilludens, 343, 377
Ondansetron, 188
Opiates, 187
Orellanine, 249–251, 257–258, 378
  detection of, 253–254
  toxicity of, 251
Orellanine poisoning, 179, **242–263**
  clinical features, 252–253
  history of discovery, 243
  incidence, 243–245
  laboratory findings, 253
  management, 256
  mushrooms causing, 246–247
  pathology, 256–258
  prognosis, 254–255

Pain management, 187
Panacea, 42–44, 77
Panaeolus africanus, 326
Panaeolus ater, 326
Panaeolus cambodginiensis, 326
Panaeolus castaneifolius, 326
Panaeolus fimcola, 326
Panaeolus foenisecii, 318, 326, 330, **382**; color plate 13
Panaeolus microsporus, 326
Panaeolus papilionaceus, 321
Panaeolus retrugis,   326
Panaeolus sphinctrinus, 326
Panaeolus subalteatus, 326
Panaeolus tropicalis, 326
Panic reaction, 108–111, **181**, 347
Paralytic ileus, 218
Parasites, 52
Parkinson, John, 6
Partial veil, 58–69, 190
Paulet, Jean Paul, 38
Paxillus involutus, 129, 138, 173, 177, 372, **382–384**; color plate 30
Penicillin, 210, 227
Pesticides (*see* Chemical contaminants)

Peziza, 267
Phaelepiota aurea, 141, 372, 385
Phallic nature of mushrooms, 6, 7, 25, 26
Phallolysins, 208, 213
Phallotoxins, 207–208, 215
Phallus impudicus, 6, 7
Phanerochaete chryosporium, 73
Phenobarbital, 188, 333
Phenol, 358
Pholiota hameji, 73, 83
Pholiota highlandensis, 369
Pholiota limonella, 369
Pholiota nameko, 120
Pholiota squarrosa, 146, 155, 291, **367–369**; color plate 27
Phytophothora infestans, 54
Pileus, 58, 60
Pilobolus, 54
Piptoporus betulinus, 83
Plasma exchange transfusion (plasmapheresis), 230, 256, 384
Plasmopara viticola, 54
Platelets, 381
Pleurocybella porrigens, 385
Pleurotus, 87
Pleurotus ostreatus, 21, 73, 83, 84, 154, 362
Pleurotus spodoleucus, 83
Pleurotus tuber-regium, 15
Pleuteus cervinus, 105
Pleuteus nigroviridus, 325, 326
Pleuteus salicinus, 325, 326
Pliny, 32–33, 36–37, 44, 48
Plutarch, 31
Pneumanitis, 119–120
Poisoning, mushroom
   in amateur mycologists, 159–160
   ancient remedies, 38–40
   criminal use (murder), 40–41
   incidence, 152–156
   management (general), **183–189**
   in professional mycologists, 162
Polyporus indigeus, 24, 200, 203, 243, 244
Polyporus myllitae, 26

Polyporus umbellatus (see *Grifola umbellata*)
Porcini (see *Boletus edulis*)
Poria cocus, 18
Prishvin, Mikhail, M., 21, 22
Protein content of mushrooms, 65
Pseudohydnum gelatinosum, 146
Psilocybe, 54, 105, 111, 148, 164–165, 185, 206, 212, 318, 319
   distribution and habitats of, 324
Psilocybe argentipes, 321
Psilocybe aztecorum, 326
Psilocybe baeocystis, 324, 330
Psilocybe bonetii, 326
Psilocybe caerulescens, 326
Psilocybe caerulipes, 326
Psilocybe cambodginiensis, 326
Psilocybe coprinifacies, 326
Psilocybe cubensis, 326
Psilocybe cyanescens, 324, 326; color plate 11
Psilocybe cyanofibrillosus, 326
Psilocybe fimetaria, 326
Psilocybe mexicana, 326, 327
Psilocybe pelliculosa, 326
Psilocybe quebecensis, 326
Psilocybe semilanceata, 324, 326
   spores of, 194
Psilocybe sempervira, 326
Psilocybe serbica, 326
Psilocybe strictipes, 326
Psilocybe stuntzii, 326; color plate 10
Psilocybe subaeruginosa, 326
Psilocybe zapatecorum, 326
Psilocybin/psilocin, 182, 192, 318, 383
   legal status of, 334–335
   toxicity of, 328
   toxicology of, 327–328
PSL syndrome (*see* Muscarine poisoning)
Puffballs, 18, 19, 57, 69, 105, 119
   in bee keeping, 47
   in cautery, 46
   as styptic or hemostatic agent, 18, 46–47
   See also Lycoperdon.

Pyridoxine, 269–270, 275, 276–277

Radioactivity, 125–127
Ramaria, 103, 371–372
Ramaria flavobrunnescens, 372
Ramaria formosa, 371, 372
Ramaria gelatinosa, color plate 32
Ramaria pallida, 372
Raverat, Gwen, 7
Raw mushrooms, 71–72, **146–147**,
     149, 174, 277–278, 291–
     292, 356, 365, 382, 384
Red-backed vole, 148
Rees's *Cyclopaedia*, 45
Regulations, 20, 21, 37, 112
Renal failure, 173, 178–180, 187,
     274–275, 384
   with *Amanita smithiana*, 180, 244,
     378–379
   in amatoxin poisoning, 217– 218,
     220
   in orellanine poisoning, 180,
     244–245, 251, 253–255
Rhodophilus rhodopolium (see *Entoloma*
     rhodopolium)
Rhodophilus sinuatus (see *Entoloma*
     lividum)
RNA polymerase, 208, 214, 232
Rocket fuel, 269
Romans, 30–38, 42
Rozites caperata, 105, 385
Russia, **21–23**, 126, 157–159, 200
Russula, 126, 138, **369**
Russula brevipes, 138, 369
Russula cyanoxantha, 369
Russula emetica, 369; color plate 26
Russula sardonia, 369

Salish, 18, 19
Saprobes, 52
Sarcosphaera, 267
Sayers, Dorothy, 302–303
Scleroderma, 49, 103, 105
Scleroderma aurantium, 372
Scleroderma cepa, 372
Sclerotium, 18, 29, 56

Secondary compounds, 82, 143,
     147–148, 353–354
Seizures, management of, 188
Selenium, 125
Sewage sludge, 124
Shaman, 19
Shelley, Percy B., 9
Shiitake (see *Lentinus edodes*)
Shiku, 2
Siberia, 297–298, 300, 303, 308
Silymarin/silibinin, 199, 210, 223,
     227–228
Slime molds, 51
SLUDGE syndrome (*see* Muscarine
     poisoning)
Sodium sulfate, 186
Soma, 296–297
Sorbitol, 185, 186
South America, 23–24
Spermatozoa, 287
Spore identification, 193–194
Spore print, 60, **190**
Squamish, 19
St. Anthony's fire, 59
Stanley, H. M., 15
Staphylococcal food poisoning, 113
Steroids, 119, 230, 256, 384
Stinkhorn (see *Phallus impudicus*)
Stropharia coronilla, 372
Suetonius, 34
Suicide, 169, 210, 251
Suillus, 117, 120, **370**
Suillus luteus, 86, 370, 384
Sumptuary laws, 31
Switzerland, 20, 124, 152, 361, 370
Symbionts, 52
Syrup of ipecac, dosage, 175, 184
Szechwan purpura, 381

Teonanacatl, 322
Terfezia, 48
Termitomyces, 14, 16, 42, 53, 73
Testes, 287
Thiotic acid, 210, 230
Thomas, Lewis, 76
Tinder, 45, 46

Tiramania pinoyi, 129
Toads, 8, 328
Toadstool, 4, 10, 11–12, 22, 45, 164
Tolstoy, L. N., 23
Totems, 19
Toxicity, variation in, 139–142
Toxins in edible mushrooms, 142–146, 147
Trametes suaveolens, 47
Trametes versicolor, 83
Transaminases, 179, 189, 219, 221, 231, 331
Trattinick, L., 68
Trehalose, 66, 116, 142
Tremella mesenterica, 12
Tricholoma, 370–371
Tricholoma flavoviriens, 105, 109, 258
Tricholoma georgii, 21
Tricholoma gigantium, 17
Tricholoma inamoenum, 371
Tricholoma irinum, 372
Tricholoma magnivalere, 25, 105, 378
Tricholoma matsutake, 25, 83
Tricholoma muscarium, 304, 308
Tricholoma pardinum, 153, 370
Tricholoma pessundatum, 371
Tricholoma populinum, 18
Tricholoma sulphureum, 372
Tricholoma ustale, 154
Tricholoma venenatum, 371
Tricholoma zelleri, 371
Tricholomic acid, 308–309
Tricholomopsis platyphylla, 372
Truffles, 8, 16, 20, 21, 32, **48–50**, 57, 139, 148
Tryptamine poisoning (*see* Hallucinogenic syndrome)

Tuber melanosporum, 63
Tuckahoe, 18, 24
Tulostoma, 19
Turkey, 210–202
Tyromyces sulphureus, 79

Universal veil, 58–59

Varthema, Ludivicio de, 48
Vasopressors, 187
Veil, 58–59
Verpa, 143
Verpa bohemica, 267, 268, 291
Verpa conica, 267, 268
Virotoxins, 208
Vitamins in mushrooms, 67
Voltaire, 35
Volva, 58–59, 190, 191, 211
Volvariella volvacae, 17, 73, 198, 362

Waimiri Atroari peoples, 23
Wasson, R. G., 3, 5, 8, 22, 34, 35, 88, 296, 299, 301, 322–323
Water content of mushrooms, 64
Weiland newspaper test for amatoxins, 191–192
Wells, H. G., 302
Witches, 4, 8, 9, 12
Wolfporia cocus, 18

Xanthochrous hispidus, 79

Yanomama, 23–24

Zhu ling, 84
Zuni, 18